Extending

Equipment's Life Cycle

The Next Challenge for Maintenance

(7th Discipline on World Class Maintenance Management)

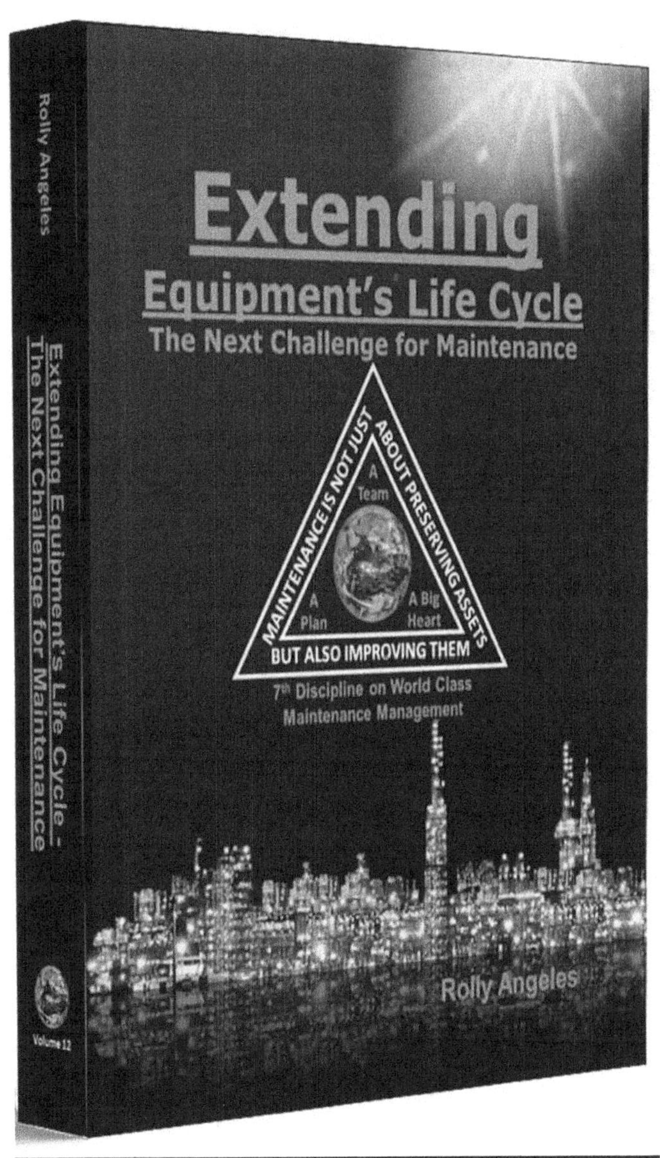

By Rolly Angeles

Extending Equipment's Life Cycle – The Next Challenge for Maintenance
(7th Discipline on World Class Maintenance Management)
Copyright © 2022 Rolly Angeles

10 9 8 7 6 5 4 3 2 1

Printed in the Philippines by **Centralbooks**
Head Office: Phoenix Bldg. 927 Quezon Avenue, Quezon City, Philippines 1101

The National Library of Philippine Catalogue

For international reproduction of this book, will be printed by amazon.com

Kindle Asin Amazon	B0B84HH3CK
Paperback Amazon	979-8842668830
Hardcover Amazon	**979-8802362006**
Paperback Ingramspark	979-8885260077
Hardcover Ingramspark	979-8885260084

Published by:

Central Book Supply Inc.,
Phoenix Bldg. 927 Quezon Avenue, Quezon City, Philippines 1101

This book was designed and produced by:

RSA Reliability and Maintenance Consultancy Firm
Sta. Rosa, Laguna, Philippines 4026

Website: http://www.rsareliability.com
Email: rollyangeles@rsareliability.com

First Printing: August 2022

World Class Maintenance Management Concept

Figure A: Original Concept of World Class Maintenance

By: Rolly Angeles

Figures

Note: For simplification purposes, all pictures, graphs, charts, tables, and drawings in this book will be referred to as figures, which are consecutively indexed below.

Chapter 8: Considering Redesign to Improve Reliability

Chapter 9: Addressing Equipment's Design Weaknesses

Chapter 10: Using Life Cycle as Part of the Overall Maintenance Strategy

Chapter 11: ISO 55000 on Asset Management

Chapter 12: The Conclusion

Table of Contents

Table of Contents

About the Author

Rolly Angeles is a seasoned technical and international reliability and maintenance trainer and book author. His portfolio of reliability and maintenance training includes maintenance management and reliability courses on Total Productive Maintenance (TPM), Planned Maintenance, Autonomous Maintenance, Lubrication Strategy, Tribology, Oil Contamination Control, Condition-Based Maintenance (CBM), Predictive Maintenance, Reliability-Centered Maintenance (RCM), Root Cause Failure Analysis (RCFA), Planned Maintenance, World Class Maintenance Management (WCM), Meaningful Measures of Equipment Performance, and more.

Rolly is a graduate of Mechanical Engineering from Mapua Institute of Technology in the Philippines, batch 1985, and passed the Licensure Board Examination the following year in 1986. With more than 30 years of solid experience, he had worked in various industries from shipping, woodworking, foundry, cast-iron machining, assembly lines, semiconductor manufacturing, and the mining industry. From 1994 to 2002, Rolly worked as a TPM Senior Engineer at Amkor Technology Philippines, a Multi-National company engaged in the manufacture of integrated circuit products, and spearheaded Amkor's Planned Maintenance Organization, composed of maintenance managers and engineers. He was responsible for the dramatic reduction of their machine's unplanned breakdowns in their TPM journey as well as RCM implementation on their Facilities Air Handling Units (AHU) and their sub-station equipment. Here is where he gained hands-on experience and understanding of both TPM and RCM, respectively. His last corporate employment was in 2002, where he worked as a technical training specialist at Lepanto Consolidated Mining Industry. In 2005, Rolly retired early from the industry and decided to establish his own consulting business, **RSA Reliability and Maintenance Consultancy Firm**, where he dedicates his time and passion to work as an independent reliability and maintenance consultant. His email is rollyangeles@rsareliability.com and his website is https://www.rsareliability.com. Rolly has written the following books in a series that is all about his passion for Reliability and Maintenance.

• Volume 1: World Class Maintenance Management – The 12 Disciplines
• Volume 2: Maintenance – Roadmap to Reliability
• Volume 3: Reliability – A Shared Responsibility for Both Operators and Maintenance
• Volume 4: Cutting–Edge Maintenance Management Strategies
• Volume 5: Problems and Solutions on MRO Spare Parts and Storeroom
• Volume 6: Lubrication Tactics for Industries Made Simple
• Volume 7: Decoding Reliability-Centered Maintenance Process for Manufacturing Industries
• Volume 8: RSA Reliability and Maintenance Newsletter Vault Collection, Subscribers Edition
• Volume 9: Investigating Equipment Failures through Root Cause Failure Analysis
• Volume 10: Maintenance Indices – Meaningful Measures of Equipment Performance
• Volume 11: Implementing Preventive Maintenance for Industries the Right Way
• Volume 12: Extending Equipment's Life Cycle – The Next Challenge for Maintenance

Acknowledgment

This is to acknowledge all these good maintenance people from different races, cultures, and industries that have attended my reliability and maintenance training in the past. Training is always a two-way process and I have to admit that I also learned a great deal from my students. It is not only the students learning from the teacher, but it can also be the other way around.

I would like to thank my family, most especially my three kids: Marie Vic, Kathleen Kay, and Christian Joseph, my wife, Marites, and my dear and first granddaughter, Kalie. I thank you for the love and support they have provided me during the good and tough times.

Most of all, I would like to thank God Almighty for allowing me to complete this book. The struggles, challenges, pressures, opportunities, blessings, and wisdom in my life made me a better person. I would like to dedicate this book to Him as a way of reaching out to maintenance and reliability people in industries in search of better ways to maintain their equipment and assets. I believe that as maintenance myself, we all have a duty and responsibility not only to preserve and sustain our equipment and assets but to preserve the environment and planet as well. It is my hope that this book provides the readers with the guidance, learning, principles, knowledge, and wisdom about maximizing the use of equipment and assets as well as the possibility of extending its life span.

Life Cycle Costing Quiz Part 1

1. TPM pillar that deals with the Vertical Start-up of Equipment is known as;
   a) Maintenance Prevention (MP) Design
   b) Life Cycle Management
   c) Early Equipment Management or Initial Flow Control Activities
   d) Quality Maintenance

2. Activities on Planned Maintenance Phase 2 is all about
   a) Equipment Restoration
   b) Stabilizing MTBF
   c) Periodically Restore Deterioration
   d) Addressing Equipment Design Weaknesses

3. The subject of Initial Flow Control Activities involved two major subjects, which include;
   a) Quality and Safety
   b) Maintenance and Reliability
   c) New Product Development and New Equipment Purchases
   d) Operations and Maintenance

4. Life Cycle Costing is simply the sum of the following;
   a) LCC = Initial Cost + Running Cost
   b) LCC = Maintenance Cost + Labor Cost
   c) LCC = Running Costs + Decommissioning Costs
   d) LCC = Design + Fabrication Cost

5. According to Ernest Rabinowich, there are 3 main causes why equipment can no longer be used which include obsolescence, accidents and;
   a) Corrosion
   b) Surface Degradation
   c) Mechanical Failures
   d) Fractures

6. According to Ernest Rabinowich, 70% of why equipment can no longer be used is because of;
   a) Corrosion
   b) Surface Degradation
   c) Wear and Tear
   d) Accidents

7. The running cost of the equipment will begin during the following;
   a) Design Phase
   b) Fabrication Phase
   c) Acquisition, Installation, and Commissioning Phase
   d) Operations to Decommissioning Phase

8. The Vertical Startup time for the equipment will begin during the following;
   a) Design Phase

b) Fabrication Phase
c) Acquisition, Shipment, Installation, and Commissioning Phase
d) Start of Operations to Decommissioning Phase

9. Increasing the reliability of the equipment can be done through;
    a) Improvements, Redesign, and Modification
    b) Applying the Correct Maintenance Tasks
    c) Implementing Predictive Maintenance Consistently
    d) Implementing both Predictive and Preventive Maintenance

10. The biggest contributor to the Life Cycle's Running Costs is;
    a) Taxes and Depreciation
    b) Raw Materials
    c) Maintenance Costs
    d) Manpower Costs

11. To protect the rights of the industry regarding the new redesigned items, the industry can apply the following;
    a) Copyright
    b) Trademark
    c) Patent
    d) Intellectual Property Right

12. Implementing TPM Planned Maintenance is done two folds, the first is to reduce unplanned breakdown and the second is to;
    a) Sustain the equipment so it will not revert back to its old condition
    b) Promote Preventive Maintenance Tasks
    c) Address Human Error
    d) Repair and Troubleshoot the equipment

13. According to James Reasons and Anthony Hobbs, the majority of human error in maintenance occurs during the following;
    a) Reassembly Process during Overhauls
    b) Disassembly Process during Overhauls
    c) Parts Replacement
    d) During Inspection Process

14. According to lubrication experts and tribologists, the most destructive contaminants usually fall under the size of;
    a) 40 to 50 microns
    b) 50 to 59 microns
    c) 35 microns and below
    d) 60 microns and above

15. If an item or part is modified or redesigned and the lifespan increased from 30 to 120 days, the original parts inside the storeroom will definitely become;
    a) Fast Moving Parts
    b) Slow Moving Parts

c) Non-Moving Parts
d) Obsolete Parts

16. According to R. Keith Mobley, maintenance-induced errors constitute around how many percent of equipment-related problems, production interruptions, quality defects, and failures?
   a) 17%
   b) 83%
   c) 90%
   d) 40%

17. According to the author of this book, an example of a Product Life-Cycle with a finite or limited lifespan includes;
   a) Food
   b) Pharmaceuticals
   c) Soft drinks
   d) Electronic Appliances

18. According to the author, addressing equipment basic condition is the first step toward extending equipment lifespan. In TPM, the pillar responsible to address this activity will be;
   a) Early Equipment Management
   b) Both Planned and Autonomous Maintenance
   c) Solely Autonomous Maintenance
   d) Solely Planned Maintenance

19. The concept of Poka-Yoke is usually designed to address;
   a) Component Failures
   b) Human Errors
   c) Instrumentation Failures
   d) Electronic Failures

20. The TPM Pillar that will address breakdown losses will be;
   a) Focused Improvement Pillar
   b) Planned Maintenance Pillar
   c) Autonomous Maintenance Pillar
   d) Initial Flow Control Activities Pillar

21. Implementing the Initial Flow Control Activities can improve and optimize the following;
   a) Design Phase
   b) Fabrication Phase
   c) Operation to Disposal Phase
   d) Vertical Start-Up Phase

22. Equipment or machines that experience a lot of errors, assists, and minor stoppages can be addressed through;
   a) MTBF Analysis
   b) MTBA Analysis
   c) MTTS Analysis

d) MTTF Analysis

23. This is an equipment loss that will not result in downtime;
    a) Design Speed Loss
    b) Changeover and Conversion
    c) Breakdown Losses
    d) Minor Stoppages

24. When a redesign is performed to address design weaknesses, those original parts that are stored inside the storeroom become;
    a) Obsolete Parts
    b) Slow Moving Parts
    c) Non-Moving Parts
    d) Replacement Parts

25. Thirty-one participating countries were present in the development of ISO55000 asset management structure, which of these countries is not included in the lists
    a) Australia
    b) Germany
    c) Republic of the Philippines
    d) The United States of America

I. Life Cycle Costing Quiz Part 2

1. In purchasing new equipment, the decision should always be to go to the lowest bidder to save costs for the industry.
    a) True
    b) False

2. Initial Flow Control Activities (IFCA) is one of the pillars of TPM (Total Productive Maintenance) which aims to reduce the vertical start-up time of the equipment.
    a) True
    b) False

3. The vertical startup time in IFCA or Initial Flow Control Activities begins at the design of the equipment.
    a) True
    b) False

4. For lubricating oil, a reduction of moisture from 2500 to 156 ppm can extend the life of the equipment by a factor of five times
    a) True
    b) False

5. According to IFCA or Initial Flow Control Activities, what we can control on the entire lifecycle of the equipment is only the running cost. Design, fabrication, installation, and commissioning are fixed and already given, which means that they cannot be improved.
    a) True

b) False

6. Human Errors can be eliminated completely.
   a) True
   b) False

7. Integrating Precision Maintenance with Preventive Maintenance can reduce Infant Mortality Failures, especially during Preventive Maintenance.
   a) True
   b) False

8. If a redesign or modification has been performed on the equipment to address a design weakness and the equipment has three similar pieces, the modification should be replicated or fan out to all equipment even if other equipment does not exhibit the same problem.
   a) True
   b) False

9. Industries that passed ISO5001 certification will assure them to continue operating and be in business for the next decade.
   a) True
   b) False

10. The sole objective of maintenance is to sustain and preserve their equipment and assets.
   a) True
   b) False

11. If the equipment is dedicated and can only produce one product and the product is not selling well in the market, it is recommended to increase the life cycle of the equipment.
   a) True
   b) False

12. The term fail-safe means that the equipment is free from any form of failures and breakdowns.
   a) True
   b) False

13. In the aviation industry 80 % of accidents are caused by human errors.
   a) True
   b) False

14. Implementing both Preventive and Predictive Maintenance will help extend the Life Cycle of the equipment.
   a) True
   b) False

15. The last stage in the study of Life Cycle Costing is about decommissioning or retiring the equipment.
   a) True

b) False

16. The tragedy at Union Carbide in Bhopal India is a clear case of Management Cost Cutting Schemes.
   a) True
   b) False

17. Having an EAM or Enterprise Management Software is ensuring compliance with ISO55001 certification making sure equipment and assets are well managed,
   a) True
   b) False

18. Establishing the basic equipment condition is the sole responsibility of the maintenance function of the organization.
   a) True
   b) False

19. Extending the equipment's life cycle will begin by addressing the very basic equipment condition.
   a) True
   b) False

20. In TPM, the pillar that can impact OEE the most will be the Focused Improvement or the Kobetsu-Kaizen teams.
   a) True
   b) False

Answer on Appendix A, page 284 of this book.

<u>Preface: Going Beyond Equipment's Life Cycle</u>

Maintenance is ensuring that physical assets continue to do what the users want them to do. In its simplest term, maintenance is all about sustaining and preserving our equipment and assets so that it remains available whenever the users need them. However, this is just one side of the maintenance function. The objective of any maintenance program in industries is to take care of the Equipment's Total Life Cycle at the most reasonable cost with compliance to quality, safety, and the environment. To perform this we need the following;

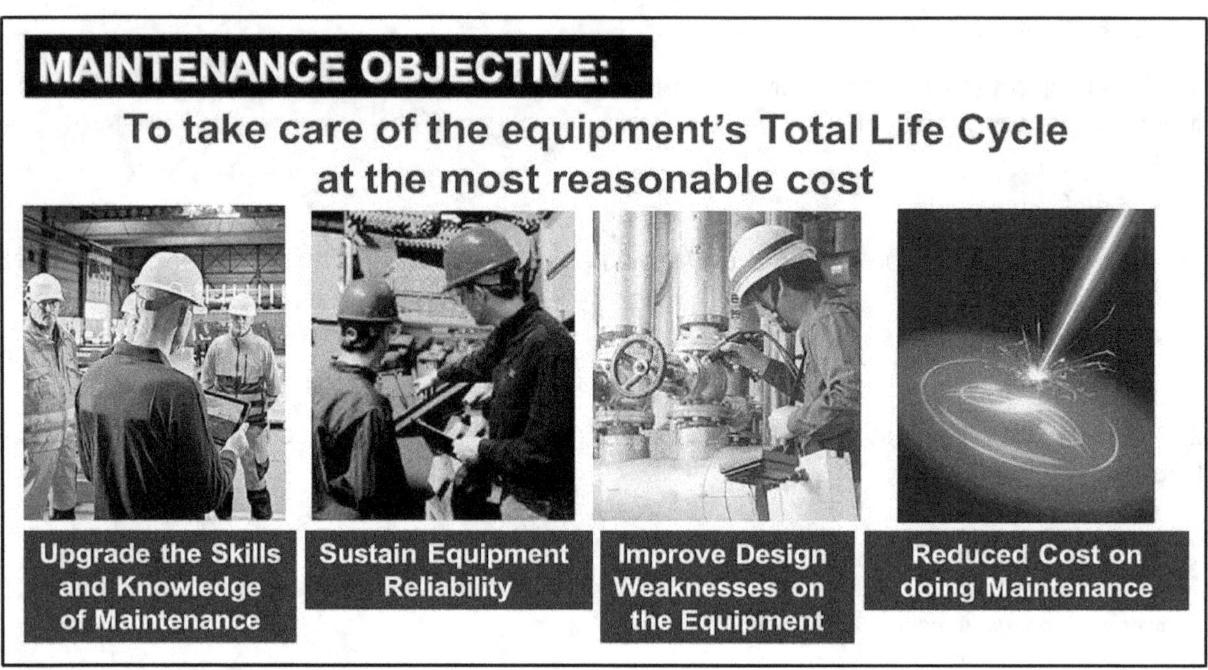

Figure A: Objective of Maintenance

• Upgrade the skills and knowledge of our maintenance people
• Sustain equipment reliability by doing the correct maintenance
• Improve design weaknesses in the equipment
• Reduced the cost of doing maintenance

Although the last objective, which is reducing cost, will only happen by sustaining and improving the equipment and this can only be possible if maintenance is equipped with the right skills and knowledge to execute their jobs correctly. Reducing cost is just the effect of doing the correct maintenance on the equipment. This should not be the primary objective of maintenance. In fact, in many cases reducing maintenance costs can be detrimental to the health and the reliability of the equipment.

The discussion of Life Cycle Costs involves two subjects, which include the Product Life Cycle and the Equipment Life Cycle. Although this book is focused more on discussing the Equipment Life Cycle and most importantly what we can do to possibly extend it. This book covers the 7th Discipline on World Class Maintenance Management which is all about the

Extending Equipment Life Cycle and how it can be part of the overall Plant's Maintenance Strategy.

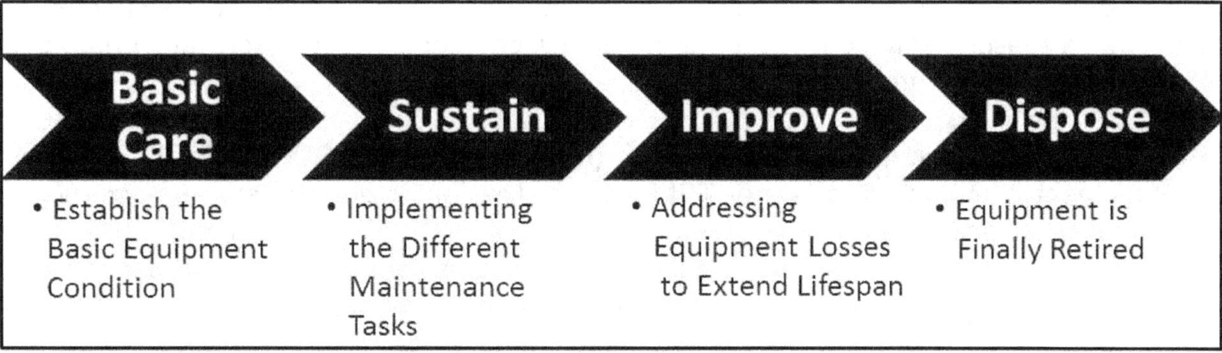

Figure B: The Basic Concept of Extending Equipment Life Cycle

Equipment has a finite or dictated life and usually, this is estimated by the Vendor or OEM, but what if there is a possibility to extend the lifespan by 20, 30, 50, percent, or even 2, 3 folds, or even more? Then this will be of great benefit to industries as they do not need to purchase additional new machines in the future. There are many reasons why equipment does not reach its useful life. There are those, which can be managed, and those that are simply beyond the control of the maintenance function such as the Market Demand. This means that if the product manufactured has limited demand from the public, then there is no need on using the equipment unless the equipment is non-dedicated which means that it can produce other products to market.

The study of the Life Cycle will begin at the Design Phase of the equipment until it is finally retired and disposed of. There are different costs involved in the different phases of an equipment life-cycle. Once the OEM has completed both installation and commissioning, it will now be endorsed to the plant and the start of operations begins which now becomes the responsibility of both operations and maintenance people. The operators had been taught how to correctly operate the equipment, while the maintenance complies with the different tasks needed to sustain and preserve their equipment. Preventive, Predictive, On-Condition Tasks and Failure Finding Tasks are adopted as part of the sustaining program. However, during operations, different equipment losses will be experienced which will affect the uptime of the equipment and this is where the challenge begins.

This book covers three parts. The first will include the Vertical Startup of the equipment. This will start during the acquisition of the asset and what can be done to optimize its start-up. The second part will be the running costs, which include the Life Cycle of the equipment. Finally, the feasibility of extending the Life Cycle of the equipment until the time the equipment or asset will be disposed of. The concept of Life-Cycle covers the Total or Overall Cost incurred on the equipment during its entire lifespan from the beginning up to the time the equipment will be retired and disposed of. What we all want is to operate and maintain the asset with the least amount to own over its entire life span, which is the main objective of this book. This book covers 12 chapters, which are summarized as follows.

Chapter 1: The Concept of Equipment's Life Cycle Cost begins with what Life Cycle is all about and why industries need to adopt it. The three phases of the Equipment Life Cycle, which include the OEM Phase, the Vertical Start-Up, and the time it will be operated until it will be decommissioned, are explained in this Chapter. Also discussed in this chapter are the various reasons why equipment can no longer be used. A brief explanation regarding the different stages of the Product Life Cycle is likewise included.

Chapter 2: Initial Flow Control Activities (IFCA) also called EEM or Early Equipment Management is one of the pillars of TPM that involves optimizing the start-up activities of new equipment and assets. IFCA covers the vertical start-up of the equipment, which starts from the acquisition up to the completion of the commissioning time. This chapter also covers the six conditions for implementing IFCA and what we want to achieve. An actual case study of IFCA is also covered in this chapter. Finally, this chapter explains why IFCA is important for industries to implement for newly purchased equipment.

Chapter 3: Understanding the Equipment's Running Costs begins when we start operating the equipment. This chapter starts with the concept that purchasing cheap may actually be expensive and expensive may turn out to be cheap in the long run. The message of this chapter is that purchasing based on the initial or acquisition costs is only one side of the story since there are other costs involved when the equipment starts to operate until the time it will be retired or decommissioned. This cost refers to the running cost of the equipment. This chapter also provides details on what the running cost will include. One unique maintenance indicator will be the Percentage of Maintenance Cost to RAV or Replacement Asset Value. This indicator provides insights into whether a piece of equipment needs to continue running, retire, or the possibility of modifying it. This indicator refers to the cost of maintaining the asset, which is measured against its value.

Chapter 4: Sustaining Equipment Against Failure explains the different stresses involved once the equipment is loaded. These different stresses will affect the different parts of equipment, which can subject them to failure. For mechanical parts, once the stress finally exceeds the strength of the material then that item will be expected to fail, or fracture. Since these failures are inevitable, the need to create sustenance is essential and this is where the different maintenance tasks are performed to sustain and preserve our equipment and assets from these stresses, which can eventually lead to failures and breakdowns. What is important is that all maintenance tasks performed on the equipment should address a particular failure or breakdown.

Chapter 5: Mitigating Human Errors in Maintenance provides the readers an understanding of human errors. This chapter explains if it is possible to totally eliminate human errors, or not. Another topic of interest to the reader is how can we reduce and manage human errors committed in industries. Also provided in this chapter are cases where human errors committed in maintenance can be devastating and can cost human lives. Maintenance Induced and Non-Maintenance Induced Errors are likewise explained in this chapter. Human errors are part of being human, this chapter provides the readers with what we can do to reduce, mitigate, or how we can manage them in maintenance.

Chapter 6: Reducing Equipment's Running Costs briefly explains that there is no distinct and universal standard on what maintenance cost includes as it varies from one industry to another. Maintenance cost is a universal and common measurement and indicator for all types of industries, unlike other KPIs. Once we understand what to include in the running cost, there are several correct practices that we can adopt to reduce them. These strategies include adopting an MRO Spare Parts Management, implementing the correct Preventive Maintenance, application of Predictive Maintenance, and implementing a Plant-Wide Lubrication strategy in the plant.

Chapter 7: Built for Reliability explains why both Reliability and Life Cycle are intrinsically important and connected which means that they are inseparable. While others may define reliability in terms of failure and MTBF, a deeper meaning is provided in this Chapter by a dear friend of mine, R. Keith Mobley. One of the things that must be understood is that reliability is not just for the reliability people but it is each and everyone's responsibility in the plant. Top Management and C-Level people should be the drivers of reliability. This chapter also explains what can be achieved if the three forces of industries which include Safety, Quality, and Reliability activities can be consolidated and integrated.

Chapter 8 Considering Redesign to Improve Reliability begins by comparing the east and the west approach to improvements as well as what redesign and modification are all about when it is used, and what it includes. Another interesting topic discussed in this chapter is the contradicting beliefs of both RCM and TPM on improvements. This chapter also discusses the different losses that can be experienced on the equipment and how to conduct Focused Improvement activities to address the different losses that can occur on the equipment and assets. Each of these equipment losses will require a different strategy on how to deal with them. Lastly, this chapter also explains other losses that can be experienced beyond the equipment losses besides breakdowns.

Chapter 9: Addressing Equipment's Design Weaknesses provides a detailed Step-by-Step approach on how to address equipment design weaknesses using Phase 2 of Planned Maintenance. The standard 4 Phases of Planned Maintenance include Phase 0 which is the Preparatory Phase, Phase 1 is all about restoration, Phase 2 addresses equipment design weaknesses, Phase 3, developing sustenance, and Phase 4 is all about predicting equipment lifetime. In Phase 2, one of the requirements needed is for the team to be trained on at least three or more means of analyzing equipment-related problems using some conventional problem-solving tools or how to conduct an actual Root Cause Failure Analysis investigation. An actual case study of Planned Maintenance Phase 2 is provided which includes the results obtained.

Chapter 10: The Next Challenge - Extending Equipment Life Cycle explains why Life Cycle Costing is important and should be part of the Overall Reliability Strategy. This chapter explains the different means of extending equipment life cycle such as reducing oil contamination in lubricants. Also covered are the detailed steps in monitoring equipment life cycle. Finally, the process of decommissioning is explained which is the last phase of the equipment's life cycle.

Chapter 11: ISO 55000 on Asset Management defines the need for industries to have an asset management structure in their organization so that they can manage the risks that can affect the way they do business. This will help industries find a structured approach to find the best possible solution to deal with the risks involved in their day-to-day operations. A good asset management system entails the identification, assessment, management, and mitigation of the risks involved together with their consequences. ISO 55002 provides industries a universal framework and guidelines for managing the use of their physical assets. This chapter also includes a brief explanation of the newly launched ISO 55010, which is about an alignment between the financial and non-financial functions of an organization.

Chapter 12: The Conclusion: As I have mentioned several times that training and education will serve as the very basic foundation of an effective maintenance strategy and structure. Every industry is plagued with different problems and risks that they need to address to remain in business. While many industries will opt to go for an all-out Management Cost Cutting Schemes, which can have harmful effects and repercussions as, indicated in the case study on Bhopal India and Boeing 737 Max. Finally, the readers need to understand that sustaining the equipment is not just the role of maintenance in industries but also for them to accept the challenge that equipment and assets can still be improved.

This page is intentionally left blank.

Chapter 1

The Concept of Equipment's Life Cycle Cost

> *Looking at the initial cost of the equipment is just the tip of the iceberg. Remember that what is beneath the iceberg is bigger than what is visible. This means that when we speak about Life Cycle, the initial cost is just one part of the cost, the bigger cost lies in its running cost which is the time the equipment is commissioned until the time it will be removed and decommissioned. The best way to reduce cost is to understand the equipment's Life Cycle Costs.*

1.1: The Concept of Equipment Life Cycle Cost

Life Cycle Cost refers to the total cost of the equipment throughout its entire lifespan. The US Management and Budget defines LCC as the sum of the direct, indirect, recurring, non-recurring, and other related costs of a large-scale system during its period of effectiveness. [1]Japan Institute of Plant Maintenance (JIPM) defines LCC as the systematic decision-making technique that incorporates life-cycle cost as a parameter at the design stage, performing trade-offs to ensure an economic life cycle cost for the user's system design. It can be said that In terms of production equipment, LCC can be described more simply as the design and fabrication cost, which is the initial, or acquisition cost plus the operation and maintenance cost which is the running costs of the equipment. The initial cost will always be easy to see, but the running cost is not. Failure to consider the running cost can lead to many problems. At least 80% of an equipment LCC can be conceptualized during the design stage. The goal of LCC is to operate and maintain the equipment with the least possible cost to own over its entire lifespan.

LCC = Initial Cost + Running Costs

The study of Life Cycle Costing provides us three main objectives. The first is to address the very basic equipment condition. The second is how to sustain and preserve our equipment and assets. Finally, challenging the maintenance people to improve the existing

---

[1] Suzuki, Tokutaro, **TPM in Process Industries,** Productivity Press, Portland, Originally published by Japan Institute of Plant Maintenance (JIPM), page 200

equipment by identifying and addressing design flaws so that we can maximize its entire life span, or even have the possibility of extending it. But before anything else, we need to understand the needs and requirements for purchasing new equipment in the future. There are two kinds of costs associated with the equipment. First, we have the initial costs, which is the tag price of the equipment. This can also be referred to as the procurement cost. Just like when we go to a shopping mall to buy a T-Shirt or whatever we want to purchase, these items always come up with a tag price, which is embedded in the bar code. This refers to the initial cost of the products that we are purchasing. The initial cost already includes the taxes to be paid which includes the VAT, which is quite common in most countries. This is what we pay for when we go to the cashier. When we purchase something, the receipts already covered these things. The second type of cost is called the running cost. This refers to the overall costs incurred when using the product we purchased. Knowing the initial costs will be the easy part since it's always given in the receipt, but trying to understand the running costs would be quite more challenging. In the example above, the initial cost will contain the cost of the clothes that we purchased, while the running cost will include the cost of the laundry, water bills, electricity, detergent, labor for the laundry, and other costs that can be incurred while we still keep using the shirt.

The sum of the initial and the running costs is known as the Total Life Cycle Cost. The message on Life Cycle Cost is that looking only at the initial cost of the equipment is just like the tip of the iceberg because underneath the iceberg lies other costs involved from the time the equipment was commissioned up to the time the equipment will be finally retired or decommissioned which in the long run is much bigger than the surface of the iceberg. As the equipment is used continuously, the running cost will be surpassed by its initial cost. The objective of LCC analysis is to select the lowest possible cost in operating and maintaining the equipment as well as to determine the most feasible approach from a series of alternatives, so the least long-term or cost of ownership can be achieved throughout the entire life cycle of the equipment and assets. LCC analysis can help engineers justify the equipment and process selection based on the total costs rather than the initial purchase price of the equipment. The correct way to reduce costs is to understand the equipment's Life Cycle Costs, which will consider both the initial and the running costs of the equipment. Looking only at the initial cost is just one side of the story and may not be a good decision when purchasing new equipment or even spares as we need to understand how it will also perform in the long run. This is mostly the problem with Purchasing people in industries as they seemed to go always to the lowest bidder.

Sometimes higher performance equipment costs less than a commodity type, even though the initial price is higher. To determine whether this will be the case, we should look at the entire Life Cycle Cost of the equipment, rather than its purchase price or initial cost. Life Cycle Costing is a way of analyzing equipment purchase choices. Therefore, if the decision was based on several factors rather than its initial costs, then we need to make our selection based on the least amount to own the equipment over its entire life span.

Equipment must not only be inexpensive in terms of its initial cost which is also termed the procurement cost or cost of purchase, but it must also be inexpensive in terms of its running cost, which is the cost of operating and maintaining the asset. This is what is important because this is where we can realize the true value of the asset. The goal of Life Cycle Costs is to develop equipment with the lowest possible life cycle costs that can benefit industries in the long-term in terms of their profitability. Remember that humans design equipment and humans are prone to mistakes and errors. Therefore, there is nothing like a piece of perfect equipment by design. There will always be design flaws, and weaknesses. Inevitably, failures will occur, and certain parts of the equipment will have inherent design flaws and weaknesses, which can lead to a short lifespan. It is just a matter of time before it will happen. When the equipment is placed into service, operators and maintenance will experience common problems with their assets. There would be parts that will have a short inherent lifespan or those that will fail prematurely and spares are often withdrawn from the storeroom. There are also cases of the same parts that will repeatedly fail during operations.

1.2: No Equipment is Perfect by Design

Eventually, equipment varies depending on the industry type. Industries can either be producing a product or providing a service to the public in general. There are equipment and machines that experienced repeating failures and breakdowns, while there are pieces of equipment that seldom fail. Although TPM (Total Productive Maintenance) claims to target a zero breakdown in the equipment. Technically, what TPM is referring to in this case will be to zero out all possible unplanned breakdowns on the equipment. The truth is, all equipment will fail. The thing is, we do not declare a breakdown or failure when maintenance is ahead of the breakdown. We only declare a breakdown when it happens first and then we react or repair the machine. If we replace a part or item during a Preventive Maintenance shutdown because of a wear-out mode, then this will not be declared as a breakdown, but technically the part that was replaced eventually is on the verge of failing as far as the part is concerned. In this case, we are not eliminating the breakdown totally, but we are just prolonging the equipment's MTBF or Mean Time Between Failure. If zero breakdown is really possible, then we cannot proceed with the higher Phases of TPM Planned Maintenance, which is all about Lengthening Equipment's Lifetime by Addressing Design Weaknesses as we need a failure or breakdown to occur so we can identify those parts with inherent design weaknesses as these parts are candidates for TPM Planned Maintenance Phase 2 improvement and modification. I can vividly recall our previous JIPM Consultant telling us that we need to have a breakdown before proceeding to Planned Maintenance Phase 2 so that you can identify these parts that frequently fail in the equipment. What TPM is simply claiming is having a zero unplanned breakdown is just for a temporary period. If zero breakdowns are really possible, then the fact lies that there will no longer be work for the entire maintenance population. It's as simple as that. When a piece of equipment or machine fails or breaks down, what it simply meant is that something went wrong inside the equipment, perhaps a part or a spare failed, or a software malfunction. These parts whether electronic, electrical, or mechanical parts are all subject to stress inside the equipment and it's just a matter of time before one of these parts will actually fail. The only time breakdowns will not happen is when the equipment will not be loaded.

By definition, the word failure means the inability of equipment to perform its required function. The failure of a component is viewed as terminating its life. Breakdown is a lost time due to equipment failure that may or may not cause downtime. This also refers to machine-related downtime, unplanned downtime, tooling failure, or any unplanned replacement of spares, and components. The Japanese term for breakdown is kosho. Equipment failure refers to an event in which the equipment cannot accomplish its intended purpose. It may also mean that the equipment stopped working, is not performing as desired, or is not meeting its target expectations. Breakdown and failure are almost similar, as failure will lead to a breakdown. A failure can occur, such as an increase in vibration, which is a potential failure and can eventually end up in a failed state, and in this case, the equipment finally breakdown. There are two types of breakdown. A function loss breakdown or failure of the primary function is a situation in which the failure occurs, the equipment will totally stop and will definitely halt operations. RCM termed this as a failure of the primary function. A function reduction breakdown is also termed as a loss or failure of a secondary function. This is a situation where a failure occurs, but still, the equipment is running. This means that the equipment is still capable of operating and producing the needed production. An example will be a machine that has an oil leak or sensors that is bypassed. All equipment has both its primary and secondary functions that we must be aware of. There are cases where the consequences of failure of a secondary function can be far much worse or even more dangerous than the failure of the primary function.

As time passed easily, the products, goods, and services industries provide become more diversified to accommodate the needs of the consumers, and equipment is designed to withstand the harsh environment to which it will be exposed. Equipment and machines are redesigned and become more rigid and automated as technology progressed especially in this era of digitalization, but as we use the equipment in our operations, different problems emerged from the lack of basic care, breakdowns, failures, minor stoppages, long conversion time, product defects, changing blades or cutting tools, and not to mention, the human error committed during the time the equipment was operated and maintained. The question is, will the equipment still function satisfactorily through the test of time?

Even though OEMs and vendors improved the performance of the equipment they manufactured from its initial inception during the design stage to how they operate currently in industries. The thing is there will always be some inherent design flaws and weaknesses in the equipment. Equipment, machines, and assets are not perfect by design. In fact, there is no such thing as a perfect design. Perhaps if Jesus Christ, Buddha, or Allah designed the equipment, then there will be no argument that it will be perfect.

In today's industries, there will always be the temptation to purchase equipment and spare parts based on the lowest or cheapest possible cost, especially if management is under cost-cutting schemes. One common problem is that Purchasing people will always go for the lowest bidder at all costs. This might not be such a good idea since purchasing based on the initial costs will just tell us one side of the story. The true cost can be seen and felt based on the equipment's entire Life Cycle Cost. In my experience, this is the problem with most procurement and purchasing departments I have experienced during my employment days since they make the decision to purchase parts and new assets based on the lowest possible cost thinking that they are saving a few bucks for the industry. The savings generated by these department claims are insignificant, since both operations and maintenance encounter problems as a result which generate a much bigger cost which can generate a loss of profit and revenue for the entire industry. The problem, in this case, is that these people always look at the initial cost, and not at the entire overall cost of the problems the part may provide the user in the end. This is the problem with buying cheaper parts and assets. Equipment that fails due to this will likely yield a much larger amount of cost after a given time compared to slightly higher-priced equipment. The true cost of the equipment can be felt from the time the equipment was commissioned in the plant until the time it is finally retired or disposed of. What we are saying is that cheap can be expensive in the long run, and expensive can be cheaper in the end. What is important for the decision-makers is to create a balance between the two.

1.3: Why Life Cycle Costing is Important for Industries

Although this book covers the subject of equipment Life Cycle, this can also be used in making decisions regarding future expenses by the industry. The concept of Life Cycle Costing simply tells us two things. There will be expenses before using the equipment which is the initial or acquisition cost, and there will also be expenses that will be incurred once the equipment will be used until the day it will be finally decommissioned which are called the running costs which in the majority of cases will exceed the initial or acquisition cost of the equipment or asset. Just like human beings, when people die. whether we like it or not, there will still be cost involved. Life Cycle Costing does not only limit us to purchasing new equipment and assets, but this can also be used in making decisions to whom to award the bidding whether we are purchasing, new equipment, spare parts, Xerox machine for offices, purchasing a car, or whatever the industry will be purchasing in the future. The message of Life Cycle Cost is before making any decision on purchasing anything that the industry needs, the decision to purchase should not only be decided in the short term but also on the long-term benefit. This means that purchasing anything will also involve future costs besides the cost of the item that was purchased.

If an industry purchased a Xerox machine, besides its initial cost or tag price, there will be additional costs in the future. These costs may include the ink cartridge, toner, electricity bill, paper, repairs, and depreciation until the Xerox machine will be finally disposed of. What is important for the industry is to estimate the cost of expenses the Xerox machine will incur in the future. Providing an estimate on what will be the future cost may not be actually the same as the actual cost, but the benefit of estimating the future cost will allow the user to create a balance between making a decision on which item to purchase and which vendor to award this particular item. While in the majority of industries. Purchasing people can be tempted to award the bid to vendors that can provide the lowest cost, a better approach is awarding the bid to vendors that can provide the lowest possible cost in the long term. This is important since we also need to consider the long-term benefits of our decision. Considering only the tag price or initial cost of an item being purchased may only benefit us for a short-term period, which can end up actually spending more in the long term.

PERSONAL CAR TOTAL RUNNING COSTS

No.	New Car Initial Cost (Tag Price)		CAR A		CAR B		CAR A	
1	Car's Initial Costs		PHP 648,000.00		PHP 578,000.00		PHP 725,478.00	
2	Car's Fuel Mileage		19.6 km / liter		12.5 km / liter		14.5 km / liter	
	CARS RUNNING COSTS IN PHP	Expenses	Monthly	Year	Monthly	Year	Monthly	Year
	Gasoline Fuel (PHP 78.00 per liter) at 60 km/day	500.00 / day	PHP 7,163.25	PHP 85,959.00	PHP 11,232.00	PHP 134,784.00	PHP 9,682.78	PHP 116,193.36
	Toll Gate (Back and Forth)	350.00 / day	PHP 10,500.00	PHP 126,000.00	PHP 10,500.00	PHP 126,000.00	PHP 10,500.00	PHP 126,000.00
	Car Wash (2 times a month)	300.00 / month	PHP 300.00	PHP 3,600.00	PHP 300.00	PHP 3,600.00	PHP 300.00	PHP 3,600.00
	Periodic Maintenance PHP 4000.00 / Quarter	4,000 / quarter	PHP 1,333.33	PHP 15,999.96	PHP 1,333.33	PHP 15,999.96	PHP 1,333.33	PHP 15,999.96
	Insurance PHP 10,000 per year)	10,000 / year	PHP 833.33	PHP 9,999.96	PHP 833.33	PHP 9,999.96	PHP 833.33	PHP 9,999.96
	Car Registration	1500 / year	PHP 125.00	PHP 1,500.00	PHP 125.00	PHP 1,500.00	PHP 125.00	PHP 1,500.00
	Car Emmission Test	500 / year	PHP 41.67	PHP 500.00	PHP 41.67	PHP 500.00	PHP 41.67	PHP 500.00
	Driver's License (5 year Validity)	675 / 5 years	PHP 11.25	PHP 135.00	PHP 11.25	PHP 135.00	PHP 11.25	PHP 135.00
	Depreciation @ 15% per year	15% / year	PHP 8,100.00	PHP 97,200.00	PHP 8,100.00	PHP 97,200.00	PHP 8,100.00	PHP 97,200.00
	Others for Car Features	1,500 / month	PHP 1,500.00	PHP 18,000.00	PHP 1,500.00	PHP 18,000.00	PHP 1,500.00	PHP 18,000.00
	TOTAL RUNNING COSTS MONTHLY		**PHP 29,907.83**	**PHP 358,893.92**	**PHP 33,976.58**	**PHP 407,718.92**	**PHP 32,427.36**	**PHP 389,128.28**

Figure 1.1: Using Life Cycle Costing to Select Which Car to Purchase

In figure 1.1, let us assume that we want to purchase a new car for our family and we went to three Car Dealerships of known brands, we also estimated the running cost that will be incurred once we start using the car. Let us assume that the car will be paid in cash as indicated in the initial cost, which is the total value of the car when paid in full without any monthly amortization. If the car will be amortized, then the cost will be much higher. In this case Car A will be having the lowest monthly cost at PHP 29,907.83 compared to Car B and C. Although there might be other factors that we will be considering such as the design of the car, ergonomics, seating capacity, and others. The table indicates that in terms of Life Cycle Costing, Car A will provide the least amount to own over its entire life span, which we can consider in the decision-making process on which car to purchase for your family.

Similarly in industries, this should be the practice when purchasing anything, not only for new equipment or spare parts but also for other assets the industry plans to purchase in the future. The bottom line is that considering the Life Cycle Cost will benefit the industry in the long term instead of making decisions on purchasing at the lowest initial cost since we are only looking at one side of the story.

1.4: The Initial and Running Cost

There are cases where when industries purchased new equipment, the initial cost will be higher, and this is where the negotiating skills of the purchasing will be needed to reduce the initial cost of the equipment. If the projected running cost is expected to be higher than the initial cost, then this will be a collaborative effort between the engineering, maintenance, reliability, and other cross-functional technical people from the entire organization to conduct improvement during its operations phase to redesign their equipment to lower the equipment's running costs during its entire life span. This is done by identifying design flaws and weaknesses in the equipment. Once these design flaws are identified modifications will be done to further lengthen its lifespan.

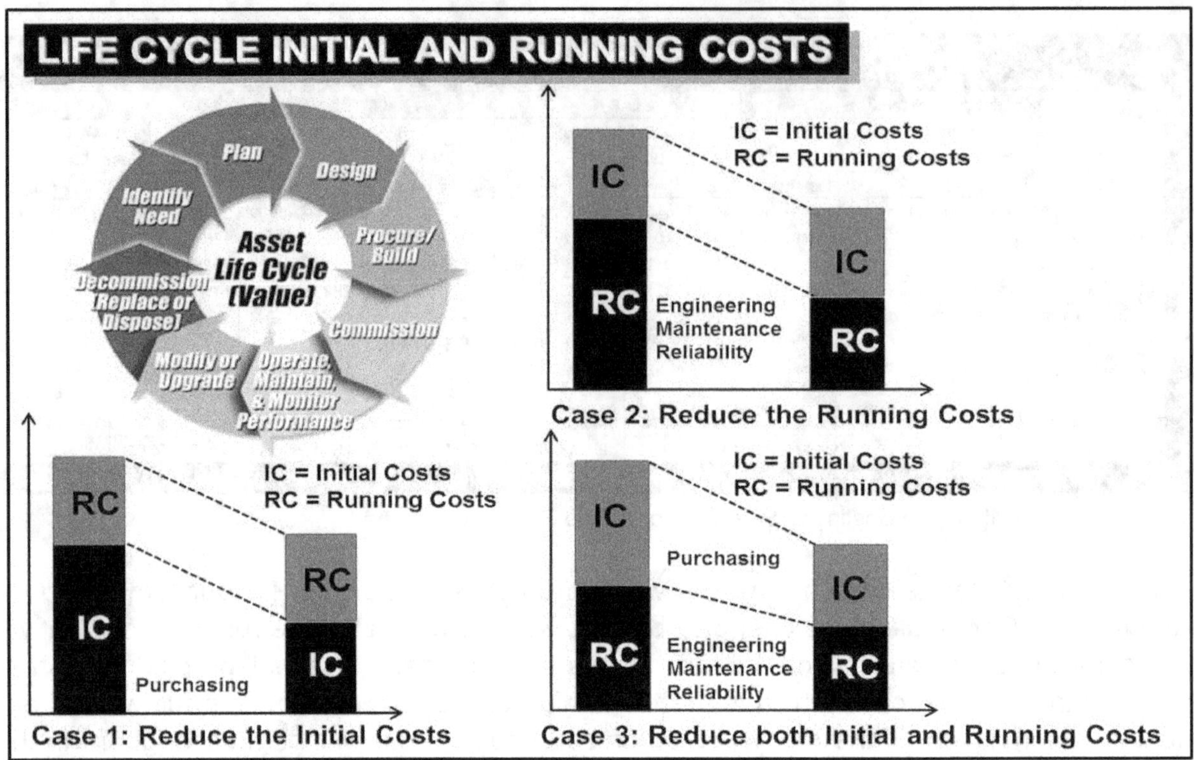

Figure 1.2: Strategy for Lowering Life Cycle Costing

[2]**Case 1: Reducing the Initial Costs:** This approach concentrates on reducing the initial costs of the equipment. If the initial cost of the equipment is much too expensive, the running costs will be less prioritized. Priority will be on trying to reduce the initial costs. If it is typically obvious that the initial cost of the equipment is high or expensive, it is for some reason that the chances are that the running cost of the equipment will be lower. We need to ask the OEM what makes this equipment different from the others or explore other plants that also used this type of equipment. This happens all the time with me when negotiating the cost of my training seminars. Purchasing people are good at negotiating the initial cost of the equipment. Remember that the initial cost will be a one-time event. In contrast, the running cost will continue for its entire duration and lifespan.

Case 2: Reducing the Running Costs: There are several cases where it is best to emphasize reducing the running costs over the initial costs. In complicated automated equipment, higher quality generally means greater risks of defects, breakdowns, erratic start and stop, minor stoppages, and other equipment problems. This also leads to additional costs on maintenance resources and spare parts. As a result, the running costs, such as downtime loss and maintenance costs, exceed their initial projections. In this case, more emphasis should be given to how equipment should be cared for and sustained. Maintenance should have an initial projection or estimate in this case regarding the running cost of the equipment since this is expected to be much higher. A piece of equipment may have a low initial cost, which can end up with a high running cost.

---

[2] Fumio Gotoh, ***Equipment Planning for TPM, Maintenance Prevention Design,*** (Productivity Press, Portland, Oregon, USA, Dover Publications Inc., Mineola, New York, 1991), Page 16

Case 3: Reducing Both the Initial and Running Costs: This approach gives equal emphasis to both the initial and running costs. Obviously, this amounts to a holistic LCC reduction approach. This case can apply to large equipment with also a large projection on what will be involved in the running costs. Although we cannot project the exact amount of the running costs that can be accumulated on the equipment during its entire lifespan, we can provide an estimate based on records from our previous equipment.

Considerably, operators, maintenance, and engineers' job are to consolidate their efforts in identifying these flaws in the equipment and improve their design weaknesses. However, improvements should not always be taken for granted. There are cases where modifying or redesigning the equipment is not always the right thing to do. Let me explain a bit more about this. If the equipment or product would be phased out in the next few months and the equipment would be disposed of or decommissioned, it is not recommended to proceed further with improving or modifying the equipment. Likewise, if a major breakdown or accident occurs in the equipment where the cost of rebuilding and restoring the equipment would be similar to the cost of purchasing a piece of new equipment, then the default is to no longer redesign or modify the equipment since economic wise, it will not be further feasible to do so. Again, the decision is for the equipment to be decommissioned and put out of service.

When purchasing new equipment, there is often a temptation to look at the initial cost or the tag price of the equipment. In hindsight, we must also look at the cost of operating, maintaining, and decommissioning the equipment. The bigger cost lies in how the equipment is maintained. The study of Life Cycle cost involves two parts. First, what to do with the equipment that industries currently have, and secondly, the equipment that industries will decide to purchase in the future. In these cases, the study of Life Cycle Costs must not be ignored, as it will contribute a crucial role not only in operations and maintenance but also for the whole organization itself.

1.5: The 3 Phases of Equipment's Life Cycle

Phase 1 **OEM Phase**	Phase 2 **Vertical Start-Up Time**	Phase 3 **Operations to Decommissioning**
• Design Phase • Fabrication • MP Design for Repeat Order Machines	• Price Negotiation • Pre-Buy Off • Shipment • Transportation to Plant • Installation • Commissioning / Debugging • Final Buy Off	• Start of Operations • Normal Operations • Extended Life Cycle • Decommissioning

Figure 1.3: The Three Phases of Equipment Life Cycle

Although if we consider the entire life cycle of the equipment, this will start from the design phase until the equipment is finally decommissioned. There are two points to consider first if the industry will be purchasing a repeat order of the equipment that they already have, which means ordering the same model and type, and second, if the industry will be purchasing entirely a piece of new different equipment for the very first time. The entire equipment Life Cycle can be broken down into three phases, which can be shown in figure 1.3.

1.5.1: Phase 1: Design and Fabrication

This phase is mostly owned and controlled by the OEM or vendor as it includes the design and fabrication phase until the equipment is now ready to be sold in the market. For industries that will be ordering a repeat order of their existing equipment, the IFCA team usually discusses some improvements and modifications performed in their existing equipment to the OEM or vendor so that the previous problems experienced on the equipment will no longer pose a risk when they purchase additional repeat equipment. These modifications, improvements, and redesign are usually in the form of MP Design improvements, which will be explained, later in the next chapter.

Design Phase: Everything will start in the design phase. Today's equipment design should be ergonomically easy to use and easy to operate. If automation is included, it should be user-friendly. Not only should equipment be easy to operate, but it should be intrinsically safe to operate as well. Previous problems on other equipment models are evaluated and modified on the new design so that the old problem will not emerge once more. If the equipment or machine will be used for different products, then the conversion or change-over time should be minimum. The challenge with today's machines and equipment should be as follows;

• Design for higher Reliability
• Design for Greater Safety
• No Damage to the Environment
• Longer Equipment Lifetime (LCC)
• Better Cost Effectiveness
• Real-time data and Analytics
• Ergonomically Friendly
• Wireless Technology and the use of smart sensors

Fabrication: Once the equipment's design is completed, then the fabrication phase begins where the different parts of the equipment will be sourced out from different suppliers while others can be fabricated by the OEM themselves. Different parts and items will be sourced out by the manufacturer to different vendors until the equipment will finally be completed and ready to be sold out to the marketplace.

1.5.2: Phase 2: Vertical Start-Up

The Vertical Start-up time usually starts when the purchaser generates a Purchase Order for additional new equipment or a repeat order of new equipment, which will be the same model that they are currently using. The Vertical Start-up time will include price negotiation with the vendor, shipment and transportation of the equipment, pre and post-buy-off, installation, and commissioning which includes debugging the equipment for flaws, bugs, errors, and problems during the pre-operation phase.

Equipment Purchase: When equipment will be purchased by the industry, it is common for purchasing to select a minimum of around three or more vendors who will submit their proposal. Although the initial cost will be provided, the specification of the equipment will also be included and evaluated by the engineering function or technical group. There are several factors to consider on whom to award such as operator's training, equipment warranty, MRO spare parts, maintenance cost, and others. Usually, the OEM will define both the capital and insurance spares that are usually recommended which majority end up as non-moving items. Capital Spares are vital spares for critical equipment. The stock-out cost for such spares and unit costs is very high. The number of items that may be consumed during the entire lifespan of the equipment can be 1 or 2 units. A decision has to be made whether to stock the item or not. An insurance spare includes those parts that will be used to replace a failed part in a piece of equipment whose penalty cost for downtime is very high. By definition, this is the insurance against such failures for which the downtime costs are very high. These parts do not become obsolete until the equipment is finally retired from service. Most of the time these parts are big and classified as non-moving parts. Usually, the equipment's initial cost will include the cost of the equipment, the spares recommended by the OEM that needs to be stocked, the warranty cost, the cost of shipment or transportation, and the cost of training for operators, as well as the installation and commissioning costs. In other cases, the OEM may recommend that they will also be responsible for the equipment's maintenance during the warranty period, or even its entire life span.

Freight and Shipment Cost: This will include all costs from the shipping point to its final destination. This can include the cost of transporting the equipment to the plant, packaging costs, customs declarations, taxes, insurance, the density of the machinery, fuel surcharge, toll fees, and other relevant costs in transporting the equipment to the industry or plant.

Installation and Commissioning: Commissioning a piece of equipment means carrying out all the necessary tests needed to be required according to industry standards. It is the process of ensuring that all systems and components of a building are designed, installed, tested, debugged, operated, and maintained according to the operational requirements of the owner or user. This may also refer to as the start-up of the equipment for the first time to verify that it will be running as intended according to its design specifications. Installation is the process of placing, positioning, building, and connecting equipment in its proposed and design nominated location following design specifications. A commissioning checklist is usually provided to check that all systems are functional before endorsing the equipment to operations and maintenance. The commissioning checklists should also include the activities as well as the duration spent on each task.

1.5.3: Phase 3: Operation to Decommissioning

Once the vertical start-up has been completed by the OEM, and the equipment has passed its final buy-off. The equipment is now ready to be operated until the time the equipment will be decided to be retired and decommissioned in the plant. All modifications, improvements, and redesign performed on the equipment to further extend its lifespan will also be included in this final phase. All in all, these three phases will cover the entire life cycle of the equipment.

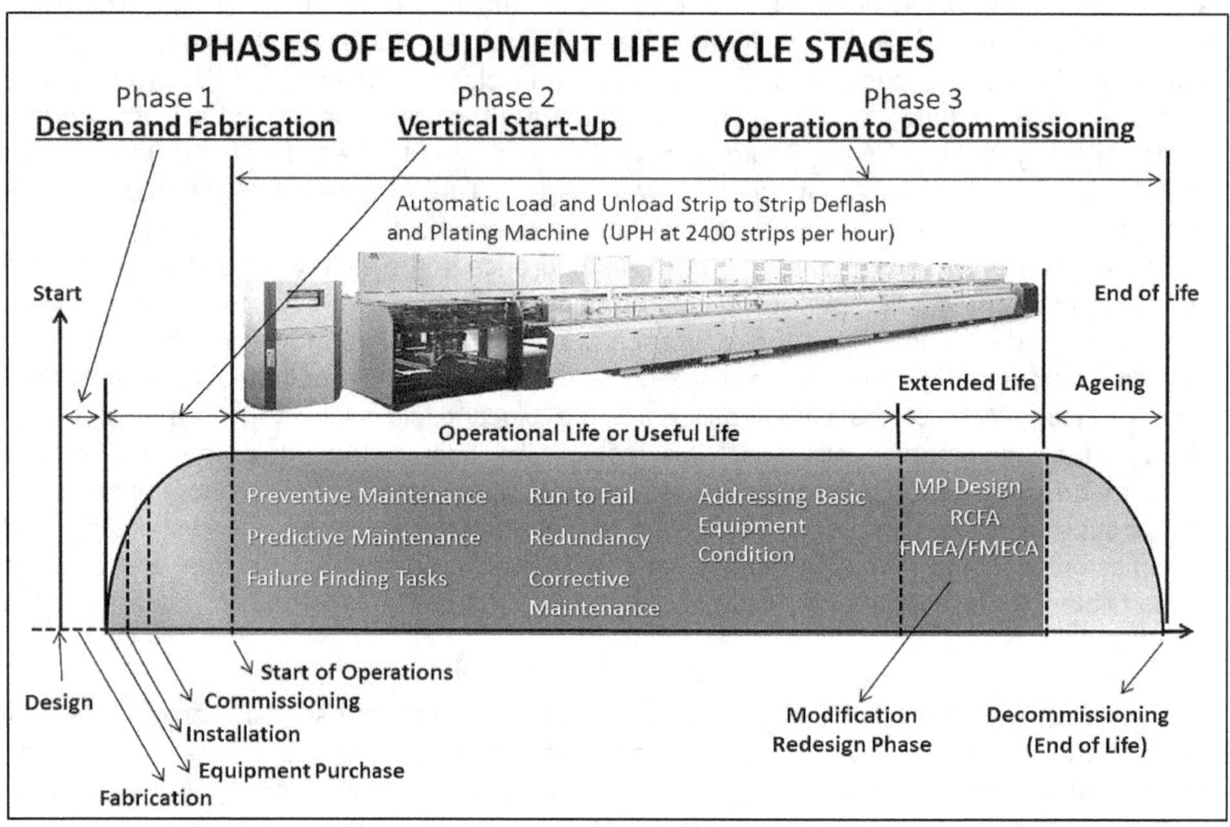

Figure 1.4: The Complete Equipment Life Cycle Process

Start of Operations: Once the installation and commissioning activities are still ongoing by the OEM, the vendor now provides the necessary training needed for the operator to operate the equipment. The production people are finally advised that the equipment is now operational and they can start operating. The equipment will now be operated based on the needs of the operations to generate some revenue. An equipment manual is usually provided for basic guidance where the maintenance derives the different maintenance tasks needed to sustain the equipment. However, during the initial months, the equipment is still protected by the warranty and in some cases, a contract will be added that the OEM provides the necessary maintenance needed on the equipment. At this point, maintenance will carry out the necessary maintenance tasks to preserve and sustain their equipment and assets.

Decommissioning: This will be the final phase, which refers to the removal or retirement of a piece of equipment or asset from active use. At this point, the equipment will now stop operating and the decision-makers finally decide that it is time to retire the asset for good. The equipment will be dismantled and removed from the production floor and will no longer be used. This can be said as the last phase in the equipment's lifetime.

There will also be cases where part of the decommissioning process will be cannibalized where some common parts that can be used by other equipment will be removed and kept by maintenance. These parts, items, or components removed will be kept stocked by the maintenance themselves, which can be used in the repair, replacement, and restoration of other equipment in the plant.

1.6: Why Equipment Can No Longer be Used

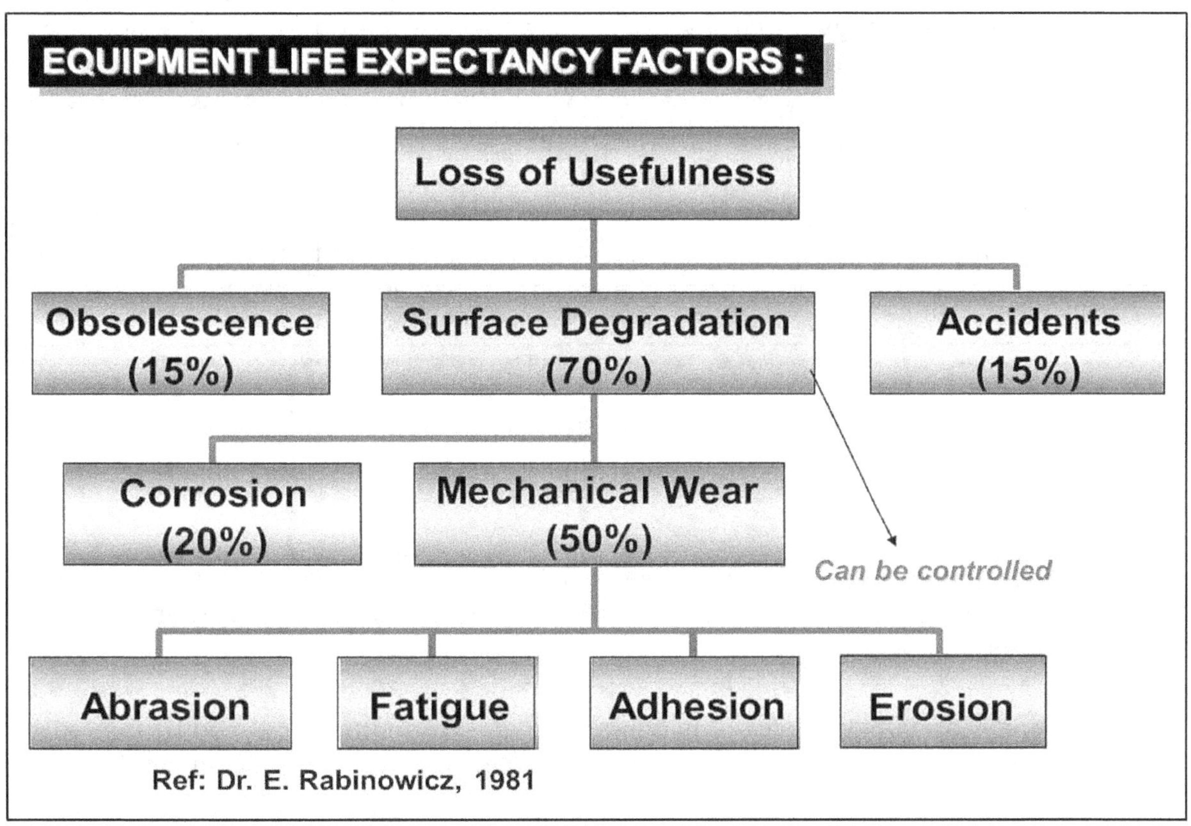

Figure 1.5: Why Equipment Can No Longer be Used

According to Dr. Rabinowicz, there are three main reasons why equipment can no longer be used. Dr. Ernest Rabinowicz is a Professor Emeritus of Mechanical Engineering. He worked at MIT for 43 years before retiring in 1993. He was also known for his work on the subject of tribology, the study of three subjects, lubrication, friction, and wear of interacting surfaces such as bearings. In 1998, he received the Tribology Gold Medal Award from the Institution of Mechanical Engineers in England. According to Dr. Ernest Rabinowicz, the three main reasons why industries can no longer use their equipment include obsolescence, surface degradation, and accidents. First, is obsolescence, where either the equipment or the product being manufactured on the equipment is considered as obsolete or will no longer be used. This accounts for 15% of why equipment can no longer be used. The second is accidents where a major breakdown has occurred in which the cost of rebuilding or rehabilitating the equipment is almost equal to buying new equipment, so the management's decision is to no longer use the equipment. Accidents also account for 15% of why industries can no longer use the equipment. However, the bigger percentage of why industries can no longer use their equipment is due to surface degradation, which accounts for 70% of why industries can no longer use their equipment.

However, the good news is although surface degradation contributes to a higher percentage, it can still be managed and prolonged. Surface degradation can be composed of corrosion, which constitutes around 20%, and mechanical wear, which accounts for 50 % of surface degradation. Wear in this case, will include abrasion, adhesion, fatigue, and erosion. Out of these four types of wear, the most common type of wear will be abrasive wear. In fact, erosion, fatigue, and adhesion wear can also be possible to cause abrasion inside the equipment. There are two kinds of abrasion, which includes two-body abrasion and three-body abrasion. Two-body abrasive wear occurs when the asperities of two mechanical components come in contact with each other. One surface, which is usually harder than the second, cuts material away from the second. However, this mechanism very often changes to three-body abrasion as the wear debris then acts as an abrasive material between the two surfaces. A three-body abrasion means that there is a third party, usually a hard contaminant that is trapped between the clearances of two mechanical parts.

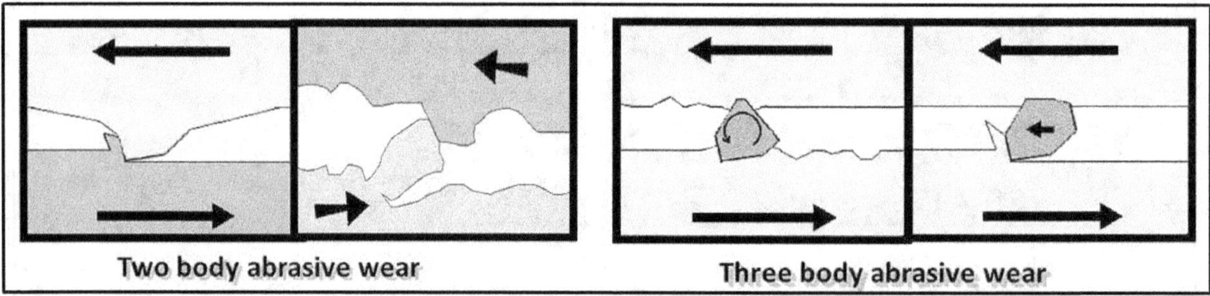

Figure 1.6: Two and Three-Body Abrasion

By definition, wear can be defined as damage to any solid surface caused by the removal or displacement of material using a mechanical action of a contacting solid, liquid, or gas. This process can cause significant surface damage, and the damage is usually thought of as a gradual form of deterioration. Wear can also be defined as the undesired removal of these metal fragments from a contacting surface through some means of mechanical action. This means that when the capability of equipment falls below its design performance standard, it is because the stress causes the asset to deteriorate and eventually wear out. Let us assume that a new piece of equipment was purchased brand new. The parts inside the equipment are new. Once the equipment completes its installation and commissioning process, it will now be used for operations. Inside the equipment are mechanical parts that have some sort of movement. Let us say that a movement of the piston will be on an up and down motion. If we put this in slow motion, once we turn on the equipment, the piston will move away from the centerline to an upward motion and then back again to its centerline and then to downward motion, then back again to the centerline once more. This is equivalent to one cycle. If the equipment is continuously running, this means that the piston is moving up and down on a continuous cycle. The piston is moving continuously reaching 100 cycles, 1,000 cycles, 10,000 cycles, 1,000,000, and so on. After reaching an (n) number of cycles, there are very tiny bits of metal fragments that will break off from the piston caused by this continuous cycle. These tiny fragments of metal that have fractured from the piston can be trapped between the piston and the walls of the piston block, thereby abrading the metal surface and creating another batch of metal fragments. These tiny metal fragments will be mixed with the oil, which then acts as a catalyst and speeds up the oxidation process of the lubricant. Since the oil is circulating, these metal fragments join the flow of oil. Some of these metal fragments will be trapped by the oil filter, but in the majority of cases, most oil filters are rated nominal and have a pore size of 30 to 40 microns. Hence, smaller contaminants may not be captured by the oil filter itself. These contaminants can be trapped in some tight clearances on other mechanical parts, which repeats the process of generating more metal fragments. This is how the wear process usually happens inside our equipment and machines.

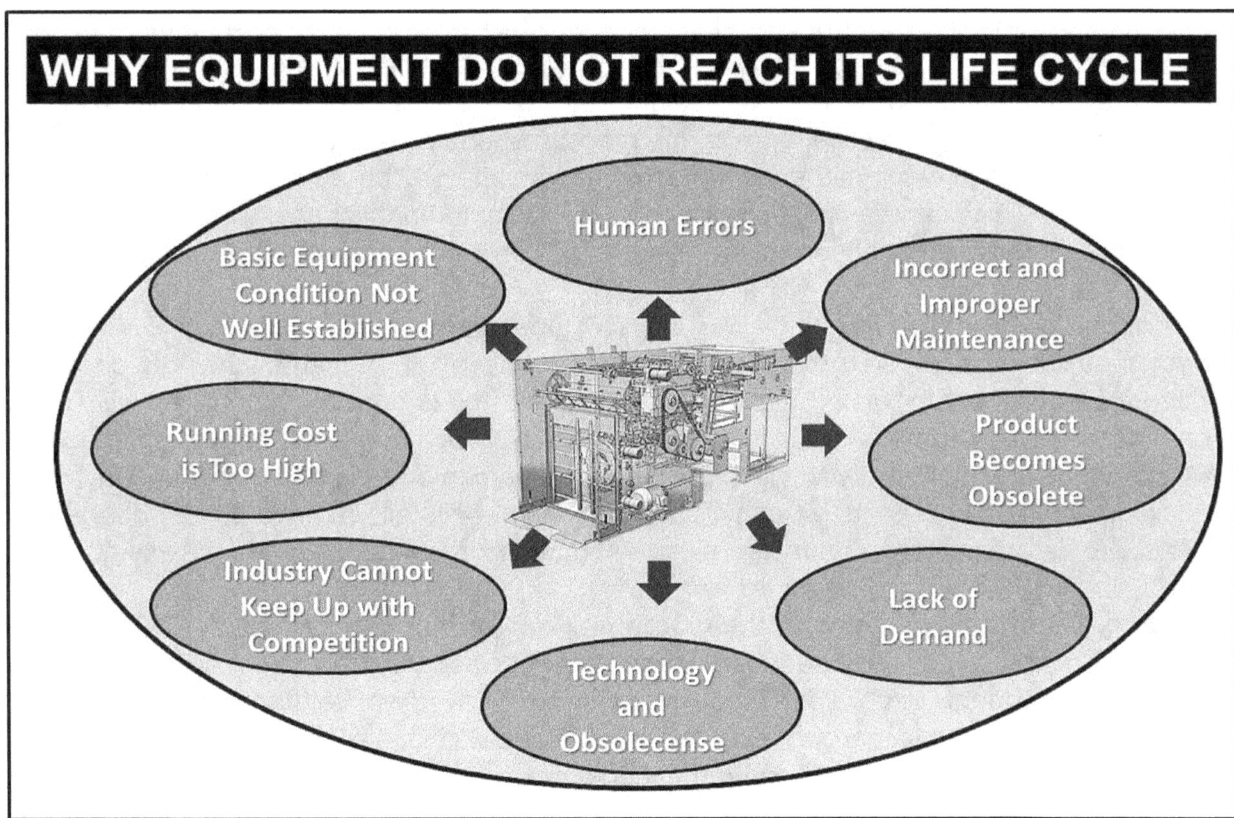

Figure 1.7: Why Equipment Do Not Reach Its Life Cycle

Industries are diversified but one thing common about all industries is that they have their equipment, machines, and assets to maintain. These machines and assets can be system-based and linked to each other, process-based, or the machines can be independent of one another just like in manufacturing industries. Most manufacturing industries will have their own facilities or utility equipment that provides them the power and energy needed to run their manufacturing machines and pieces of equipment. Industrial equipment and machines usually contain both ancillary and auxiliary assets. Both ancillary and auxiliary assets provide support to the main equipment or machine. The main difference between ancillary and auxiliary is that ancillary refers to providing something additional to a main part or function whereas auxiliary refers to offering or providing help. The word ancillary implies that there is a primary function or activity, but ancillary refers to providing necessary support for this primary function. Auxiliary equipment is defined as peripheral equipment that may be an integral part of the extrusion process to improve or optimize the extrusion process efficiency and ease of operation. Ancillary equipment includes support machines, which are used with the main equipment to create a complete function. These can be any means of devices such as piping, fittings, flanges, valves, pumps, meters, gauges, and others.

Although Rabinowich's concept on why equipment can no longer be used is still valid and holds true, there are also other reasons why equipment can no longer be used which can include the following:

Technology and Automation: Industries may decide to retire or decommission their equipment due to technology or automation especially since we are now in the age of digitalization, where equipment is designed to be fully automated to cope with the digitalization age. In this case, equipment, machines, and assets will contain PLCs, PWBs, sensors, and many electronic parts. Machine downtime will be automatically recorded by the machine, which can be directly linked with the system. In this case, what is important is to provide the code for every downtime experienced on the equipment.

Running Costs is Too High: If the costs of operating and maintaining the equipment become too high compared to other equipment in the plant, management may decide to retire the equipment for good. A piece of equipment that is not well maintained will simply end up in costly repairs and maintenance. An indicator called the Percentage of Maintenance cost to RAV, which is expressed as a percentage, can be one of the factors in deciding whether to continue operating the equipment or totally remove it from operations. The lower its value the more effective we are in maintaining and preserving the asset.

Industry Cannot Cope Up with their Competition: In manufacturing industries, there is always competition as there are other industries that can manufacture your products. Some industries can manufacture your products which are way cheaper than your product cost and this can lead to either stopping the manufacture of the product or the competition cannot be beaten. In this case, either the industry can choose not to proceed and the worst may end up in the industry going bankrupt.

Lack of Market Demand: As the demand for the product or services the industry provides to their clients decreases, then it may no longer be feasible to continue manufacturing their product. It is also possible that the product demand is already in its declining mode. In this case, if the machine is not dedicated and can produce other products, then it may continue to operate. However, if the machine is dedicated, then the machine's life cycle may be over as there is no more demand for the product being manufactured and no more reason for the equipment to continue operating.

The vendor or OEM is No Longer Around: Although Dr. Ernest Rabinowich mentioned obsolescence, this refers to either the equipment or product. In this case, it is not the equipment that is obsolete but the vendor. The vendor or OEM may have shut down their operations or may have moved to other businesses as well. Although one solution, in this case, is to source other vendors who can fabricate the part for as long as a mechanical drawing of that part can be provided and for the purchasing to source other vendors who can provide the needs of the equipment.

Basic Equipment Condition is Not Addressed: This means the equipment should be clean, well-lubricated, all bolts secure and contained, and no form of leaks.   The life of an electric motor varies from five months for a dirty motor to twenty years for a clean motor and all we are speaking about is dirt. Catastrophic breakdowns can greatly be reduced if Basic Equipment Condition is in place. Lacking and lose bolts often lead to excessive vibration which can produce secondary damage on parts affected by the vibration itself. Most maintenance organizations find it hard to advance to any continuous improvement, some had advanced to Predictive Maintenance stage yet they find it hard to control the failures and breakdowns. In most cases, the parts' lifespan had not been reached. Why? Because the basic equipment condition had not been well established.

Human Errors: This can happen anywhere at any given time. This may occur during the design, fabrication, installation, commissioning, operations of the equipment, or how the equipment is being maintained. This will be discussed in detail in Chapter 5 of this book.

1.7: Life Cycle for Current and New Equipment

When equipment is used in production for a considerable amount of time, operations and maintenance experience problems and failures on the equipment. They notice that some parts of the equipment tend to fail easily or break off, while others simply have a short lifespan. Maintenance and engineers challenge themselves on identifying these parts and analyze how to prolong the lifespan of parts with inherent design weaknesses.

Document the Improvement: Most of the time, improvements and modifications are done to the equipment to improve its performance, however many of the improvements are not well documented, or simply the document has been lost or does not exist. Worst is the case when the document relating to the modification is taken home by the engineer who has resigned from the plant. What is important is to have control over the improvements done on the equipment. If you have not done this at all, it is best to start consolidating these improvements and assign a control number for each improvement done on the equipment. The engineering department should be responsible for this. Improvements done by engineering, maintenance, or other functions should be controlled, documented, and kept by the Engineering Department. It is likewise important for any department responsible for any improvement to have a copy of the improvement documents. All original improvement documents should be forwarded to the Engineering Department for control purposes.

Calculate the savings for the year: If we purchase equipment based on the Life Cycle Cost, we might not save on the initial or procurement cost of purchasing them, but the savings will be felt on the running cost, which includes operating and maintaining the equipment. In hindsight, modifications and redesign will be feasible if there are some economic value and savings to be generated as a result of the improvement itself. It is not feasible to modify an asset if there will be no trade-off in terms of cost reduction. The part subject to modification can have a higher cost, but in the long run, the savings will be felt in how we maintain the equipment since it will be more economical to use and have a longer lifespan.

Discuss the modification with the equipment manufacturer before purchasing additional equipment: As modifications and improvements are done from time to time by the Autonomous Maintenance, Planned Maintenance, and the Engineering Group, some of the parts have been modified to lengthen their lifespan. If the parts are supplied by the equipment manufacturer, an MP (Maintenance Prevention) design improvement and specification of the modification should be discussed thoroughly with them. Although not all improvements should be carried out in this fashion, make a list of the improvements generated and prioritized as to which of them should be discussed with the equipment manufacturer, most especially if the industry has plans for purchasing additional equipment in their plants. Remember that the industry patronizing the equipment are the customers and this will be both beneficial to the user and the OEM. On the other hand, if the equipment sold to your plant will also be sold to other plants, the subject of the patent should be discussed by the industry, the person responsible for the modification, and the OEM or vendor. Although this is not covered in this book.

1.8: The Importance of Life Cycle Costing in the Maintenance Strategy

If we consider a piece of equipment; from the moment it was conceptualized and designed. There were already costs involved from the manufacturers and vendors such as evaluation costs, design costs, labor costs, revision costs, engineering costs, and so on. Once the equipment's design was approved, the fabrication begins and another batch of costs will commence. We have the cost of fabrication, material costs, procurement costs, assembly costs, evaluation costs, shipment costs of these parts, and so on. When an industry purchases the equipment, the vendors and its contractors will transport the equipment to the site and once it reaches the destination, then the start of both installation, and commissioning costs begin as well as utilities required to run the equipment. In this case, the industry will have to shoulder the procurement costs, installation costs, transportation, shipment costs, contractor costs, debugging costs, and warranty costs until the equipment can finally be endorsed to the plant and is now ready for operations. From this moment onwards, the plant will now operate the equipment and will encounter the cost of operating and maintaining the equipment such as labor costs, overtime costs, energy costs, downtime costs, spare parts costs, consumable costs, lubrication costs, operator's training costs, repair and maintenance costs, and other costs that will be incurred during the entire life span of the equipment. Finally, when the equipment is decommissioned after (N) number of years, we still have to go with the burden of shouldering the costs of decommissioning, or disposal, which includes transportation costs, and spare parts costs that have to go away since it now becomes obsolete and so on. Just like human beings, when a person dies, there are still costs, which include the costs of the coffin, cremation, cost of cookies, and coffee, as most visitors will be taking them anyway while finally saying goodbye to the deceased. In the Philippines, you can even hire crying ladies. They will just cry. The more you pay, the louder they cry. Hence, during all the stages of the life cycle, there will be cost involved in each of these stages.

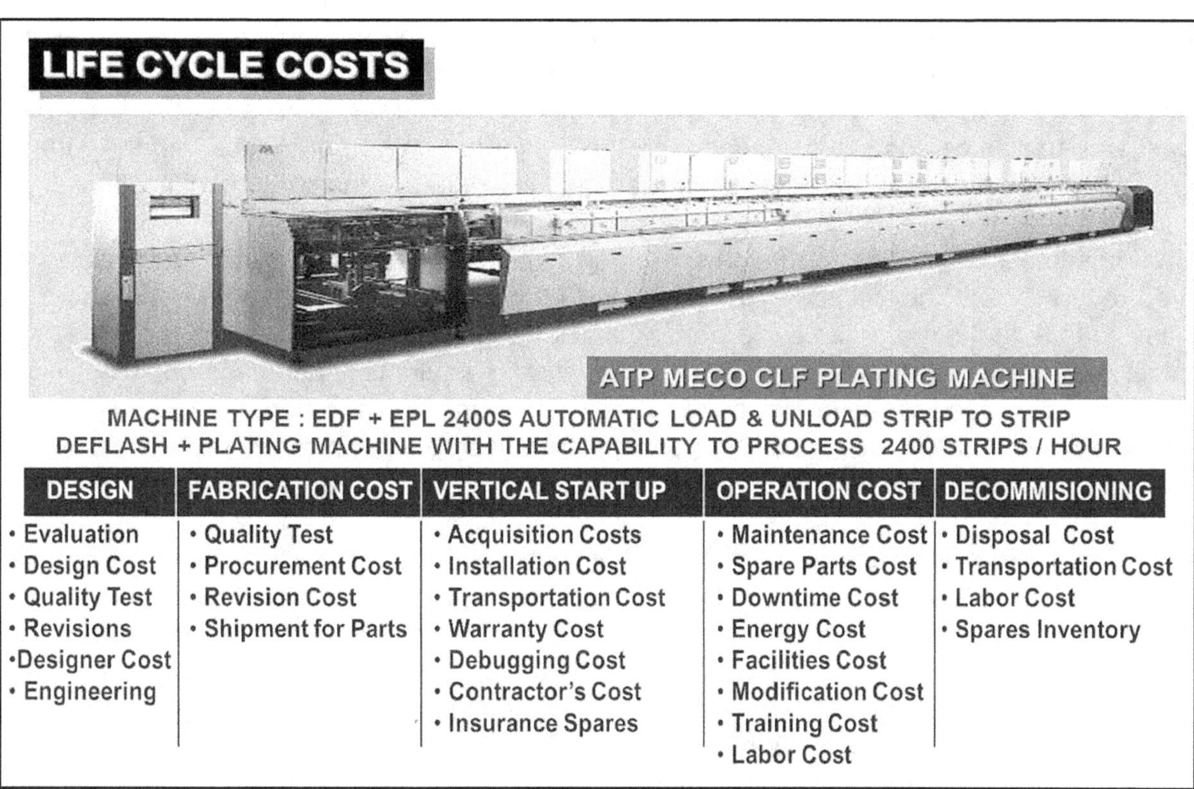

MACHINE TYPE : EDF + EPL 2400S AUTOMATIC LOAD & UNLOAD STRIP TO STRIP DEFLASH + PLATING MACHINE WITH THE CAPABILITY TO PROCESS 2400 STRIPS / HOUR

DESIGN	FABRICATION COST	VERTICAL START UP	OPERATION COST	DECOMMISIONING
• Evaluation	• Quality Test	• Acquisition Costs	• Maintenance Cost	• Disposal Cost
• Design Cost	• Procurement Cost	• Installation Cost	• Spare Parts Cost	• Transportation Cost
• Quality Test	• Revision Cost	• Transportation Cost	• Downtime Cost	• Labor Cost
• Revisions	• Shipment for Parts	• Warranty Cost	• Energy Cost	• Spares Inventory
•Designer Cost		• Debugging Cost	• Facilities Cost	
• Engineering		• Contractor's Cost	• Modification Cost	
		• Insurance Spares	• Training Cost	
			• Labor Cost	

Figure 1.8: What the Life Cycle Costs Will Include

What this means is that when an industry is making decisions about purchasing equipment for their plant's operation, there are actually two choices to make in which the first choice will be to consider only the initial costs of the equipment which is the tag price, or procurement costs. In this regard, the decision will be based on the vendor with the lowest bid on the equipment. However, a better choice will be to consider the cost of owning the equipment in its entire lifespan where one has to understand about the costs of maintaining and operating the equipment. Decisions about purchasing or acquiring new equipment and spares should not just be solely based upon the tag price or procurement cost of the machine, or equipment itself but rather on the entire Life Cycle Costs. Some of the questions that need to be considered will be the following:

• How will the equipment perform in the long run?

• What will be the estimated cost of operating and maintaining this equipment?
• What are the critical parts of this equipment?
• How frequently should these parts be replaced in a given period?
• How much is the cost of spares and downtime in a given month?
• How long will the equipment be down in the event of a breakdown?

Purchasing cheaper equipment or even spare parts may not be the best decision so far as this can actually mean spending more in the long run. A true and meaningful cost reduction effort can be achieved not by purchasing the cheapest item but rather by having a thorough understanding of what Life Cycle Cost is all about. Many plants are playing a dangerous game, called cost cutting. If your industry is involved in this, just beware as this can backfire on industries with devastating and unfathomable consequences. For industries doing this, remember that you will have your day of reckoning. What I believe is that the best way to reduce costs is to understand that it is more important to look into the running cost of the equipment or machine rather than the initial costs of the equipment itself.

Sometimes a higher-performance spare part or item will be more feasible than a localized type, even though the price is slightly higher. To gauge whether this will be the case, we need to look at the entire life cycle of this part or item, rather than the purchase price or its initial cost. Life Cycle Costing is a way of analyzing equipment purchase choices. If the decision was based on several factors rather than its initial costs then we would make our selection based on the least amount to own the equipment over its entire life span which indeed is the better decision.

On Purchasing New Equipment: What is important in purchasing new or additional equipment for the plant is to consider reducing the installation and commissioning time of the equipment. In TPM, they called this the vertical set-up time for new equipment. Remember that the longer the new equipment is being commissioned in the plant, the plant is not making any revenue or profit from it. As the equipment runs continuously in the plant, problems appear, and modifications, redesign, and improvements are done to prevent these problems from recurring. These modifications should be properly documented and prioritized based on their importance and criticality so that when the plant decided to order additional similar equipment in the future, these modifications and improvements should be discussed with the equipment designers and OEM to be included in their next purchase. The reason behind this is to expect fewer problems since these problems have already been experienced by both operations and maintenance. These improvements and modifications done in the equipment are prioritized and discussed with the equipment vendors and OEM if the new design can be included in their next purchase of similar equipment to avoid problems previously experienced.

If the industry really wants to reduce its operations and maintenance costs, we need to focus not on the Initial cost but on the Life Cycle Cost. This is where true and meaningful savings can be realized on maintenance. The role of maintenance is to sustain reliability so that the equipment can perform what it is intended to do.

1.9: Different Stages of Product Life Cycle

The study of the Life Cycle Cost involved two subjects, the product, and the equipment. Although the emphasis of this book is most likely on the equipment life cycle, a limited discussion will be made available on this book for the Product Life Cycle. The study of the Product Life Cycle will be limited to manufacturing industries and those that produced goods and products only. In manufacturing industries, their product can be considered the lifeblood of their existence. For industries to survive, depending on what the manufacturing industries manufacture, some products can be considered to have a limited lifespan and will no longer be available after some period. Other products will have an unlimited lifespan especially when it is a consumable and a need for human life such as food, pharmaceuticals, drinks, and the like. A good example of this will be electricity and water as this will be a lifetime expense for every family that they need to pay every month. Another example of this is the Coca-Cola soft drink originated in 1886 by an Atlanta pharmacist, John S. Pemberton from 1831 to 1888, at his Pemberton Chemical Company. His bookkeeper, Frank Robinson, chose the name for the drink, which then later became the famous drink that became the Coca-Cola trademark. This means that when a product will be launched, then it just needs to be upgraded and marketed well by these industries. This lifespan is known as the product life cycle. The product life cycle refers to the sales of the product over time, from its launch and throughout its entire lifespan and decline. When a product reaches the end of its life cycle, it is often replaced by an updated version of the product or a totally different product. Sales and Marketing will play a key part in promoting the industry's products.

Figure 1.9: Products with Non-Finite Lifespan

Figure 1.10: Products with Finite Lifespan

However, more importantly, for products with a finite or limited lifespan, it is their customer that will spell out the difference why these companies still remain in business. A good example of this will be electronics and technology. To let the customer remain faithful and loyal to their clients, industries must need to satisfy these things.

• Those that can produce the lowest possible costs
• Those that can produce the highest quality product
• Those that can provide the best customer satisfaction
• Those that can produce the fastest delivery of them all
• Those that can make decisions to adapt and innovate to rapid changes in the market trend

For industries manufacturing products with a limited lifespan, there are actually several stages in the product development, which include the following:

Phase 1: Conceptualization to Development Stage: This is where people involved in product research and development start to conceptualize the new product on how it will be accepted by the customers. Before the product hits the marketplace, you will be refining your concept, testing the product, and creating a launching strategy. The important point in this stage is to answer why people need this product and what makes this product better than the others. There are cases where the product will be tested to target people and generate feedback so the product can be adjusted according to the needs and satisfaction of the customers and consumers even begin to produce the product. This stage will include a brief sketch of a prototype of your product. This is also considered part of the company's investment as no revenue is generated at this stage. The important thing is to understand if the new product will be valued and needed by their customers.

Phase 2: Product Launching: Once the new product has been launched, product manufacturers need to introduce their product to the marketplace. The Sales and Marketing people will play an important role in this case. What is important for the organization is to find ways on how people will know that a new product has been launched and how their product is better than their competition. It is also important to understand the demand for the product. Other industries may use advertisements such as social media and TV Commercials to introduce their new product to the public. Again, industries will invest heavily in this phase when introducing their new product to the market.

Phase 3: Market Growth: At this phase, the product is now accepted and purchased by the public, and industries have now started to earn their revenue and profit. Marketing and advertisements will continue at this stage and demand has now been available to the consumers. Industries also think of innovative ways how to deal with their existing competition and what makes them better. This also helps industries to strengthen their brand and position in the marketplace.

Phase 4: Market Saturation: At this point, the saturation point of the product has been reached perhaps a new competitor emerged in the market with better products, or a new model was once again introduced to the market. Sales and Marketing will find ways to deal with the situation. Industries may adjust their product prices to stay competitive.

Phase 5: Decline and Obsolescence: If the sales of the product decline and consumers are reaching for a newer product or new competition, then it is most likely that this will now be the end of the product. Manufacturing of the product will now be stopped as it will no longer be produced by the industry. This will not be considered the end of the Product's Life Cycle. Either a new product will be once again introduced into the market and again we shall begin on Phase 1 of the Product Life Cycle.

Figure 1.10: Big Companies that are Already Closed Today

There are also cases where the decline in the product can cause the company to close down. Although there may be various reasons why the industry closed and goes bankrupt. If we review the history of the top Fortune Companies in the US, only around 52 US companies have made it to the Fortune 500 companies from 1955 to 2019. This means that only 10.4% of these companies have remained operational until today after 64 years since their inception in 1955. More than 89% of these companies have either gone bankrupt, merged, or acquired by other industries, or have fallen out from the original Top 500 Fortune Companies in the United States. Many of these companies were big shots during their time in 1955. Still, today, they are now unrecognizable, forgotten, or merely gone. Two of the main reasons why industries closed are due to technology and competition.

In general, the most fruitful success strategy is, to begin with, leadership tools, starting with hiring the right people, envisioning the change in place, adapting to the changes in the marketplace, defining their roles in the organization, developing their objectives, and closely monitoring their indices and measurement. But the most important part is listening to the voices of the customers and consumers. Here are just a few of the big names that are no longer with us here today.

Polaroid went bankrupt in 2001. The CEO of Polaroid, Gary DiCamillo said, kids today do not want hard copies anymore. We made this major mistake.

Nokia was not ready for the booming of the Internet. Other manufacturers started to understand that data and not a voice was the future of communication. Despite this, Nokia still focused on the hardware system instead of its software since management feared that the users might go away if they changed too much.

Sony Walkman was booming, and a must-have for all teenagers in the 1990s. I used to have one of these during my time. It was cool to carry such a thing with you, but it was killed by MP3, which was then killed by Smart Phones today. I just cannot say about tomorrow.

IBM's nickname was Big Blue for the International Business Machines Corporation, which was very popular in those days and in the financial press since the 1980s. Big Blue was an American multinational technology company that had its breakthrough in the 60s. In the early 1990s, IBM failed to adjust to the personal computer revolution by focusing more on their hardware instead of their software solutions.

Compaq was one of the biggest sellers of PCs in the entire world in the 1980s and 1990s. They produced some of the first IBM PC-compatible computers, being the first company to legally reverse engineer the IBM Personal Computer. Compaq ultimately struggled to keep up its price against its competition with Dell and was acquired for $ 25 billion by HP in 2002. The Compaq brand remained in use by HP for lower-end systems until 2013 when it was discontinued.

BlackBerry was a big hit in 1998. They outsmarted their competition by introducing a keypad in their mobile phone, but after a few years, the entire mobile industry was focusing on the touchscreen display, while BlackBerry was more concerned about protecting what it already had. Failing to adapt to changes, in 2017 the CEO John Chen announced that BlackBerry was out of the smartphone manufacturing business

1.10: Balancing the Product and Equipment Life Cycle

Although the coverage of this book is more on Equipment Life Cycle, there should be a balance between both the Product and Equipment Life Cycle, especially for industries that produce goods and products, such as manufacturing. The goal of any industry is to produce products to generate some revenue for the industry. Revenue refers to the income a company earns by providing services or selling products in a given financial year. On the other end, profit refers to the amount realized by a company after subtracting the expenses it incurred when providing a service or goods from the total revenue. This means that the subject of the Equipment Life Cycle can only be of value if the industry is generating revenue and profits from the products or services it renders. If the product or service provided is no longer needed by its consumers, then it is plain useless to maximize or even extend the Life Cycle of the equipment unless the equipment can produce other products beyond the obsolete ones. Extending Equipment's Life Cycle can only be feasible if we answer yes to the following questions;

• Will the product or services rendered by industries still be needed by their consumers?
• If the product becomes obsolete, can the equipment still be used to manufacture other products?
• Will the equipment still be used and needed even if the product gets obsolete?
• Will the equipment be used for a longer time?

This means that extending the Equipment's Life Cycle can only be feasible if the demand for the product or services is consistent on the market, if not, then the equipment will be decommissioned or retired together with the product.

Chapter 2

Initial Flow Control Activities

> *Initial Flow Control Activities (IFCA) is one of the pillars of TPM that can reduce the vertical start up time of the equipment. The vertical start-up time begins at the time the equipment will be acquisitioned by the industry until the commissioning of the equipment had been completed. Remember that the longer the equipment is being commissioned in the plant, the more industries do not make money from it.*

2.1: The Concept of Initial Flow Control Activities

As new products emerged and become more diversified, what the industry needs is to have a structured and systematic approach to New Product Development and equipment starting with the shortest possible time, with minimal debugging, and can be used by operations with minimal flaws and problems. The study of Initial Flow Control Activities, which is also called Early Equipment Management, focuses on two main subjects;

• Early Product Management (EPM): For new products that will be introduced in the market.
• Early Equipment Management (EEM): For new equipment that will be purchased in the future.

Typically, when the equipment will be newly commissioned, there will be many hidden defects and problems that emerged which prolong the installation and commissioning of the equipment. The equipment will be debugged during the commissioning process until all flaws, bugs, and defects are finally corrected on the equipment. Although industries that produce products will be focusing on both, those that provide services will be focusing on Early Equipment Management. The keyword here is being early. This means that all problems or any design flaws should be anticipated during the design and development of the new product or equipment.

IFCA or EEM is one of the pillars of TPM. This pillar is being deployed once the main TPM pillars which include Autonomous Maintenance, Planned Maintenance, Focused Improvement, and Training and Education have been kicked-off and are already in progress. This is usually Step 8 on the TPM 12 Developmental Steps as in figure 2.1. IFCA or EEM has something to do about purchasing new equipment, improving the vertical set-up time in the shortest time possible, developing easy-to-manufactured products, and extending the life cycle of equipment through MP (Maintenance Prevention) design modifications and improvements. What is important is to have a detailed and structured process so we can identify the flaws and problems easily at the very beginning.

For New Products

• Is the product easy to manufacture?
• If equipment is designed to produce different products, is the machine's set-up time short?
• Is the new product easy to set up?
• What quality problems does the product experience mostly?
• Are Quality conditions easy to set?
• Does the Quality conditions should resist variation?
• Is it possible to standardize the conditions to avoid a wide range of variations in conditions?
• Are changes in quality conditions easy to detect and easy to correct?
• Is the cost of manufacturing the product economically feasible?
• Is the cost of the product competitive?

For New Equipment Purchased

• Are the flaws, and problems identified during the installation and commissioning process identified?
• What was done to address these bugs, flaws, and problems during installation and commissioning?
• Was the corrective action temporary or permanent?
• What was the total installation and commissioning time?
• How many days does it take before the start of operations?
• What problems seem to occur during the start of operations?
• Were there any infant mortality failures experienced during the start of operations?
• Is the equipment ergonomically friendly compared to the old design?
• Were all the processes and activities during commissioning well documented?
• Were all the process and activities duration documented?
• What specific activities during installation and commissioning consumed the most time?

Typically, when a plant purchased new equipment, problems with defects and breakdowns emerged. Equipment is modified until the problems have been finally addressed; however, as equipment continues to operate, a new set of problems occur. The goal of IFCA or Initial Flow Control Activities is to observe these problems at the very beginning. The overall goal of IFCA or EEM is to design equipment with fewer problems if possible. Designers of equipment must design equipment that can hold tolerances over time and build equipment that can resist problems in the future. One of the main activities for EEM or IFCA is to create a loop between problems experienced in the past with existing equipment and correct them through redesign, modification, and improvements. Once these modifications have been successful in eliminating these defects and problems in the equipment, they are being discussed with the OEM and vendors to be included in the next purchase of the same equipment in the future. Although the OEM was responsible for the design and fabrication in most cases, it will be both operations and maintenance that have experienced these problems. Modifications and improvements that have been done on their existing equipment with success are discussed thoroughly with the designers of the equipment so that when future equipment of the same make and model is re-ordered, then-current problems experienced will no longer be a problem. As we have discussed previously, Life Cycle Costing design refers to a design approach to minimizing both its initial and running costs.

TPM MASTER PLAN			Preparatory Stage				Implementation Stage								Stabilization			
Item	Details of Activities		2022				2023				2024				2025			
			Q1	Q2	Q3	Q4	Q1	Q2	Q3	Q4	Q1	Q2	Q3	Q4	Q1	Q2	Q3	Q4
1	Top Management Formally Announce TPM	Plan	█															
		Actual																
2	TPM Introductory Education and promotional campaign	Plan	█	█														
		Actual																
3	Creation of TPM Promotional Organization and Office	Plan	█															
		Actual																
4	Established Basic TPM Policy and alignment of goals	Plan		█														
		Actual																
5	TPM Master Plan of Completion	Plan			█													
		Actual																
6	TPM Kick-Off	Plan				█												
		Actual																
7.1	Maximize Production Effectiveness Implement Focused Improvement (Kobetsu-Kaizen Pillar)	Plan					10 cases	20 cases	30 cases	40 cases	50 cases	60 cases	61 cases	62 cases	63 cases	64 cases	65 cases	66 cases
		Actual																
7.2	Implement Autonomous Maintenance	Plan					Step 1			Step 2 - 3			Step 3- 4					Step 5 - 7	
		Actual																
7.3	Implement Planned Maintenance	Plan					Phase 0		Phase 1		Phase 2			Phase 3				Phase 4	
		Actual																
7.4	Training Skills and Education	Plan					Planning		Step 1		Step 2 - 3			Step 3- 4				Step 6	
		Actual																
8	System for Initial Flow Control Activities (Early Equipment Management)	Plan									Step 1 - 2			Step 3 - 4				Step 5	
		Actual																
9	System for Quality Maintenance Aim for Zero-Defects	Plan									Step 1 - 2			Step 3 - 4				Step 5 ➡	
		Actual																
10	Administrative/Office TPM	Plan							Step 1		Step 2 - 3			Step 3- 4				Step 5 - 7	
		Actual																
11	Establish Effective EHS System Aim for Zero Accidents, Zero Pollution	Plan									Step 1 - 2			Step 3 - 4				Step 5 ➡	
		Actual																
12	Challenge the TPM Excellence Awards 2nd Category	Plan																	█
		Actual																

Figure 2.1: TPM 12 Developmental Steps

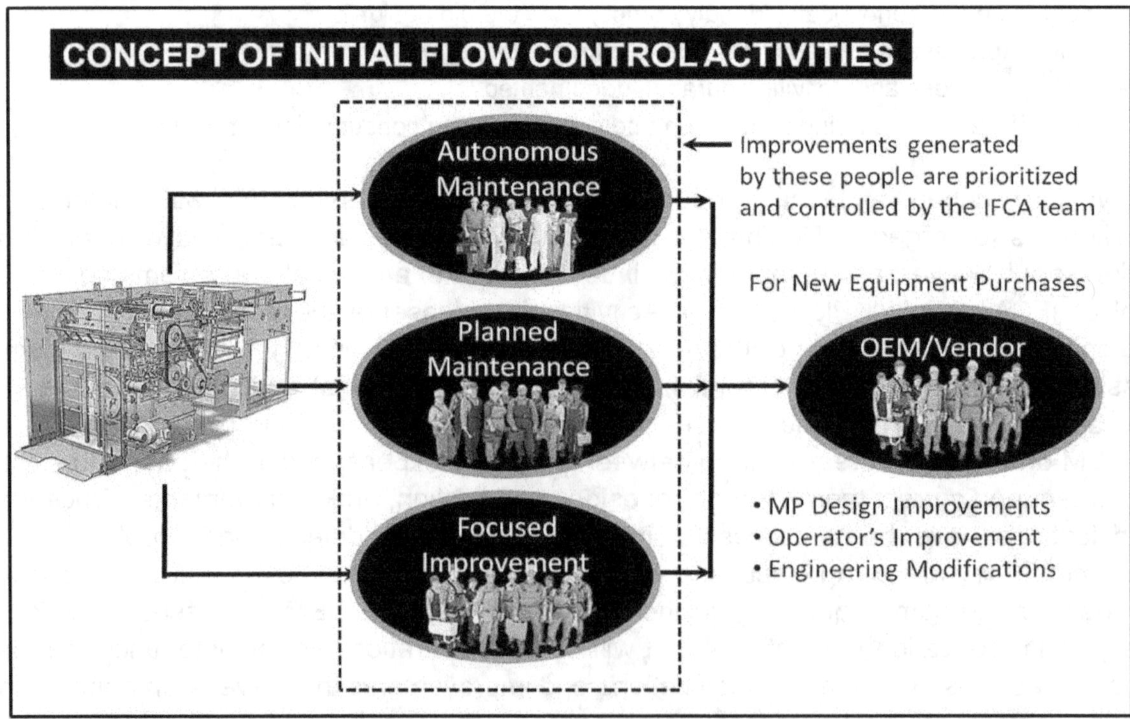

Figure 2.2: Creating an Improvement Loop to OEM

2.2: Patents for Improvements

One of the things that need to be addressed by industries will be the improvement patent. First, let us explain briefly, what this patent means. A patent is the granting of a right by an authority to the originator of the invention. Just like in publishing this book, copyright is placed at the beginning of this page, which grants the author the right to publish this book. In the case of a patent, this provides the inventor the exclusive rights to the patent process whether this is a design or a new invention. The purpose of the patent is to protect the right of the inventor so that it will not be copied, replicated, and sold by other organizations. They are a form of right for the industry. A patent can also be declared as a legal right to an invention given to a person or entity. A patent usually has a lifespan of 20 years in the United States but might differ from one country to another.

In the US, the government involved in approving patents is the United States Patent and Trademark Office (USPTO), which is part of the Department of Commerce. This agency provides the application and approval of patents. In the Philippines, the government agency involved in this case is the Intellectual Property Office of the Philippines or IPOPHL. For Philippine Patents, the term of protection is 20 years from the date of filing with no possibility of renewal. A Patentable Invention is any technical solution to a problem in any field of human activity, which is new, inventive, and useful. An Invention may relate to a product (e.g. machine, device, an article of manufacture, a composition of matter, a microorganism), process or method, (e.g. a method of use, method of manufacturing, a non-biological process, a microbiological process), computer-related inventions, and improvement of any of the foregoing.

Most patents have a validity of 20 years in the U.S. starting from the date it was applied at USPTO, but there can be exemptions to extend the term of the patent. A patent may be given to a person or entity who has invented any new and useful process, machine, manufacture, or any new and useful improvement thereof. They exclude others from reproducing, using, or profiting from it without the expressed permission of the patent owner. The granting authority issues a patent in exchange for permission to publish details about the invention, such as how it's made and what it's used for.

Assuming that a design weakness was found on a pick and place of a pneumatic machine and was redesigned by Charlie where he extended the pick and place arm to 3 millimeters so the product can be carried by the pick and place without falling which caused frequent minor stoppages. In the process of IFCA, these modifications will be discussed with the designers of the equipment so that when this pneumatic equipment will be reordered in the future, what we want is for the revised design to be incorporated so we can eliminate the previous problems experienced in this equipment. Now here's the thing, although the redesign came from Charlie's brain, he may not actually qualify to apply for the patent since Charlie does not own the machine, and second, there would be a conflict of interest on the part of Charlie since he is just an employee unless this will be granted by the company which may seem to be unlikely. This means that if the new design will be used by the OEM designers for their other clients, the industry where the redesign originated can only grant permission if they will be paid a certain amount of royalty or may totally refuse the OEM as this will only be exclusive to the industry who invented the modification. If the OEM used Charlie's design with their other customers, the industry has the right to sue the OEM since the modification or redesign has been patented.

Although applying for a patent for the modification or redesign may not be 100% fool-proof in protecting the rights of the inventor, other people may copy the design and change the material or may have a slight difference in measurements. Hence, when discussing the improvements to be carried out for repeat order equipment, the industry needs to discuss this matter with the vendor to avoid problems in the future.

2.3: The Six Conditions for Equipment's Initial Flow Control Activities

There are six conditions to consider for factory-friendly equipment, which include the Development Phase, Reliability, Economics, Availability, Maintainability, and Safety. Each of these beginning letters will spell out the word **DREAMS**. Just like when the US sent their first professional basketball players to compete in the 1996 Olympics in Barcelona Spain, which they called the Dream Team composed of only the best players in the NBA. The US Dream Team was composed of an impossible line that includes the following players, Magic Johnson, Larry Bird, Charles Barkley, Karl Malone, John Stockton, Patrick Ewing, David Robinson, Clyde Drexler, Scottie Pippen, Christian Laettner, Chris Mullin, and "**The Goat**" Michael Jordan (Goat means Greatest of All Times). Their head coach was Chuck Daly. He led the Detroit Pistons to win their two consecutive National Basketball Association (NBA) Championships in 1989 and 1990. Industries likewise would want equipment to possess such quality and characteristics.

[3]**Development**: According to Fumio Gotoh, author of Equipment Planning for TPM, Maintenance Prevention Design, today's strength in Research and Development is a key determinant of corporate competitiveness. True competitiveness comes from the synergetic combination of product development capability and rigorous development of equipment and related fabrication methods to produce high-quality products at the lowest possible cost. This means that newly designed equipment should not only be inexpensive in terms of its acquisition or initial cost but also in terms of its running cost to ensure profitability for the industry and to remain competitive.

Reliability: By definition, reliability is a measure of how long the item performs its intended function for a specific period under specified operating conditions. It is the probability that an item will operate without failure throughout a specified interval. According to Bazovsky, the modern concept of reliability in popular language is simply the capability of equipment not to break down or fail in operation. This means that when a piece of equipment works well and performs to do the job for which it was designed, such equipment is said to be reliable. One of the key challenges for equipment designers is building equipment that is reliable and will not break easily during operations, especially in this era of digitalization and

---

[3] Fumio Gotoh, *Equipment Planning for TPM, Maintenance Prevention Design,* (Productivity Press, Portland, Oregon, USA, Dover Publications Inc., Mineola, New York, 1991), Page 8 to 11

automation. As technology and automation are usually integrated into today's equipment, designers must have a forecast of what problems can occur in the future and how to address these issues early during the design phase. New equipment must be reliable and should not fail or break easily during operations. Equipment designers must understand the current problems of their existing equipment and machines as well as what can be done so that these problems will not occur with the new equipment. The inherent reliability has something to do with the design of the equipment. We cannot force the equipment to deliver beyond its design and capability.

Economics: Industries purchase equipment for one good reason, which is to generate revenue so that they can sustain their operations. Even if the equipment is reliable, if the fabrication cost and the running cost will be too high, then the equipment may not turn out to be profitable for the industry. Equipment designers must create a balance between both the initial and running cost of the equipment for the industry to generate some revenue from their investment. The overall principle for manufacturing industries is that their products must be sold out for a profit. This means that equipment designers and OEMs should design their equipment at the lowest possible life cycle cost.

Availability: Availability is when a machine is available for work less all the downtime divided by the total available time. This refers to the time the equipment was available for production at a given time whether it will be used or not by operations. It is the proportion of time the equipment is available for use for its intended purpose. It can be measured as the total Uptime divided by the Total Available Time where the total time will be equal to the summation of both the uptime and the downtime. Problems that contribute to availability will include breakdowns, set-up, and changeovers, cutting blade change, start-up losses, and all Planned Downtime. This means that availability drops when the equipment is left idle as a result of these equipment losses just mentioned. These problems should be carefully studied and addressed during the design phase. Even if the initial cost of the equipment is acceptable but if the machine will take more than an hour to convert from one product to another, then this will be a total loss of revenue for the industry. Equipment design for flexibility means that losses to availability will be minimal. Machines should be flexible and should be easy to convert from one product to another if the industry demands a changeover from the equipment. A change-over usually exists when a piece of equipment is non-dedicated and can able to produce different products.

Maintainability: New equipment should be easy to operate and maintain. If a piece of equipment is too expensive and difficult to maintain, then this will be detrimental to the economic criteria of the equipment, while if too much time is spent on maintaining the equipment, then this will likewise affect the availability of the equipment. Although one-way to improve the maintainability of the equipment is to modify or redesign the asset. These modifications will require additional investment. As John Moubray has written in his book on RCMII quotes, which comes first redesign or modification? [4]According to the book of John Moubray on RCM II, quote, which comes first redesign or maintenance? Reliability, design, and maintenance are inextricably linked. This can lead to a temptation to start reviewing the

---

[4] Moubray, John, *Reliability-Centred Maintenance II,*(Butterworth Heinemann., 1997), Page 189-190

design of existing equipment before considering its maintenance requirements. In fact, the RCM process considers maintenance first for two reasons. First, most redesign and modification takes from six months to three years from initial conception to commissioning depending on the cost and complexity of the new design. Hence, maintenance has to maintain the equipment as it exists today and not what should be there in the future. Second, most organizations are faced with many more apparently desirable design improvement opportunities than are physically or economically feasible. In this case, when maintenance can no longer address a particular failure mode, then this is where redesign or modification takes place.

Safety: Equipment should always be safe to operate. The safety of the operator will always be the priority in this case. Equipment includes defenses and protective devices that not only warn us of a problem but also ensure the safety of the operator. An example of this would be a protective device designed for the safety of the operator, which may be a proximity sensor whose function is to automatically shut down the equipment if the operator accidentally opened the machine's door. In this case, the machine will stop automatically.

2.4: Vertical Start-up Time for Equipment

Designers of equipment are not involved during the equipment's day-to-day operation as well as maintaining the equipment during its entire lifespan. They assume that their equipment will perform as intended, but that is not likely the case. MP Design improvements generated by Autonomous Maintenance, Planned Maintenance, Focused Improvement, and Engineering group help provide a communication loop and feedback for the equipment designer to avoid repeating future mistakes and improve the design of the new equipment. Autonomous maintenance will be delighted to provide visual controls for its equipment as well as reduce the time to clean and inspect the equipment.

For Planned Maintenance, improvements and modification of parts that they encounter which fail prematurely and frequently can be adopted, while for the Engineering Department, any improvement and project done to improve the cycle time of the equipment can likewise be added to the lists. Improvements to be discussed with the equipment vendors should be planned and prioritized carefully by the IFCA team. The goal of the equipment designer should always be to improve the design of future equipment. Likewise, many problems also arise during the equipment commissioning and start-up process. What is important in the start-up process is that before commissioning the equipment, equipment manufacturers should have a detailed process and plan of activities on how to go about every detail on commissioning the equipment. Time the completion of each process should be noted and list down all the problems and difficulties encountered during the commissioning process from wiring the equipment, laying out the foundation, piping installations, alignment of the equipment, and all sorts of activities and adjustments. These records can be used as a way of improving the vertical set-up time of commissioning new equipment from the time the equipment has been transported to the plant until the time it is ready to be endorsed and used by operations. It is important that during the installation and commissioning stage, each activity is noted down together with its duration. If flaws, bugs, and problems arise, what has been done to correct them as well as what activities prolong the commissioning process should also be well documented. Once the equipment is ready to operate, the documentation for the vertical-set up time should be completed and reviewed as this will serve as a guide and procedure when there will be a repeat order of the equipment thereby reducing the total vertical start-up time of the new equipment. A final checklist and inspection should be ready before handing the equipment back to operations.

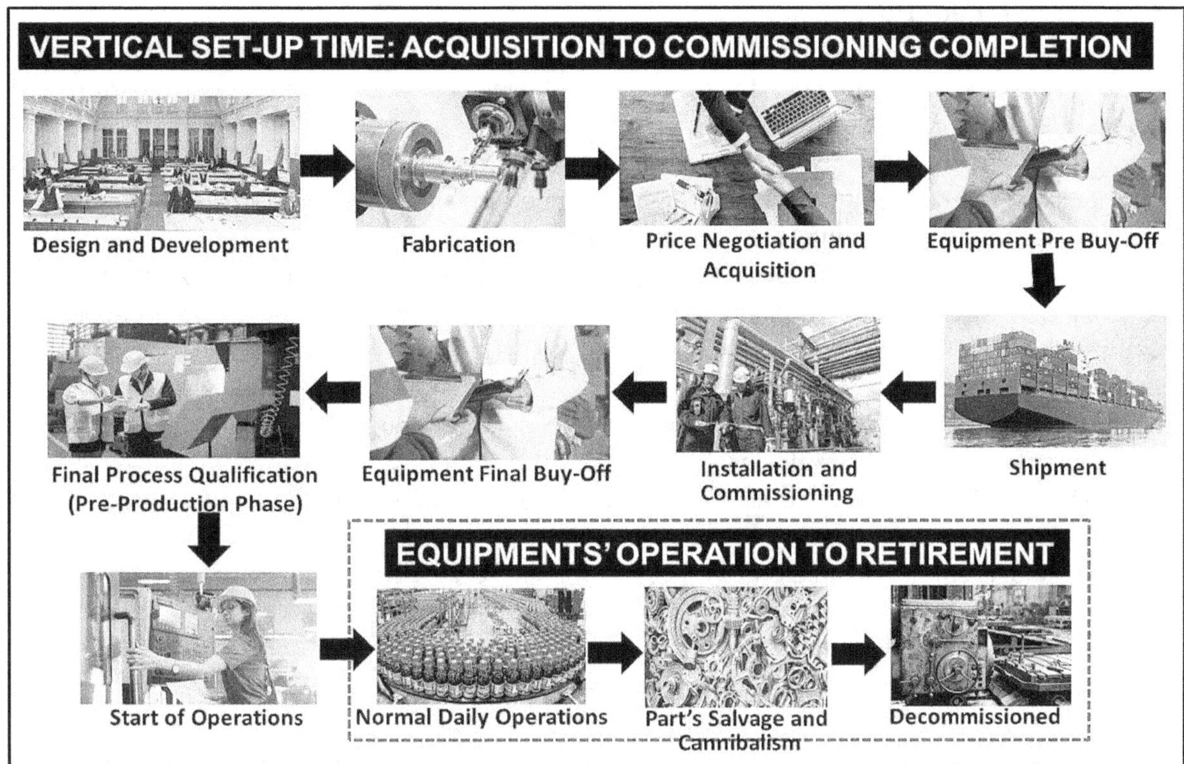

Figure 2.3: Equipment Vertical Set-Up Time

If industries purchase additional equipment in the future in support of their production needs and goals, they should ask for some documentation regarding the vertical start-up time from their equipment manufacturers or simply ask how fast they can commission the equipment in the plant. Their previous data should be used as a way of having some benchmark on the process. The goal of equipment manufacturers should be to continuously improve the vertical set-up time of equipment being commissioned. If the previous commissioning time has been reduced by 3 days, then this is an added revenue to the industry. Remember that the longer the equipment is being commissioned in the plant, the more industries do not make money from it. Profit only starts when the equipment has been handed over to operations, and production starts to generate products. Both equipment manufacturers and industries should consider the vertical set-up time as a record. This is like running a 100-meter sprint and improving the personal best time in completing that race. Improving the vertical start-up time of the equipment is a continuous effort and is not a one-time event. The best record so far achieved in the vertical start-up time is challenged continuously by the team.

2.5: What We Want to Achieve on IFCA

Initial Flow Control Activities is one of the pillars of TPM, which focused on new product development, and the early stages of equipment design to the start-up stage until the equipment is ready to be operated.

For New Equipment Purchased
• For new machines and equipment to be purchased, what we want is to achieve a shorter vertical start-up (Installation and Commissioning) time from the initial development period to the start of full-scale production.
• Fewer problems as experienced from the existing equipment and machines since improvements, and modifications are done from Autonomous, Planned Maintenance, and Focused Improvement, which are looped backed to the OEM and incorporated into the new design.
• Achieve the lowest possible Life Cycle Cost, which is the costs incurred during the entire life span of the equipment from the initial start-up, which is the time the machine was purchased, installed, and commissioned to its running cost, which is the overall cost of operating and maintaining the equipment.
• Equipment that is flexible to respond to abrupt changes in the market demand to help the company remain competitive in the marketplace.
• Possibility of extending the Overall Life Cycle of the equipment through modifications and MP Design Improvements.
• Equipment that is safe to operate and with built-in Intrinsic Safety Design.

For New Product Development
• To produce products that are easy to manufacture and produced.
• Products of high quality that provide total customer satisfaction.
• Be able to generate products whose cost to manufacture will be competitive in the market compared to existing products and their competition.
• Develop a system for gathering feedback information based on the problems encountered on existing products.
• We also want the new products, which can be easily converted to the equipment if the new products will be processed on existing types of equipment. This means lower set-up or changeover time.
• Products that will not become obsolete easily.

2.6: TPM Certification Requirements for IFCA

For those industries aiming for the TPM Awards for TPM Excellence – Category B, Category A, and Award for TPM Excellence – Category A. Initial Flow Control Activities will be Chapter 8 of the TPM Book Report. This will be submitted as part of the TPM preliminary assessment criteria. Although previously, JIPM was responsible for assisting industries that would challenge the JIPM-TPM Excellence Awards, there are now several certified groups from across the world that can assist industries in this regard. Here is a message from the JIPM (Japan Institute of Plant Maintenance) Official Website regarding these changes.

Previously, it will be the JIPM that will provide the initial and final TPM assessment, however, changes were made, and the following amendments were made to the Japanese legal system. JIPM transformed itself on April 1, 2012, from a Shadan-Hojin, a public interest corporation under the control of the Ministry of Economy, Trade, and Industry, to a Ko-eki Shadan-Hojin, a public interest incorporated association approved by the Cabinet Office. The TPM Awards, which started in 1964, have been certifying Assessment Agencies since 2007 and, in cooperation with these certified agencies, have been operating the TPM Award system for enterprises and factories outside Japan. Due to the change in its structure, JIPM will directly accept applications for evaluation from the 2013 TPM Awards onward. Agencies that have served to date as assessment agencies will include the following:

• CETPM (Centre of Excellence for TPM, Germany)
• CSD (Corporate Synergy Development Center, Taiwan)
• KSA (Korean Standard Association, Korea)
• SMMT (The Society of Motor Manufacturers and Traders Limited, UK)
• TPA (Technology Promotion Association Thailand-Japan, Thailand)
• Confederation of Indian Industry in India (TPM Club India)

These organizations are going to assume the role of Associate Agency during the evaluation for the JIPM-TPM Awards. We recommend that enterprises and factories that would like to be evaluated consult with the nearest Associate Agency as required. The requirements for IFCA will be included in Chapter 8 of the Overall TPM Report that they will submit to their assessors. This will require both activities done to improve new products and new equipment, which will include the following.

Chapter 8: Development Management Activities
8.1: Product Development Management
    8.1.1: Outline – Concepts, Aims, and Schedule
    8.1.2: Designing easy-to-make products in the development stage
    8.1.3: Product Development Management System
    8.1.4: MP Information, its collection, and use
    8.1.5: Designing Recyclable Products and Manufacturing Systems
    8.1.6: Results achieved and Future Plans

8.2: Equipment Development Management
    8.2.1: Outline – Concepts, Aims, and Schedule
    8.2.2: Integrating Product Development Management and Equipment Development Management
    8.2.3: Equipment Development Management Systems
        – Status Analysis, Capital Investment Plans, Economic Comparisons, Development and control of equipment budgets
    8.2.4: MP Information; its collection, storage, and use
    8.2.5: Results and Future plans

8.3:Individually developed Management – Examples, and Effects

2.7: Case Study for IFCA for CLF Plating Machine

When implementing IFCA or EEM, two IFCA groups will be formed. This will be for the new product development and for the new equipment and assets that will be purchased in the future. Although discussion of IFCA for new products will only have a limited discussion in this book since the main topic for this book is the entire Life Cycle of the equipment. During my days working as a TPM Engineer in one of the largest semiconductor industries in the Philippines, I was also assigned to handle both the Planned Maintenance and Initial Flow Control Activities pillar of TPM. These two TPM pillars are not only connected but are entirely linked and integrated with one another.

Before Implementing Initial Flow Control Activities for Equipment

• Experienced many abnormalities on the equipment specification and design stage due to insufficient feedback system and the absence of MP (Maintenance Prevention) System, hence the same problems encountered before are being carried over to repeat order of equipment of the same type and model.
• Other maintenance and reliability people discuss directly with equipment vendor and OEM without prior notice and knowledge to purchasing and other people involved resulting in additional cost due to modifications involved in the equipment.
• Long vertical start-up time, which includes the time to purchase, install and commission the new equipment. Since there is no tracking and comparison between the existing previous equipment and those that are newly purchased, the delivery time or lead time varies even for a repeat order of equipment and machines.
• Experienced presence of non-checked items during the vendor or OEM presence due to insufficient buy-in checklists and standards.
• Installation of new machines on the production floor with insufficient training for operators, maintenance and process people resulting in excessive vendor support which comes at a fee even for small and minor problems encountered such as bulb replacements.
• Price changes due to non-standardization of specifications due to the variation in packages even when this is a repeat order of the same equipment or machine with the same model.
• We have no idea how to verify those parts under warranty and non-warranty price.
• Experienced long lead time on MRO Spare Parts for new machine upgrades due to modifications.
• Incomplete equipment information and specification during ordering especially if there are revisions and amendments of P.O. (Purchase Order Specs). Due to modifications and revisions in the current model, vendor overpriced their equipment even for a repeat order of the same model.
• Difficult to find new equipment that will satisfy our requirements and based on the specifications provided or cases where the specification is overkill, which involved too many requirements and additional functions of the equipment.
• Incomplete equipment delivery since this has not been well communicated with the vendor.
• Improvements from existing machines are kept individually by the person who was responsible for the improvement. When the person responsible for the improvement resigns from work, the document disappears.

To address the problems above, we decided to implement Initial Flow Control Activities for newly purchased equipment and machines to address specific problems especially on repeat order machines and equipment purchases as well as for purchasing new model machines and equipment. Our objective of establishing IFCA for new equipment purchases is to reduce the time from the initial development stage to production scale and achieve significant improvements in our overall equipment performance and be able to develop equipment and machines with the least cost that fulfills all the required functions needed. We generate the following protocols.

- Generate a system for capturing and having centralized equipment improvements through MP Design forms and create a control system per station so we can discuss and prioritize these improvements in the next purchase for repeat order equipment and machines.
- We piloted one station Central Lead Finish (CLF) and selected the Meco Machine and monitor its Vertical Start-up time. Also part of the plan is to monitor other equipment, which is being co-developed with other vendors and suppliers.
- We asked the vendor to provide training programs and seminars on proper operations and maintenance from the vendors themselves, which should be done when the equipment is being commissioned in the plant.

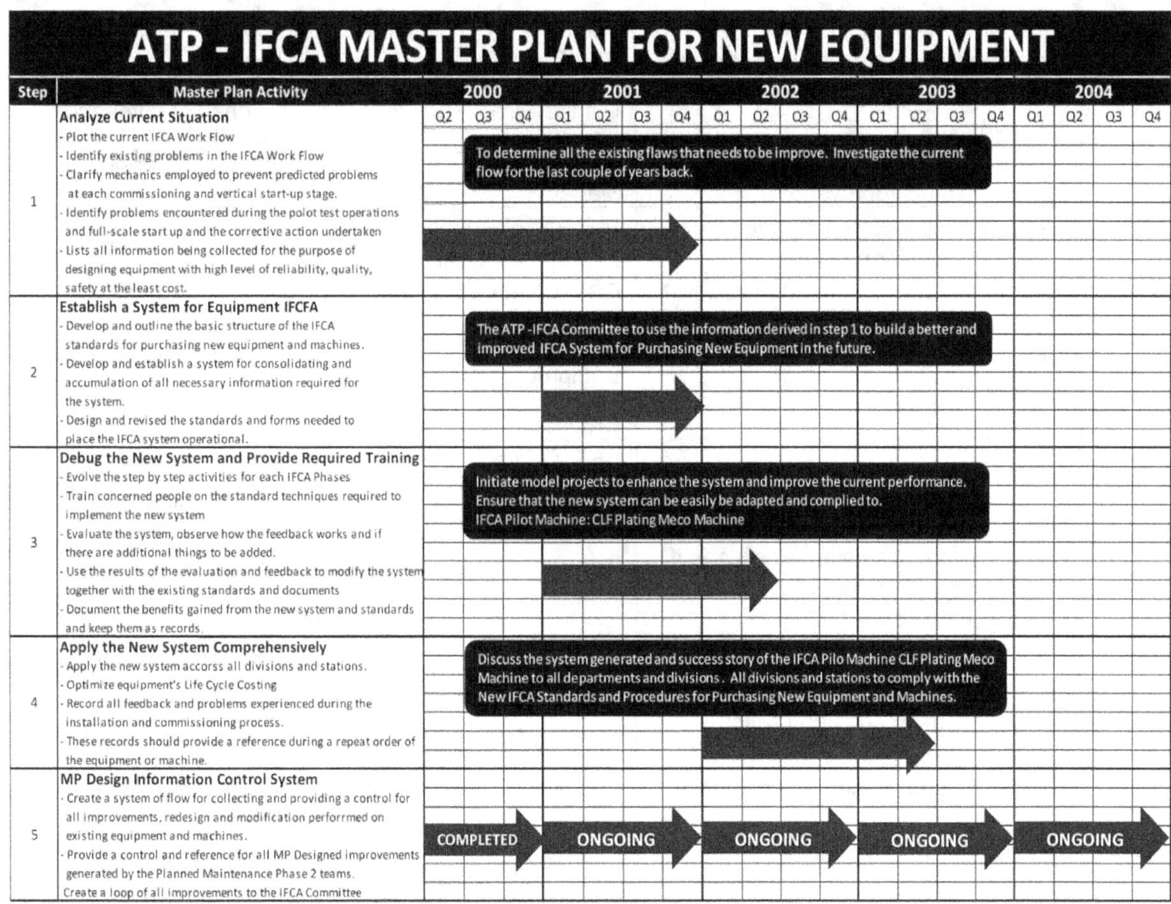

Figure 2.4: IFCA Master Plan for New Equipment Purchase

Mission: To establish a system for Initial Flow Control Activities for equipment and capture the improvements, modifications, and innovations done on the existing equipment through MP Design forms.

Vision: To reduce the overall time to develop new equipment to the time that it is qualified to run for production. Achieve significant improvement on equipment performance of new machines due to the improvement made on existing equipment and provide higher quality assurance by building equipment that is flexible and can fulfill all the required functions at the lowest possible cost.

We developed our IFCA Master Plan for the equipment as well as generated a system for collecting and controlling all MP Design Improvements so that existing problems experienced by both operators and maintenance can be eliminated if not mitigated in future purchases, especially for repeat order equipment and machines.

The IFCA team for equipment developed this step-by-step activity for acquiring new equipment purchases to minimize problems and be able to generate a feedback looped system between the user, maintenance, OEM or Vendor, and the EST or Equipment Selection Team.

Steps	Detail of Activities	Team	Owner
1	Review improvements, and modifications and generate feedback.	PCM, TDT, EST, PM, ME, PE, IE, OPTN, and QA	Equipment Selection Team
2	Generate MP Design Improvement for approval.	Planned Maintenance	Planned Maintenance
3	Review PM MP Design generated and approval of the improvement	QA Manager, Equipment Manager, TPM Office, EST	Equipment Selection Team
4	Fan-out improvement to equipment with similar problems only	ME, PM, Focused Improvement Team	Equipment Selection Team
5	Set meeting with OEM and review MP Design Improvement	EST, CEPD, Originator, Purchasing	Equipment Selection Team
6	Generate Purchase Requests for New Equipment	Equipment Selection Team	Equipment Selection Team
7	Purchase Order Generation	Originator	Originator
8	Monitoring of Equipment Order	EST, Originator, CEPD, Purchasing	Equipment Selection Team
9	Pre-Buy-Off on OEM or Vendor. If Pass, then the next step is delivery, if Fail, then the equipment will be debugged	Equipment Selection Team	Equipment Selection Team
10	Newly purchased equipment will be delivered to the site	Equipment Selection Team	Equipment Selection Team
11	Conduct training and certification for operators, ME, and PE	Training Department	Training Department
12	Equipment certification and final buy-off	Equipment Selection Team, QA, and PE	Equipment Selection Team
13	Equipment Performance Monitoring	PCM, ME, and PE	PCM, PE, and ME
14	Submit final report on Vertical Start-Up time for the new equipment	ME	Equipment Selection Team

Figure 2.5: ATP-IFCA Flow for Acquiring New Equipment

We started our IFCA on Central Lead Finish (CLF) plating area. Finally understanding the concept of Life Cycle Costing, our purchase decision for new equipment is not only decided on the initial or acquisition cost of the equipment but other aspects were taken into great consideration and detail such as the cost of maintaining the equipment, MRO spare parts cost, amount of space consumed by the equipment, operator and maintenance training cost. Provided in this report is a cost comparison between two plating machines currently used in operations. A comparison between this two plating equipment was reviewed by the Planned Maintenance and Equipment Selection Team (EST) to determine which equipment is better.

Figure 2.6: Comparison Between Technic and Meco Plating Machine 1

Comparison	Technic	Meco
Model	Technic SP 800 CMB Continuous Moving Belt	EDP + EPL Strip to Strip Planing Machine
Products	Molded Lead frame	Molded Lead frame
Length of Machine	25 meters	26.1 meters
Deposit Thickness	300 to 600 micro inch	300 to 600 micro inch
Chemical Used	Leo Ronal	Leo Ronal
Controls	Programmable Logic Control (PLC)	Electro Plating Control Software (EPCS) Windows-based check alarms, temperature, and pumps. LAN and Field Bus System ensuring minimum electrical wiring
Load and Unload System	X-Y COMM Touch Screen	Windows Software computer auto loading and unloading system and universal load and unload frame size
Plating System	Not Applicable	Menu-driven auto rectifier setting ramp up and ramp-down
Others	Changing adjusting, rotating, and extended motion for load and unload intelligent motor	UPS System CAP 2 KVA
Costing	Technic	Meco
Acquisition Cost	$ 980,000.00	$ 1,180,000.00
Maintenance Cost/machine	$ 2,500.00 monthly average	$ 2,000.00 monthly average
Total Machines	4 Technic Machines	6 Meco Machines
Monthly Maintenance Costs	$ 10,000.00 monthly average	$ 12,000.00 monthly average
Yearly Maintenance	$ 120,000.00 yearly average cost	$ 144,000.00 yearly average cost

Figure 2.7: Detailed Comparison Between Technic and Meco Plating Machine 1

Technic Machine: Technic Model SP 8 Continuous Moving Belt (CMB) system strip-to-strip plating machine where the strips are delivered by a continuous belt from the offload to the work in process cells and unload. The strips are automatically loaded by operating the belt finger that provides a gripping mechanism to hold on to the product. ATP-CLF Plating Station currently has four Technic machines.

MECO Machine: This machine is categorized as a strip-to-strip plating machine where the strips are placed in a magazine carrier that loads them into the machine loading station. The strips are fed one by one into the conveyor belt. The strip being gripped in the belt will pass through different chemicals in the process of the entire machine. After the plating process is over, the strips are taken off the conveyor belt. ATP-Plating has six Meco Machines operating.

REDUCTION IN VERTICAL START-UP FOR NEW EQUIPMENT			
STAGE 1	**STAGE 2**	**STAGE 3**	**STAGE 4**
Developmental Stage	**Pre Buy Off and Shipment**	**Final Buy Off and QA Certification**	**Equipment Performance Monitoring**
Design stage to fabrication. Discuss MP Design to OEM	To be carried out by the Equipment Selection Team (EST)	Installation, Commissioning, Training and Certification	To be monitored by PCM, PE and ME

Figure 2.8: Reduction in Vertical Set-up time

Stage	Stage 1	Stage 2	Stage 3	Stage 4
Machines	Meco 1, 2 and 3	Meco 4	Meco 5	Meco 6
With IFCA in place	No IFCA	No IFCA	with IFCA	with ICFA
Development Stage	4 months	4 months	3.5 months	3.5 months
Fabrication Stage	8 months	7 months	6 months	6 months
Total Time	12 months	11 months	9.5 months	9.5 months
VERTICAL START-UP TIME				
Pre-Buy Off Stage	14 days	14 days	7 days	7 days
Shipment	60 days	55 days	45 days	45 days
Installation and Commissioning	45 days	30 days	28 days	25 days
Buy-Off at ATP	7 days	7 days	7 days	7 days
Process Qualification	7 days	5 days	3 days	2 days
Total Time	133 days	111 days	90 days	86 days
Production Dates	March 1, 1996, April 15, 1996 April 25, 1996	September 28, 1998	August 25, 1999	February 6, 2001

Figure 2.9: Reduction in Vertical Set-up time for CLF Meco Machines

The IFCA Pilot team was successful in reducing their vertical-set-up time in purchasing additional repeat order equipment for additional Meco Machines. The team has dramatically reduced the vertical set-up time, which generates additional revenue for the industry.

2.8: Why Implementing IFCA is Important for Industries

Although the application of Life Cycle Costing is important during operations, implementation of IFCA or EEM is important during the pre-operational stage of the equipment. It is important to note that the pre-production phase can also be controlled and managed if IFCA or EEM is in place. The advantage of implementing Initial Flow Control Activities is to understand the problems experienced during the Vertical-Set uptime of the equipment. What seems to be the problems experienced during the installation and commissioning period and what was done about them. The vertical start-up time will be recorded and will serve as a benchmark if there will be repeat orders of the same equipment type and model in the future. This means that the shorter the vertical start-up time for the equipment, then the earlier the industry can start generating revenue and profit. This means that if we can reduce the vertical start-up time for the new equipment from 6 months to 3 months, then this is an additional 3 months of revenue for the industry. Previous problems experienced during installation and commissioning will be recorded and reviewed. Those activities that contribute to lengthening the installation and commissioning process will be analyzed with which the overall goal is to reduce the vertical start-up time of the new equipment so that it can be used by operations at the earliest possible time.

In figure 2.9, the Total Vertical Start-Up time was reduced from 133 days to just 86 days for Meco 6. This means that Meco 6 had been commissioned much faster compared with the previous 5 Meco Machines allowing the machine to start up early by 47 days compared to Meco 1, 2, and 3 and start generating revenue for the industry.

Chapter 3

<u>Understanding the Equipment's Running Costs</u>

> *Reducing costs has always been the focus of most maintenance managers and perhaps, we need to learn from the lessons of history. The cost must be studied not just based on its initial cost but on the entire life cycle cost of the equipment. Considering only the initial cost is just the tip of the iceberg, and underneath the iceberg lays the true cost far greater than the initial cost of the equipment.*

<u>3.1: Why Cheap Is Expensive</u>

During my first work in 1986, where I was assigned as an oiler on board a cargo ship named MV Sea Raider, our common destination was from Manila to Singapore and vice-versa. During our maiden voyage, we carry some traders who will purchase different items and sell them in the Philippines. During our first trip to Singapore, the crew decided to go out to the city on Orchard Road in Singapore where I decided to come along. There were many cheap clothes you can purchase at a bargain. I was looking at a pair of white Bell Bottom pants that were displayed and they looked really awesome I stared at them for a few minutes and decided to purchase them. I wore it during the evening in time for dinner and the crew was mesmerized by its look as it covered the 1.5 inches heel of my leather shoes entirely. It was really cheap at 4 Singapore Dollars. (1 USD = 2 SGD = 18 PHP) or 144 PHP. The following morning I washed them since I planned to go out once again that evening, finally, after ironing and wearing them, the pants shrunk by several inches. I was so appalled by the experienced that I decided to throw that thing (the white bell-bottom pants) in the sea and I never ever bought any clothes from that day forward in Singapore.

A couple of decades ago, I decided to paint my roof and hired a painter to do it. The painter advised me to purchase some well-known brand of primer paint, so I went to the paint shop to purchase some paint. There were varieties of paints to choose from. Some of them were even shown on TV commercials, but those shown on TV were priced much higher compared to some unknown brands of paints. The difference in paint is quite a big margin since low-cost paint costs around Php 245.00 per gallon or roughly around USD 5.21. The cost of some leading brand paint shown on TV was around Php 390.00 or USD 8.30 USD. I've decided to purchase the cheaper paint since I will be needing around 4 gallons of it. When I reached home, I contacted the painter, and he went to our house. I contracted him and paid him daily to paint the roof. He said that I should have purchased the more reliable brand so that it can last longer, but I guess the paint was already there, so he just started painting and completed the work, which lasted for a week. A year later, I was having some problems viewing my TV reception since the antenna's orientation changed as we were still using an aerial antenna during that time as the cable was scarce and quite a bit expensive. I decided to go to the roof to try to fix the orientation of the antenna; to my surprise, when I saw the roof, almost all the primer paint faded off. I might have saved on the initial costs of the paint, but since I needed to repaint it again, the cost of the paint has now doubled and is much more costly than the leading brand of paint. If I considered both the economic part of the paint, perhaps I would be saving more if I purchased the paint with an initial higher cost. I think that the painter knows more about what the life cycle is than I do. That was a very poor decision on my part.

In industries, if we conduct a survey and ask each maintenance manager in every industry which of these two things is more important, cutting costs on corners, or preserving equipment's reliability? Almost everyone will agree that preserving equipment's reliability will be more important than reducing costs. If failures and breakdowns are reduced, then the cost of maintenance will definitely go down. However, this is not reversible. In fact, many cases state that cutting costs and purchasing cheap spares and items for maintenance will create more problems with the equipment than you can ever imagine. If we conduct a survey on what most industries are doing, I believe that cost cutting is the name of the game most industries are playing which may not be a good strategy since this can backfire and produce more harm than good to their equipment and assets.

In a survey I started in 2009 regarding the Top 10 Problems on Preventive Maintenance, lack of training on the maintenance function ranked first. Training budgets are often slashed or rejected by top management. The common perception is that we have been doing things this way and training just adds up to the cost. Maintenance often realized a reduction in their maintenance budget, which happens every year. Headcounts are reduced as industries think they are truly saving money by finding ways to reduce the benefits of their employees. Most of these maintenance people will be laid off permanently while others will be forced to retire as these maintenance people reached their retirement age, sad to say. At the same time, industries replace them with fresh new graduates with absolutely no experience in the field whatsoever and these new guys find their way inside committing many mistakes and errors to perform their job which later on will be called experience.

The budget on spare parts is cut, and purchasing will source localized parts that are cheaper or award the part to the lowest bidder unknowingly that the material used is different and have less strength of material. All of these are just some of the few strategies of industries that are into the cost-cutting scheme. Purchasing people are busy looking for vendors and suppliers that can provide them with the lowest possible cost on their spare parts. On the other side, maintenance people are always in a never-ending battle against production people in getting the equipment for scheduled Preventive Maintenance work, where in most cases, the schedule will be deferred at the very last minute despite all the planning and preparations. The budget for maintenance training is often cut-short, and maintenance can barely attend training since all of them are too busy fire-fighting. In short, maintenance is left in a never-ending battle against reactiveness with sudden random breakdowns everywhere. In the end, not only maintenance is definitely going to lose this war but the whole organization too. The pressure is building on the maintenance side and as pressure increases, the maintenance people are in a state of confusion where the end game always boils down to chaos. At this point, the sentiment of most of these maintenance people is that how they wish management hire additional maintenance manpower as there are too many breakdowns realized during operations. To add to the burden, the parts needed are unavailable in the storeroom. With this, maintenance is left with two options, which are either to prepare for an emergency purchase or look around for any idle equipment and cannibalize the parts so the production can continue.   This is the life of a typical day of maintenance people in a reactive plant. It does not have to be this way.   If industries are dead serious about saving costs then there is a much better way, which is about applying the concept of Life Cycle Costing.

The study of Life Cycle Costing does not just limit us to developing new products and purchasing new equipment. It can also allow us to make decisions as to whom to award the spare parts and items needed for our equipment and assets. For industries, there is always a temptation to purchase equipment and spares based on the lowest or cheapest vendor. This might not be such a good idea since purchasing based on the initial costs will tell us just one side of the story. The true cost can be seen based on the overall life cycle cost of the part or item.   In my experience, this is the problem with most procurement and purchasing departments since they make the decision to purchase parts based on the lowest possible bidder. The savings generated by these department claims are insignificant, since both operations and maintenance encounter problems as a result. They always look at the initial cost, and not at the entire overall running cost of the equipment. This is the problem with buying cheaper parts. Equipment that fails due to this will likely yield a much larger amount of cost in a period compared to a slightly higher cost. The true cost of the equipment can be felt from the time the equipment was commissioned in the plant until the time it is retired or disposed of. What we are saying is that cheap can actually be expensive in the long run, and expensive can be cheaper in the end. It is just a matter of creating a balance and understanding what Life Cycle Cost is all about.

The initial costs of the equipment can be considered as the costs of designing, fabricating, installing, and commissioning them for the plant. That is already known as the tag price or the initial costs of the equipment. Although the commissioning costs can vary among vendors depending if they will provide additional costs if contractors will stay as part of maintaining the equipment and work together with the PM group for a certain period. If the vendors would be contracted, then we need to add this up to the costs of operating and maintaining the equipment. Commissioning costs, in this case, includes the costs of transporting the equipment to the plant, installing, commissioning, and debugging the equipment until it the fully stabilized and now ready to be endorsed to operations. Likewise, warranty costs, insurance spares, and the cost of labor or contractor in commissioning the equipment initially are involved. When the equipment is finally endorsed to operations, the cost of operating, and maintaining the equipment begins. The costs in operating and maintaining the equipment include the costs of doing maintenance such as Preventive Maintenance, repair or corrective maintenance, shutdown costs, costs of spare parts involved, consumables, downtime costs, lubrication costs, energy utilized such as power, air, fluids consumed, manpower costs, cost of training operators and maintenance, costs of doing modification from time to time, costs of operating the equipment, overheads, cost of repairs and so on. These are the costs we face once the equipment is being operated in the industry until the time the equipment will no longer be used and will be decided to be decommissioned or disposed of. Just like a person, there are still costs involved when a person dies, such as the cost of funeral services and the costs of a place in the cemetery, or the cost of being cremated. Decisions about purchasing or acquiring equipment and spares should not only be based upon the tag price or procurement cost of the part or equipment itself but rather on its entire Life Cycle costs. Some of the questions that need to be raised include the following:

• How will the equipment or part perform in the long run?
• What will be the cost of operating and maintaining this equipment?
• What are the critical parts of this equipment?
• How frequently should the part be replaced?
• How much is the cost of spares and downtime?
• What parts need to be stored inside the storeroom?
• How long will the equipment be down in the event of a breakdown?

Purchasing cheaper spare parts may actually mean spending more in the long run. A true and meaningful cost reduction effort can be achieved not by purchasing the cheapest costs but rather by having a thorough understanding of what Life Cycle Cost is all about which is looking on the long-term and not on a short-term basis. Many plants are initiating a cost-cutting scheme in their plant, which can have harmful repercussions. The best way to reduce costs is to understand that it is more important to look into the running cost of the equipment rather than looking at the initial or procurement cost of the equipment t itself.

3.2: The Running Cost Will Not Always be Easy to See

The initial cost will always be given as this is the tag price or the acquisition cost of the equipment. This can be known easily by the buyer. On the other end, the running cost will not be easy to see especially when this is a newly designed machine or piece of equipment that is not common to existing equipment in the plant. At least 80% of the equipment's LCC can be estimated during the design stages. The running cost cannot be calculated precisely during the design phase, and may not actually be the same when the equipment is being operated in the plant. But before we can estimate the LCC during the design phase questions to raise will be what cost will be included in the calculation of LCC? Estimating the LCC is a process of providing the most reasonable estimate of the equipment throughout its entire lifespan. Although JIPM Japanese consultants favor that the design of the equipment should be done in the plant where research and development, as well as engineering people, are involved. This may not be the same case for other countries as the equipment is usually being purchased as a whole unit. What is important is to ask the OEM or vendor if there is a study regarding the estimated running cost of the equipment. The running cost can be any costs or expenses that can be accumulated on the equipment being purchased during its entire life cycle until the time it will decommission.

Equipment's Initial Cost (IC) Includes:
• Acquisition cost of the equipment
• Auxilliary equipment if cost is separated from the main equipment
• Transportation and Shipment Cost
• Warranty Cost and Insurance Spares
• Installation and Commissioning Costs
• Operator and Maintenance Training

Equipment's Running Cost (RC) Includes:
• Repair and Maintenance Costs
• Consumable Cost
• MRO Spare Parts Cost
• Calibration Cost
• 3rd Party Contractor's Cost
• Facilities and Energy Costs
• Downtime Costs
• Modification Cost
• Preventive and Predictive Maintenance Costs
• Manpower Cost
• Capital and Insurance Spares
• Overall Maintenance Costs
• Equipment Depreciation Cost
• Decommissioning or Disposal Cost
• Fuel Cost
• Lubricant Cost
• Others (Specify in Detail)

3.3: What the Running Cost Includes

The running cost of the equipment will include all expenses and costs incurred during the entire life cycle of the equipment from the time it will be operated by the plant until the day when the equipment will be retired or decommissioned. Just like people, when people die, there is still a cost incurred to finally give the family its final respect to the deceased. In the Philippines, you can even hire crying ladies which you pay to cry. The more you pay, the louder they cry. Although when a piece of equipment is purchased from the vendor, this only includes the cost of the equipment. In most cases, the installation and commissioning will be an entirely different cost depending on the vendor or whenever it is also stated in the contract price of the equipment. Some of the costs that will be incurred during the entire lifespan of the equipment are already given but others can be reduced by applying the correct practices in maintaining the equipment.

By definition, maintenance costs will include any expenses incurred on the equipment, machines, or asset by the maintenance function to sustain and preserve it. It can be said as a major part of the total operating costs for all manufacturing and non-manufacturing industries. Depending on the type of industry, the Total Maintenance Costs can represent around 3 to 50% or sometimes more of the costs of products and services produced.

According to my dear friend R. Keith Mobley in his book An Introduction to Predictive Maintenance, he quotes that recent surveys of maintenance management effectiveness indicate that one-third or 33 cents out of every dollar of all maintenance costs is wasted as a result of unnecessary or improperly carried out maintenance. When you consider that US industries spend, more than $200 billion each year on maintenance plant equipment and facilities, the impact on productivity and profit that is represented by the maintenance and operations becomes clearer. The result of ineffective maintenance management represents a loss of more than $60 billion each year. This statement truly signifies that there is a lot of room for improvement in the maintenance function.

The cost incurred during the equipment may vary from one equipment to another and from one industry to another, hence, it is important to identify what will be the different costs that will be incurred when operating and maintaining an asset. For the maintenance cost, we also need to classify and code each of these maintenance costs, so we can understand and analyze the main contributor.

Equipment's Overall Running Costs

	↓ Costs / Marchine ⟶	1	2	3	4	5	6	7	8	9
Initial	Equipment Acquisition Costs									
	Auxilliary Components									
	Repair Costs									
Running Costs	MRO Spare Parts Costs									
	Consumable Costs									
	Training Costs									
	Installation and Commissioning									
	Facilities and Energy Costs									
	Manpower and Labor Costs									
	Overtime Costs									
	Warranty Costs									
	Capital and Insurance Spare Costs									
	Modification Costs									
	Other Costs (Specify)									
	Depreciation Costs									
	Disposal Costs									
TOTAL RUNNING COSTS										

Figure 3.1: Monitoring Equipment's Actual Running Costs

Although what to include in the lists of maintenance costs may differ from one industry to another. Other industries assign a cost center to determine where to charge the cost, especially for large organizations. Figure 3.1 represents what can be included in what is allocated to the maintenance costs. For example, I have clients in which the training costs for maintenance are under the budget of the training department or human resources, while for other industries, the training budget is allocated per business unit. Hence, if a plant has several business units, then I need to prepare a Sales Invoice and Official Receipt for each of these business units.

OVERALL MAINTENANCE COSTS

Code	Maintenance Cost Incurred	Code	Maintenance Cost Incurred
MRO	MRO Spare Parts Costs	BOM	PM Bill of Materials
MDT	Machine Downtime Costs	CC	Consumable Costs
LUB	Lubrication Costs	MOD	Modification Costs
CC	3rd Party Contractor's Costs	DC	Decommissioning Costs
RNM	Repair and Maintenance Costs	OBS	Cost of MRO Obsolete Parts
CAL	Calibration Costs	NMP	Cost of MRO Non-Moving Parts
EC	Energy Costs	FC	Fuel Costs
MOT	Manpower Overtime Costs	MOT	Motor Pool Costs
LAB	Labor Costs	OTH	Others

Figure 3.2: Coding Maintenance Costs

To visualize a clear representation of what the Overall Maintenance Costs will include, it is important to provide a code and category for each of these costs individually spent on the equipment as in figure 3.2. Categorizing every cost incurred on the equipment can provide us a clear understanding of where we can target our efforts to reduce what seems to consume where most of the money is spent on maintenance. What is important is to be consistent on what will be included in the maintenance costs since, in the majority of cases, the maintenance costs will be one of the biggest contributors to the equipment's running costs.

3.4: Percentage of Maintenance Cost to RAV

One of the measures and indicators of determining whether to still continue to operate the equipment or retire them is the use of this maintenance indicator called RAV or Replacement Asset Value also called the Estimated Asset Value (EAV). It is the cost of maintaining the asset, which is measured against the value of the asset. It can be said that as the percentage of the cost to replace the asset. As we continue to operate our equipment and assets, the value of our equipment depreciates. The main cause of depreciation is that our assets deteriorate over time. There will always be the subject of wear and tear for mechanical parts. Deterioration is given and will happen in our equipment and assets, and the best that maintenance can do is to prolong its process, however, when the time comes, then the lifespan of the equipment had been reached, and the equipment will be finally decommissioned which end its life cycle. With this, the cost value of our equipment will be reduced based on the amount of time we use our equipment and assets. The book value of equipment or asset refers to the book from the Accounting, or Finance Department and is reflected on the industry's financial statement. This will be equal to the Total Cost of the Asset minus its liabilities. On the other hand, the Market Value will be the value or cost of the asset according to the stock market. This will be the cost of the asset when sold to the market.

When we speak about the Percentage of Maintenance Cost to RAV, which is expressed as a percentage, the lower its value the more effective we are in maintaining and preserving our equipment and assets. This can be used by maintenance managers and decision-makers to decide whether to still continue operating the equipment, modify it, or just retire the equipment for good and purchase a piece of new equipment. This is like owning a car for 5 years since this will be one of the deciding factors on whether we still continue using the car or just sell it at the current Market Value.

For industries, we need to define clearly what will be included in the Overall Running Costs since this can vary from one industry to another. Second, we need to be consistent in only including the cost that is spent on operating and maintaining the asset during its entire period. This indicator, Percentage of Maintenance Cost to Replacement Asset Value can provide us information and decision on whether we need to maintain or replace the asset with a brand new one. The lower the value of this percentage, the better. This means that if the assets' Percentage of Maintenance Cost to RAV is at 25% or more, then your maintenance is too expensive. Mostly the target value will be at 5%, but this depends on the industry, and lowering this value means that we have lowered the cost of maintaining the asset. It would be more economical to keep the asset operating instead of selling it at a depreciated value and purchasing a new asset or equipment.

The Percentage of Maintenance Cost to RAV can also provide us information on which assets we need to replace, modify, and those that we need to retain. If these indices will be included in the maintenance function, this should be discussed with the IT to integrate this if the plant currently has an EAM or CMMS software as this will be very tedious to do individually for all equipment and assets in the plant. Although we can have a decision on whether to continue operating or sell the equipment whose percentage will be 25% and beyond which can be flagged down easily by the system. What is important is we also need to consider the non-tangible or qualitative factors such as the following:

• Will the new equipment be easier to operate?
• Will there be less movement on the operator?
• Will the new equipment be more ergonomically convenient for operators?
• Will the training for the new equipment be easier than the old one?

Case Study: An OEM is proposing to management to purchase a piece of new equipment. The cost of the new equipment is $ 50,000.00, and the maintenance cost is projected at $ 8,000.00 annually. The cost of the current equipment is $ 65,000.00 and the average maintenance cost is $ 12,000.00. The Replacement Asset Value today of this machine is $ 35,000.00. A TPM Planned Maintenance Team had recently concluded its modification of a prototype for evaluation on the current equipment and is proposing a modification cost of $ 5,000.00, which can increase productivity by 30% and is projected to reduce their maintenance cost by 60%. It is forecasted that the demand in the next 3 months would be increased by 20%. Would it be feasible to buy the new equipment as recommended by the vendor or proceed with the modification by the Planned Maintenance team?

Given:
- Cost of New Machine = $ 50,000.00
- Projected Maintenance Cost Annually for New Machine = 12,000.00
- Cost of Old Machine = $ 65,000.00
- Annual Projected Maintenance Cost for Old Machine = 12,000.00
- Annual Projected Maintenance Cost, Old Machine with Modification = $ 12,000 - (12,000.00 x 0.6)
- Projected Maintenance Cost Annually for Old Machine with Modification = $ 4,800.00
- RAV for the Old Machine = $ 35,000.00
- Modification Cost = $ 5,000.00

For the New Machine Recommended by OEM
- Percentage Maintenance Cost to RAV for New Machine = $(Annual Maintenance Cost x 100) / RAV
- Percentage Maintenance Cost to RAV for New Machine = $ (12,000.00 x 100) / $ 50.000.00
- Percentage Maintenance Cost to RAV for New Machine = 24 %

For the Old Machine without Modification
- Percentage Maintenance Cost to RAV for Old Machine = $ (Annual Maintenance Cost x 100) / RAV
- Percentage Maintenance Cost to RAV for Old Machine = $ (12,000.00 x 100) / $ 35.000.00
- Percentage Maintenance Cost to RAV for Old Machine = 34 %

For the Old Machine with Modification
- Total Maintenance Cost = [12,000 – (12,000 x 0.60)] = $ 4,800.00
- Percentage Maintenance Cost to RAV (w/ Modification) = (Annual Maintenance Cost x 100) / RAV
- Percentage Maintenance Cost to RAV for Old Machine = $ (4,800.00 x 100) / $ (35.000.00 – 5000)
- Percentage Maintenance Cost to RAV for Old Machine = 16 %

The conclusion, in this case, is to proceed with the Planned Maintenance Modification as this indicates the lowest value regarding the Percentage of Maintenance to RAV and disregards the proposal of the vendor to purchase the new equipment.

3.5: MTBF and MTTF Explained

Another important indicator we need to consider is both the MTBF and MTTF. By definition, MTBF is an average measure of reliability that the equipment will run without failing. The most common units used will be in hours. Its origin can be traced to the US Military Standards (MIL-STD-217), and since then, it has been widely used in other applications in various types of industries. The original Reliability Prediction Handbook was MIL-HDBK-217, the Military Handbook for the Reliability Prediction of Electronic Equipment, which was published by the Department of Defense, based on studies done by the Reliability Analysis Center. The MIL-HDBK-217 handbook contains failure rate models for various electronic systems and components such as integrated circuits, transistors, diodes, resistors, capacitors, relays, switches, connectors, and so on. These failure rate models are based on field data that can be obtained for a wide variety of parts and systems. MTBF can also be defined as the average time between two successive failures, and it is a measure of the time the equipment is operating until the time it encounters a failure. By empirical testing or allowing a part to fail which is a form of destructive testing, the length of performance or the functional life of a population of items can be divided by the total number of failures to achieve the MTBF of a part or component. By the word mean, we speak about an average amount of operating time between the occurrences of breakdowns or failures that require repair.

MTBF will be easy to compute if the equipment is suffering only from breakdown losses, however, if a machine is experiencing other losses, these must be excluded in calculating the MTBF value of the equipment. Several equipment losses will also constitute a downtime such as minor stoppages, set-up, or conversion, defect losses, start-up losses, and so on. What is important in calculating the MTBF value is to consider only the downtime caused by breakdown only. MTBF is an average measure of reliability that a device will run without failing. It is a reliability engineering term that means the average amount of operating time between the occurrences of breakdowns that requires repair. It means the average amount of time between two successive failures and is based on historical data or estimated by the vendor, which is used as a benchmark for reliability. MTBF trend will be, the higher the value, the more reliable the machine or asset is performing. Depending on how you intend to measure MTBF since this can be measured on a system-based level, sub-system based, component-based, equipment-based, several groups of machines, sub-assembly based, or even to the spare part level itself. The formula for MTBF will be as follows:

MTBF = Operating Time / Number of Failures
Where: Loading Time = Available Time – Non-Machine Related Downtime
       Operating Time = Loading Time – Machine Related Downtime

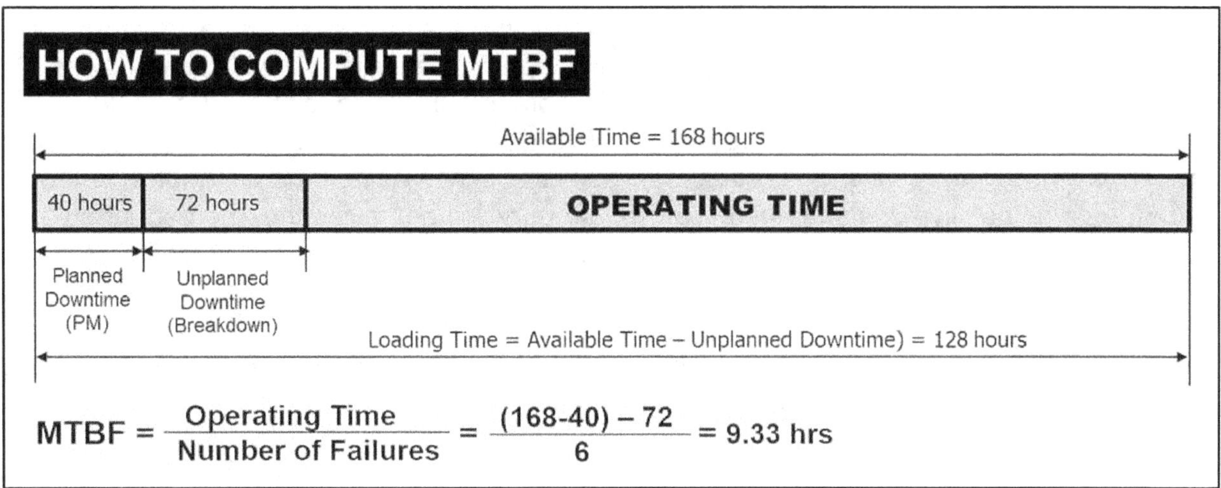

$$MTBF = \frac{Operating\ Time}{Number\ of\ Failures} = \frac{(168\text{-}40) - 72}{6} = 9.33\ hrs$$

Figure 3.3: How to Compute MTBF

Given:
• Available Time = 168 hours

• Machine Downtime due to Breakdown= 72 hours (Also called Unplanned Downtime)
• Non-Machine Downtime due to PM = 40 hours (Also called Planned Downtime)
• Loading Time = Available Time – Non-Machine Downtime = (168 – 40) hours = 128 hours
• Operating Time = Loading Time – Machine Downtime = (128 – 72) hours = 56 hours
• Frequency of Failures = 6 times

• MTBF = [(168 – 40) – 72)] hours / 6 hours
• MTBF = 9.33 hours

In figure 3.3, the MTBF is equal to 9.33 hours. So what do 9.33 hours actually mean? Most people are confused about interpreting this figure and will say that when it reaches 9.33 hours of operation, the equipment will fail at that precise period. Remember that the word "mean" refers to an average amount of time; therefore, this is interpreted as a probability or possibility of failure, but this is not the case in almost all instances since we are only getting the average time the equipment will fail. The failure can occur before or much longer than 9.33 hours. So, the correct way of interpreting this figure is by stating that there is a probability of failure when it reached 9.33 hours of operation.

Reliability predictions are used to compute the predicted failure rate or the Mean Time Between Failures (MTBF) of its system. MTBF is usually expressed in terms of hours. For example, if the system has an MTBF of 1000 hours, this means that, on average, the system can experience one failure in 1000 hours of operation. By using accepted standards for modeling failure rates of components, we can calculate the equipment's failure rate, which is the reciprocal of MTBF. The goal is to make sure that the asset's MTBF is within its acceptable limits. If the prediction analysis shows low MTBFs, this means that you can expect your system to fail more often, and you may need to take steps to improve it. By making changes to the design, for example, maybe lowering temperatures or stress levels, the MTBF can improve and expect the machine to have better performance. Only losses due to breakdown should be included in the MTBF calculation. MTBF is a good indicator that tells us that the higher the value, the better its performance and lesser breakdowns. In one week, the highest MTBF value that can be obtained in the machine will be 168 hours. For a given month with 30 days, the highest MTBF value will be 720 hours and 744 hours for 31 days. In determining if the equipment's MTBF is increasing, it is best not to compute MTBF on a cumulative basis. If the MTBF for January will be calculated, the data should be for January 1 to 31, for February, it should be 1 to 28, and so on. This way we can determine if the MTBF of the equipment is increasing or not. In case, there was no breakdown experience for the month, we will be achieving a value of infinity. In this case, we can prolong the duration of getting the MTBF, or we can assume a denominator of one to obtain the value of the numerator.

Mean time to failure (MTTF) is a maintenance indicator that measures the average amount of time a non-repairable asset operates before it finally fails. MTTF will be used for assets and equipment that can no longer be repaired which can also be said to be the average lifespan of an asset. Mean time to failure (MTTF) is a maintenance metric that measures the average amount of time a non-repairable asset operates before it fails.

Because MTTF is relevant only for assets and equipment that cannot or should not be repaired, MTTF can also be thought of as the average lifespan of an asset or item.

MTBF (Mean time between failures) for a repairable item is the average length of time between failures that requires repair. Therefore, for both MTTF and MTBF, the longer the period between failures, the more reliable the component or asset will be. MTTF (Mean time to fail) is the life test data collected, which can be used to calculate the average length of time between failure of a non-repairable component that is occurring, such as in the case of a light bulb. The difference between MTBF and MTTF is that MTBF is a key reliability metric for systems that can be repaired or that can be restored. MTTF is the expected time to failure of a system. Non-repairable systems can fail only once, hence for non-repairable items, MTTF is equivalent to the mean time of its failure time distribution. A repairable system can fail several times, while a non-repairable can fail only once. Therefore, when we speak about MTBF, we also consider the repair time since MTBF = MTTR + MTTF. The trend for both MTBF and MTTF would be the same, where both mean time indicators require, that the higher the value, the better. In most cases, MTTF can refer to spare parts or items, which cannot be repaired, or reused on the equipment or machine, while MTBF refers to repairable systems or equipment. The reciprocal of MTBF and MTTF is the failure rate, which is the probability of failure. Hence, for as long as the equipment or machine can be repaired during a failure, then we use the term MTBF. Once an asset can no longer be repaired and will be subject to decommission, then we can also refer to this as the MTTF of the asset. For human beings, MTTF represents Rest in Peace (RIP). This means that the longer an asset can operate, the higher its Mean Time to Fail. MTTF can be used to analyze which vendor to award a certain part or item as this means that the decision to whom to award will be based on the part or item with the longest lifespan.

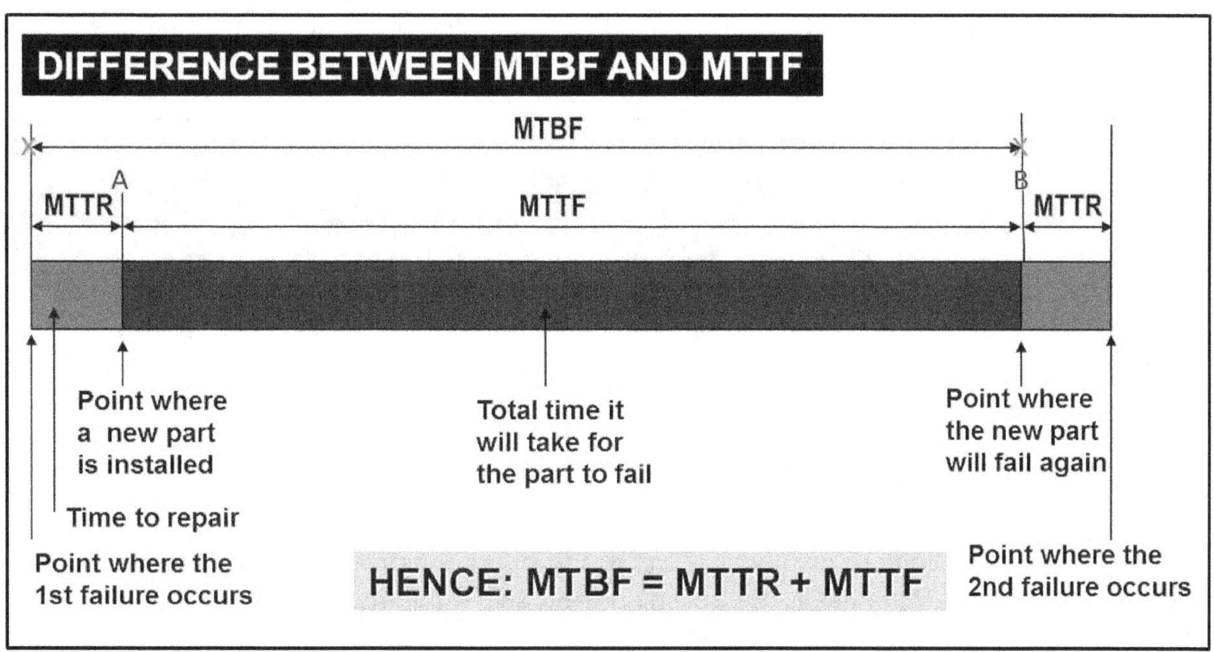

Figure 3.4: MTBF and MTTF Explained

3.6: Equipment's Life Cycle will Start with the Basics

Establishing the basic equipment condition simply means eliminating the causes of accelerated deterioration. This is when a certain part or item does not each its dictated useful life. It means cleaning to remove dirt and addressing sources of contamination, proper lubrication to prevent early wear, and understanding that bolts need to be complete and secured with the correct amount of torque. In many cases, parts do not achieve their desired lifespan due to dirt and contamination. The wrong lubricant is poured into the equipment because the guy is new. The machine fails simply because several bolts are missing. The maintenance uses the wrong tools that later wreck the equipment or equipment containing so many forms of leaks.

Once a piece of equipment or machine is loaded, automatically it will be subject to stress simply because there are mechanical movements inside. These mechanical parts that have movements will be subject to wear. In hindsight, dirt, dust, foreign material, oil contamination, inadequate lubrication, and excessive vibration can cause several parts to fail prematurely. This is called accelerated deterioration, which means that these parts did not reach their natural lifespan. Contaminants and other foreign objects can cause abrasion and can come in direct contact with rotating parts, hydraulic systems, and other critical parts of the equipment. These solid contaminants can be trapped in tight clearances between two mechanical parts thereby causing abrasion. The goal of establishing the basic equipment condition in the equipment is to expose these abnormalities, and flaws to allow these parts to reach their natural deterioration and useful life. Many failure modes can be addressed if the basics have well been established in our equipment. Establishing the basic equipment condition is a collaborative effort between the operators and maintenance. In TPM (Total Productive Maintenance), this will be parallel activities between both Planned and Autonomous Maintenance. This is no rocket science strategy, but what it takes is just a modicum of common sense.

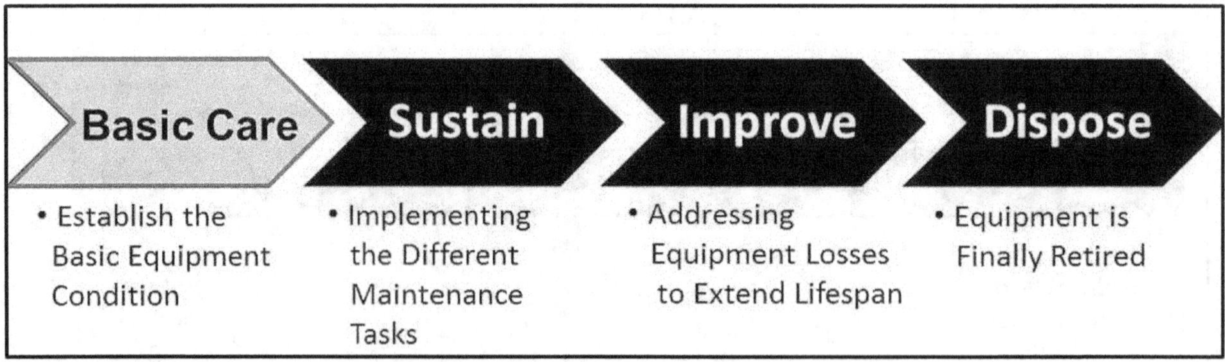

Figure 3.5: Equipment's Life Cycle Will Start with the Basics

How long the equipment or machine will run in operations depends on how we care for the equipment or machine itself. Establishing the basic equipment condition is a joint effort between both operators and maintenance people. When operators start helping maintenance in establishing these basic equipment conditions, equipment reliability starts to improve and maintenance can focus more on specialized activities and start performing improvements in the equipment. This is the importance of having Autonomous Maintenance in place for the reason that the basic things needed in our equipment are finally being taken care of. When Autonomous Maintenance is missing, the feud between operations and maintenance takes place and maintenance will always be blamed if equipment lags behind in productivity. The mindset of, "I operate you repair," must be changed. This is what is happening to most industries and sad to say that this thinking has now become a part of their culture. This means that even before we can sustain the equipment, we need to address the basic condition. If I can explain this in the simplest way possible, let us think of the equipment when it was first commissioned in the plant and how it looked currently and see if there is a difference. World-Class industries understand the importance and value of addressing the basic equipment condition which includes:

Figure 3.6: Basic Equipment Condition Includes

• Keeping the equipment clean
• Performing proper lubrication
• Completing the bolts and screws in the equipment
• Addressing leaks

Cleaning: Cleaning is the process of removing any form of any unwanted object from the equipment. It consists of removing dirt, contaminants, grime, excess oil, grease, and other foreign objects that can affect equipment parts and components. Cleaning is performed not just to satisfy the cosmetic looks of the equipment but because this dirt and contaminants shorten the life of certain parts. Detailed cleaning of components and equipment is often a no man's island because while everybody agrees that it is important, nobody wants to do it actually. World-Class Companies perform these activities. They walk the talk. Components and equipment are kept clean in every detail. Such an organization realizes that inspections cannot be done without this level of cleaning. Cleaning definitely extends the life of parts and components. Through cleaning, we can expose problems and abnormalities on the equipment that has been left ignored for a very long time. When operators clean their equipment, they touch parts, and by touching parts they are actually inspecting them. Leaks, cracks, and fractures that have been in the equipment for a very long time can be detected and exposed when operators touch and clean their equipment. Dirty and untidy equipment will tend to deteriorate more rapidly than equipment that is well-maintained and clean at all times.

Lubrication: This is another big problem for many industries. Maintenance might be unaware that many failures can be attributed to lubrication such as excess lubricant, inadequate lubrication, the wrong lubricant used, and contaminants that will definitely shorten the life of the part being lubricated. All machines contain lubricant in one form or another. The main purpose of lubrication is to reduce friction for mechanical parts moving inside our equipment. It is not only important to lubricate the equipment but also to use the correct lubricant and its correct amount. Equipment that lacks lubrication or is over-lubricated can likewise induce problems. Operators should be taught about the importance of checking the lubrication in their equipment and maintaining the correct amount of lubricant and using lubricants with the correct viscosity. So many failures can be attributed to lubrication that can be avoided if only these basic equipment conditions can be established. Maintenance must teach operators not only the necessary points in their equipment to lubricate but also what proper and improper lubrication can do with their equipment. Just like humans, lubrication is the lifeblood of the equipment and it should be maintained clean and adequate all the time. Many failures are attributed due to inadequate or excess lubrication. These things can be avoided and controlled in our equipment if we understand and learn the role that lubrication play in our equipment.

Bolting: Machinery contains bolts, nuts, and fasteners as part of their construction and they serve a particular purpose. The equipment functions properly only if fasteners are securely tightened. All equipment vibrates, but excessive vibration can be destructive as this can induce secondary damages to other parts and components affected by the vibration. When we touch the equipment, we can feel its vibration. What we feel is simply the sum of all vibrating forces moving inside the equipment. Bolts, screws, and fasteners are placed to minimize the vibration in the equipment. When vibration increases, it can result in cracks and fractures and these fractures can propagate to a point of rupture as a result of excessive vibration. Remember that it only takes one loose bolt to create a chain reaction of destruction and havoc in our equipment. As a result, other bolts become loose, and vibration increases. Match-marks are placed on critical bolts and nuts so that operator can easily detect if bolts have been loosening due to excessive vibrations for some time. Imagine driving the car and each of the tires has 3 instead of 5 stud bolts on each of your tires. Would this be all right with you? I guess not. Then why don't we treat our equipment in the same way? We can see many bolts missing or loose as a result of many activities performed by the equipment. For a start, why don't we assess all the equipment in the plant and locate equipment with missing bolts, and complete them?

Addressing Leaks: Equipment leaks may be in the form of oil leaks, air leaks, vacuum leaks, gas leaks, or any other form. The simplest method of identifying and correcting leaks is to tag them. Locate the leak and identify its source, type of leak, and severity of the leak. Set a schedule for correcting these problems on the equipment. Implementation of a simple, easy-to-use system such as this can save hundreds of thousands of dollars for a plant. Other types of leaks, which cannot be detected by the human senses, will require special Predictive Maintenance instruments such as Ultrasonic Monitoring, which can detect leaks from the compressor. Compressed air is one of the most costly utilities in plants today. The Department of Energy has estimated that around 30% of all compressed air produced in the US is lost due to leaks. An ultrasonic compressed-air leak survey can offer fast payback. One Plant averaged an electric cost of $1,760,000 to generate compressed air. The survey using Ultrasonic Monitoring identified 3,561 leaks totaling 6,340 CFM at an annualized cost of $597,000 of energy wasted. A program of leak detection and repair can have a dramatic effect on the profitability of a company such as this. In addition to being a source of wasted energy, leaks can also contribute to other operating losses. Leaks cause a drop in system pressure, which can make air tools function less efficiently, adversely affecting production.

3.7: Why Operators are Important in Maintenance

Operators are important in any maintenance strategy since it is believed that operators are the first line of defense in any failure that can occur in the equipment since they are the people closest to the asset when the equipment fails. These people will be the ones who will experience the failure first before the maintenance. Another reason why operators should be involved is that maintenance can only escape the vicious reactive cycle if operators are involved in maintenance. One of the important roles of the operator is to help maintenance work hand in hand in establishing the basic equipment condition. These operators should provide continuous communication and collaboration with maintenance because it is important for the maintenance people to understand what eventually happens before the failure itself. For those industries implementing Total Productive Maintenance (TPM), the pillar where the operator will be involved will be Autonomous Maintenance. Autonomous Maintenance has two aims. First, from a human perspective, it fosters the development of skilled and knowledgeable operators in light of their newly defined roles. Second, from an equipment perspective, it establishes an orderly shop floor where any deviation from the normal condition can be detected at a glance. Both operators and maintenance should understand that they cannot exists without each other and both should create a solid partnership because maintenance is simply a shared responsibility for both operators and maintenance.

The goal of Autonomous Maintenance is to enhance the senses of operators to detect abnormalities and signs of irregularities at an early point so that catastrophic failures can be prevented. But this is not as easy as we think, as this will take time and continuous mentorship from the maintenance people themselves. This means that for the operator to function, we need to feed them continuous knowledge of their equipment and assets. Just like in a game of chess, the white pieces move first which means that it should be the maintenance who should make this initiative happen by teaching operators regarding their equipment and assets. To be truly competent, operators should be able to spot anything out of the ordinary and immediately recognize it as abnormal. The goal of AM is to enhance and improve the people operating the equipment so that they can improve their own equipment. Operators involved in Autonomous Maintenance will generate three standards, which will include cleaning, lubrication, and inspection standards. Once these standards are in place, the operator will execute these standards on their equipment to maintain the basic equipment condition. If someone would hire a family driver and there are two candidates, where one has a valid driver's license and another one not only has a valid license but has some background knowledge about the engine, which would you prefer? I believe that this is a no-brainer.

Once an operator knows their equipment and machines intimately, they can use their senses to detect looseness, vibration, wear, misalignment, abnormal noise, overheating, and leaks. What we need are operators who can detect, correct, and prevent equipment abnormality. Ability to understand equipment functions and mechanisms as well as the possibility of detecting the causes of abnormalities. Establishing Basic Equipment Conditions is a shared responsibility by both operations and maintenance, and should be performed regularly on the equipment. Catastrophic failures and breakdowns can greatly be reduced if Basic Equipment Condition is in place. Lacking and lose bolts often lead to excessive vibration which produces secondary damage on parts affected by the vibration itself. In my third book I wrote on Reliability – A Shared Responsibility for both Operators and Maintenance, I mentioned that for as long as operators and maintenance in industries remain separated, then the maintenance can't escape the vicious cycle of reactiveness. What is important is having a partnership with operators and teaching them about the failures that can be expected and the signs. Maintenance must explain to operators the items being inspected by them and the impact of the failure if this will be neglected.

Chapter 4

Sustaining Equipment against Failure

> ***All Maintenance tasks performed on the equipment, machines, and assets should address a particular failure mode. Those tasks listed that do not address a particular failure mode are deemed useless and a waste of resources and should be totally removed from the tasks. This is how to assess whether our maintenance tasks are effective or not.***

4.1: Stress Versus Strength

When industries purchase equipment, management thinks that machines aren't supposed to break and mechanical components such as shafts, fasteners, and structures aren't supposed to fail, however, when they do fail, they can tell us exactly why. But in the majority of cases, this is not always the case since once a piece of equipment fails, then the repair process follows. When we repair the equipment, we just destroy the physical evidence. According to experts, the causes for more than 90% of all plant failures can be detected with a careful physical examination using a low power magnification scope of some basic physical testing. Inspecting the part or item that failed can tell us the forces involved whether the load was applied cyclically or single load. The direction of the critical load and influence of outside forces such as residual stress and corrosion. By knowing the physical roots we can now proceed with the human and latent cause of the failure.

In any failure analysis that involved metal fracture, the subject of both metal strength and the stress within the metal part must be considered for they cannot be divorced from one another. Stress is defined as the force per unit area often considered as a force acting through a small area within a plane. Strength in the metallurgical sense is the property of a metal part that resists the stress imposed upon the part. This means that all fractures are caused by stress, a version of the weakest link theory applies that a fracture will occur if the local stress finally exceeds the strength of the material. Stress can be said as the quantity that describes the distribution of internal forces within a given body. Strain represents the amount of deformation that occurred within a given body. Once a piece of equipment is loaded, there will be stress that will be imposed upon the part. This stress may include the movements of mechanical parts, which caused a vibration that also contributes to stress. Other forms of stress may include the temperature, shock, and the load itself. This means that there is only an (n) number of cycles before the stress finally exceeds the entire strength of the material of the part.

Vibration is accurately defined by all these common terms as swinging back and forth, oscillating, unbalanced, and shaking. The vibration occurs when a machine or machine components move from its neutral position or normal position of rest to a lower and upper extreme limit of travel and this can happen if a force is applied to the rotating machine and components. More force means more vibration and more resistance means less vibration. Vibration can also be considered to be the oscillation or repetitive motion of an object around an equilibrium position. Once the equipment is loaded, then the machine vibrates. The vibration changes as the condition of the equipment changes and can get worst over time. Vibration is a destructive force that leads to the premature failure of rotating machinery. The vibration will degrade weaker machine components until the eventual collapse of the mechanical systems. This happens because vibration forces do not only confine themselves to the faulty components. Instead, they will transmit this destructive force through connecting conduits like bearings, shafts, couplings, and belts to the other parts of the machine and even on stationary equipment. For electronic parts, environmental factors can cost moisture, excessive humidity, dust, and extreme heat, which can damage circuit boards.

Another common stress present in the equipment will be the generation of heat. To counter heat on mechanical components, lubricants are introduced. Although lubricants have different functions, the primary function of most lubricants is to create a film to avoid metal-to-metal contact. Lubricants usually have a range of temperatures they can be used. Once the temperature increase, then it creates a problem with the lubricant itself as the oil starts to oxidize. Once the oil starts to oxidize its alkalinity is reduced which can deplete the different additives present in the oil leaving the base oil unprotected. Oxidation is the breakdown of oil due to the extreme heat in the system. This can cause acidic gases and sludge to form in the crankcase. Also leads to increase viscosity, deposit formation, sludge, and increased corrosion inside the system.

The physical cause will not only be limited to mechanical failures but also to electrical and electronic failures. These electronic components have a wide range of failure modes. Electronic failures can be caused by any of the following; increases in temperature, excessive current or voltage, electrostatic discharge (ESD), ionizing radiation, handling and storage, moisture, contamination, excessive machine vibration, failure of individual components on a printed wiring board (PCB), mechanical shock, mechanical stress, poor solder contact, faulty circuit design, change in the operating environment, thermal stress, humidity, and many other causes. In semiconductor devices, problems in the device package may cause failures due to contamination, mechanical stress of the device, and open or short circuits. One of the major sources of contamination is the environment that threatens electronic components on circuit boards. Moisture can be acquired during the manufacturing or assembly process. Condensation can occur on the printed wiring board and within component packages, especially during cold temperatures where room temperature must always be regulated and monitored.

4.2: Different Types of Mechanical Stress

These stresses are the forces that can be applied to the different parts inside the equipment. Different types of forces can be applied to a mechanical object, which can either be tensile, compressive, torsion, shear, or bending forces. In layman terms, just like in boxing, the boxer can give his opponents different punches like a jab, straight punch, hook, and uppercut. Strain is the quantity that describes the amount of deformation that can occur within a material body whenever a load is applied. The stress-strain relationship varies depending on the material used, and mechanical stress applied. A moment represents the effectiveness of either a rotating, bending or twisting force applied to an object. All materials have their respective hardness. The hardness of the material is the measure of the material's resistance to plastic deformation. The most common method to determine the hardness of the material is the Rockwell Hardness Test, which can test the hardness of metals and alloys. With this method, a hardness number is determined by the difference in the penetration resulting from applying an initial minor load followed by a major larger load. Stress (S) = Force (F) / Area (A).

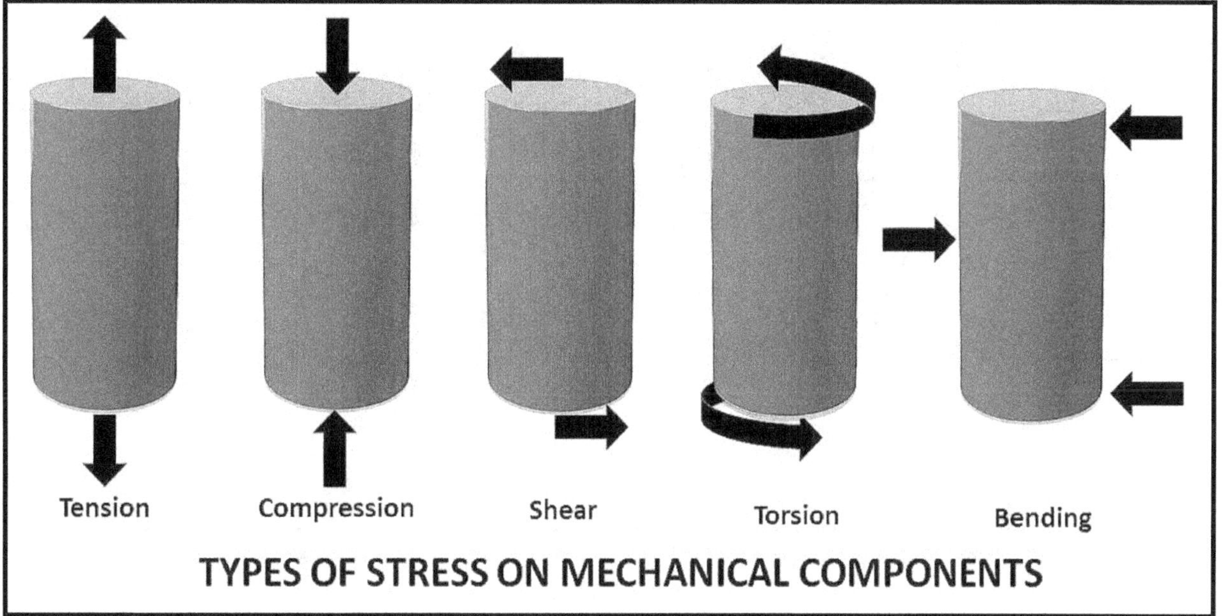

Figure 4.1: Different Types of Stress for Mechanical Components

Tensile Stress: When you have an object in a vertical position such as a cylinder, and an upward force is applied to an object, there will always be an opposite or resisting force, which will be equal to the downward force. Both the upward and downward forces will cause the object to be under tensile stress. Both the downward and upward forces will extend toward the object. This means that when we pull or stretch an object, that object will be under tensile stress. The strain will be the effect of the change of the object, which can lead to a ductile fracture if an elongation happens on the object due to heat, or it can create a brittle fracture if excessive force is applied that results in little or no deformation of an object. A good example of tensile stress can be seen when you tune a guitar in which you are stretching the strings until it reaches the correct note. If the tensile strength of the guitar string had been exceeded, the string would break. The tensile stress (or tension) applied can lead to an expansion of the object. We can also experience tensile stress when we stretch a rubber band. If we get a clay and form a shape of a cylinder, place it horizontally on the surface, and stretch it gently with both our hands, the clay elongates; as we apply more force, it will elongate more until it reaches its limit and breaks in half.

Compression: When we applied a downward force on an object, the object will have an equal upward force, which is equivalent to the normal force. Once we increase the downward force, then we also increase the upward force. These two opposing forces will cause the object to compress. A good example is placing a hardbound book (let us say the book contains 500 pages) on a table surface. If we place two additional books on top of each other, a downward force is applied to the initial book. This means that the book on the bottom of the surface is under compressive stress. The more books we add, the more compressive stress is applied. Generally, it will take much more force to compress an object compared to when pulling it.

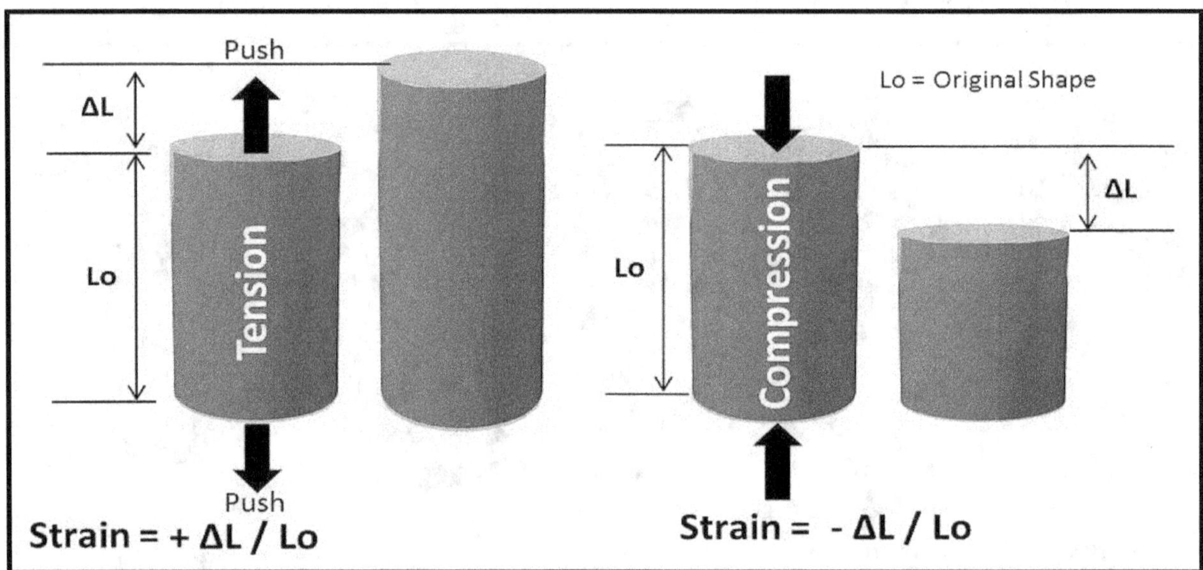

Figure 4.2: Compressive and Tensile Stress

When we talked about tensile and compressive stress, we refer to the maximum stress applied to a material. Once we exceed the maximum stress, the result will be a fracture or breakage that will ultimately lead to the failure of the mechanical part. The strain will be the change of shape of the object after compressive or tensile stress is applied, which can also be equal to the Modulus of Elasticity (E), which is equal to the stress divided by the strain of the object. For tensile and compressive stress, the force applied is perpendicular or at 90 degrees to the area.

Shear Stress: Also called tangential stress. Unlike both tensile and compressive stress, the force applied is not perpendicular but parallel to the area. Once a force is applied in parallel or in opposite directions, it can cause the object to deform. Shear stress is a force acting parallel to a surface or to a planar cross-section of an object. The shear force applied can be from an upward to a downward motion, like when we cut a loaf of bread, or from the left and right motion. Sliding is also a form of shear force. The word shear means that a force can either cut or shear through the surface or object under strain. Shear force is the sum of the effect of shear stress over a surface that results in a shear strain. The important thing to consider is that the force acting on an object is parallel. A ball or roller element sliding on the outer raceway of the bearing is a good example of where shear stress occurs.

Torsional Stress is the twisting of an object caused by the force acting on the object's longitudinal axis. The twisting effect is known as torsion. Torsional stress can lead to deformation. The twisting is called torque, which is also called the twisting moment. The most common examples of torsional stress can be seen on shafts when it rotates. Non-cylindrical objects subject to torsional stress will produce warping. When we apply torsional stress on the shaft. The longitudinal axis will be displaced, which creates an angle of twist. Depending on the material, each type of material will respond respectively to torsion; some may deform or fracture depending on the type of material used. When we wash our clothes manually and squeeze out the water, we hold the clothing with both our hands. Our right hand will move on a clockwise rotation, while the left hand will move on a counterclockwise rotation, thereby squeezing the water out from the clothes. The more force we apply, the more water is squeezed and removed. In this case, we are applying torsional stress to the clothes.

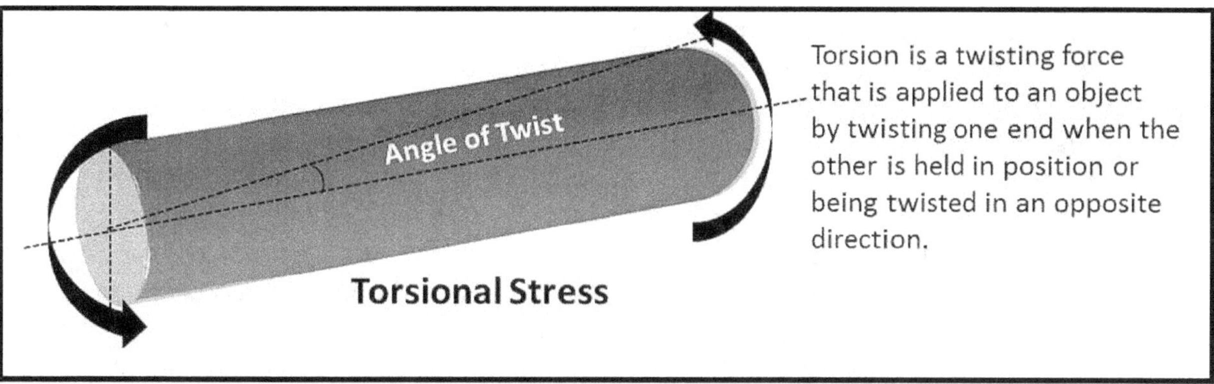

Figure 4.3: Torsional Stress

Bending Stress: This is a force an object encounters when subject to a load at a given point which causes the object to bend or warp. This type of stress usually occurs in objects when subjected to a tensile load. This is the amount of energy it takes to compromise the item from its natural shape or condition. Usually, bending stress combines both tension and compression. This can be most experienced in structural beams.

4.3: Ductile and Brittle Fracture

Fracture refers to the separation of a solid body into two or more pieces imposed upon by the stress, which is usually static and at temperatures relative to the material's melting temperature. The applied stress can be considered tensile, compressive, shear, or torsional. A fracture will involve two processes, crack formation and propagation up to its point of rupture. The mode of fracture is dependent on the mechanism of crack propagation. The fracture can either be brittle or ductile. Any fracture will involve two steps: the formation of a crack and its propagation until its final rupture.

Fractography refers to the science and art of evaluating broken things. People who study fractures are called fractographers. The study of fractography can provide many clues to the person who wants to understand why the component failed or breakdown. Fractography is often thought of as a means of getting information about how and why the part cracked by examining the fractured surface and the forces involved. More advanced fractography can also be used to shed light on many types of wear mechanisms. Fractography is getting information about how and why the part cracked by examining the fracture assembly of the failed component or part. What is important in the study of fractography is to determine the stress or forces involved, whether this is a tensile, compressive, shear, torsional, or bending force that caused the object or material to deform or fracture.

A **brittle fracture** is a fracture that involves little or no permanent deformation. Many non-metals lack ductility and are subject to brittle fracture. A brittle fracture occurs when a part is overloaded and breaks with no visible distortion. In a brittle overload failure, separation of the two halves is instantaneous but proceeds at a tremendous rate where the crack begins at the point of maximum stress. A **ductile fracture** occurs by means of plastic deformation if the yield strength of the material is exceeded. Ductile materials exhibit a substantial amount of plastic deformation with high-energy absorption before they fracture. This causes a permanent change in the shape of the part, which becomes apparent if we attempt to reassemble broken components. There is a great deal of distortion on the failed part. There are instances when a brittle fracture appears in normally ductile materials, which indicates that the load was applied rapidly and vice versa. When we heat up the material, it becomes ductile and deforms. Usually, a ductile fracture will be a slow process, while a brittle fracture will happen instantaneously and very fast. A ductile fracture will be much preferred than a brittle fracture since a brittle fracture can occur without any warning; on the other end, a ductile fracture will involve the process of plastic deformation and provide some warning that a fracture is imminent, allowing maintenance time to replace the items before it ends up into a catastrophic failure. Ductile fractures have also been termed as cup and cone fractures.

Once our equipment, machines, and assets are loaded continuously. Failures are inevitable and can never be eliminated. There will always be the subject of parts wearing out, and when these parts wear out, they actually had achieved their useful life, and therefore it failed in their technical sense. The role of maintenance is to control the timing of failure or eventually prolong the duration of the failure itself. Not all failures are created equal since every failure experienced on the equipment will have its own degree of consequences, which is important in determining which parts will undergo a thorough RCFA. The investigator must depend upon the consequences of the failure itself. Therefore, maintenance aims to control the timing of failure to select or perform a task before it happens. Making our equipment more reliable is about extending the life and the Mean Time Between Failures (MTBF). TPM says that breakdowns can be eliminated, which is unlikely to happen. What TPM is declaring is that it is simply attempting to increase the Mean Time Between Failures (MTBF) of a particular part or item. Here's the thing, when maintenance repairs the failure, conducting a root cause investigation would not be possible since all pieces of evidence that points to the failure have already been washed out. If the physical evidence has been preserved, a group of people can use some failure analysis to diagnose the mechanical cause of the failure itself. While other books can recommend performing an Ishikawa, Pareto, why-why, or other problem-solving tools to get to the root cause of the problem. This is unlikely because these problem-solving tools are really not meant to derive the root cause but only the most probable or likely cause as explained in my book on Root Cause Failure Analysis. The thing is without evidence, we can only guess, speculate, or based our judgment on the most probable or likely cause of the failure.

Figure 4.4: Ductile and Brittle Fractures

Fatigue is the phenomenon leading to fracture under repeated or fluctuating stress having a maximum value less than the tensile strength of the material. This will happen when the stress finally exceeds the strength of the material. Redesigning the material to a harder one can prolong the process of fatigue but can make the material prone to brittle fracture. The harder the materials the more they can become brittle. Fatigue fracture is progressive and can start as micro-cracks and propagates and grow into larger ones that cause destruction of the components. Fatigue fracture results from continuous cyclic slip under repetitive load applications for many thousands or millions of load cycles. These metal fragments that are removed can cause abrasive wear as well as other damage when carried by the lubricant to other parts of the mechanism. Some fracture starts as microscopic cavities and may stay microscopic as they reach their life. Secondly, microscopic fractures gradually become larger after continuous load, and the last stage will lead to a fracture or crack. Fatigue happens at stress levels well below the tensile strength of the material. The cracks take measurable time to progress across the fractured surface.

4.4: Occurrences of Failures

Besides the different stresses experienced by the equipment, another important factor we need to consider is the occurrences of failure. Failure occurs in three patterns, infant mortality failures, random failures, and age-related failures, and most of the failures we encounter on the equipment and assets are a combination of both random and infant mortality failures.

Infant Mortality Failures: These can be said as early failures or failures that happened in the beginning mostly right after a Preventive Maintenance shutdown. Usually, the main cause of infant mortality failures will be human error especially when maintenance comes in contact with the equipment during PM replacements and overhauls. Infant Mortality Failures are failures, which occur at the beginning of the operation. Others refer to them as commissioning failures, start-ups, or debugging failures. Many factors contribute to Infant Mortality Failures, which include poor design, poor quality manufactured, incorrect installation, incorrect commissioning, incorrect operation, poor set-up or conversion practices, unnecessary maintenance, slips and lapses, human errors, bad workmanship, and excessive or intrusive Preventive Maintenance. Infant mortality failures can be reduced by reviewing the current maintenance tasks performed on the equipment and removing non-value added maintenance tasks, which do not address any single failure mode as well as the integration of Precision Maintenance with Preventive Maintenance.

Random failures are failures that occur at any given period. They are also called chance failures. This means that the probability that an item will fail in any one period is likely the same as it is in any other given period. The conditional probability of failure remains constant. If the failure is random, then the part or item will not wear out. This means that if I have 100 bearings and run those all at the same time, the failure of each of these bearings will not happen in the same period. This means that the failures that occur are not failures caused by deterioration or wear out but only by chance failures. They will be characterized by a sudden breakdown without any signs of deterioration. This means that the failure can happen at any given time in which maintenance will be unaware of the failure. Although the failure is considered random, they do have a cause-and-effect relationship. A good example of these will be electronic failures, which can fail at any given time.

Wear-Out or Age-Related Failures means that the parts or items will eventually survive to their guaranteed age and wear out gradually. This means that the part or component has reached its useful life. Once it reached its useful life, then it will be replaced. Age specified may be in the form of running hours, calendar days, number of strokes, number of revolutions, number of stress applied, or any other form. The best maintenance strategy to use for this type of failure will be to identify when the majority of these parts will start to wear out and to apply Preventive Maintenance by replacing them. This is the easiest failure to address. Age-related failures refer to failures that are consistent with reaching their remaining useful life in the process. When the majority of the failures reached a certain period consistently, then we can declare that the failure of the part is age-related. For example, a study on 100 impellers indicates that around 85 % of the pump's flow rate starts to drop after two years of continuous operation. In this case, the impeller is starting to erode, and the rated capacity drops until it is no longer acceptable to production. In this case, the impeller will be scheduled for a replacement. It will be done at a point where the rated flow is no longer acceptable to the user. Hence, we can declare that the failure of the impeller is age-related.

The occurrence of failures can be compared to the lifespan of human beings, which simply states that even if there is an average lifespan for humans, not all human beings will reach their dictated lifespan as others will die either prematurely or randomly. By understanding the occurrences of every possible failure mode that can affect our equipment and assets, we can plan for the most feasible maintenance task to adapt.

Figure 4.5: Occurrences of Failures

4.5: Why Do Equipment and Machines Fail?

There are several reasons why equipment fails. Every industry or manufacturing plant acquire machinery and equipment to manufacture their goods or provide services to their consumers. Their survival depends on many factors such as uptime, cycle time to manufacture, on-time delivery, quality of goods, productivity, and product demand, all of which have a direct impact on the equipment at hand. According to dictionary.com, failure means the inability of a system or system component to perform its required function within specified limits. A failure may be produced when a fault is encountered. This is also considered a lost time due to equipment failure, a downtime loss. The Japanese term for breakdown is Kosho. Sometimes refers to as machine-related downtime, unplanned downtime, tooling failures, and unplanned replacement of spares and components. As discussed previously, there are two types of breakdowns. A functional loss breakdown is a failure or breakdown which it will result in the equipment being stopped, resulting in a total loss of function for the equipment. A function reduction breakdown means that a specific function had failed but the equipment is still running and capable of producing output or delivering, although in most cases this is not listed as a breakdown. This can also be considered a failure in secondary functions. Reliability Centered Maintenance states that in some instances losses in secondary functions may turn out to be more dangerous than the loss or failure of the primary function of the equipment. An example is a car seatbelt is defective, or the left brake light of a car is not functioning, which can cause injury to the driver. Here are some reasons why equipment and machines fail in operations.

1. Stress Finally Exceeds the Strength: Once a piece of equipment is loaded, mechanical stress occurs which can be in the form of vibration, temperature, shock, as well as the movement of several mechanical parts. These forces will have an impact on the different parts of the equipment. This means that the parts inside the equipment have a dictated life and there will be an (n) amount of cycles that these parts will run before it finally fails. Therefore, when the stress finally exceeds the strength of the material then a fracture is most likely to happen which can turn out to be a functional loss breakdown and stop the machine from operating.

2. Human Error: This will be discussed in a separate chapter. Although when we discuss the subject of human error, it can occur anywhere from the design, fabrication, installation, commissioning, or how the equipment is being operated and maintained. In aviation, human errors caused 80% of accidents.

3. Wear and Tear: Wear is usually a gradual process and usually being covered during Preventive Maintenance replacements. This means that the end life of the part or item had been finally reached. The process of wear can be defined as damage to any solid surface caused by the removal or displacement of material caused by some means of mechanical action of either a contacting solid, liquid, or gas, causing significant surface damage. The damage is usually thought of as gradual deterioration. It means that it can be a solid to solid mechanical item that will cause wear or a continuous bombarded of either a gas or fluid to a solid object that will cause the object to directly wear out for as long as the bombardment is continuous, just like in the case of a continuous bombardment of a fluid on a pump impeller. Wear may also mean the undesired removal of material from any contacting surfaces through some means of movement, or mechanical action. This means that several parts and spares have some movements. These mechanical parts can either slide up and down, left and right cycles, or rotate inside our equipment. Any part, item, or spare that moves inside our equipment, machines, or assets will be subject to wear and degradation after a definite amount of period. Wear can either be considered premature or natural while natural wear means that the life span has been reached and maximized.

4. Premature wear is also termed accelerated deterioration, which means that the part or item did not reach its natural lifespan. Other cases of random failures can be considered premature wear. According to the Jost Report, friction wear, corrosion, and lubrication cost the United Kingdom over 500 million pounds per year or roughly around 639,749,500.00 USD. He reasoned out that significant education and research are required to reduce these losses by 30% of the primary energy consumed by friction throughout the world, while 60% of the machine failures are due to premature wear.

5. Failing to Address the Basics: The process of addressing the basic equipment condition simply means keeping the equipment clean, applying the correct lubrication, tightening bolts, and equipment with no leaks. These are the very basic tasks that need to be done on the equipment. Catastrophic failures are usually the result of small problems neglected on the equipment. These small problems accumulate over time causing the parts inside the equipment to stress out easily. Establishing basic equipment conditions simply means eliminating the causes of accelerated deterioration. This is when a certain part of our equipment does not reach its useful life. It means cleaning to remove dirt and sources of contamination, proper lubrication to prevent early wear, and understanding that bolts need to be complete and secure. Parts do not achieve their desired lifespan due to dirt and contamination, or the wrong lubricant is poured into the equipment.

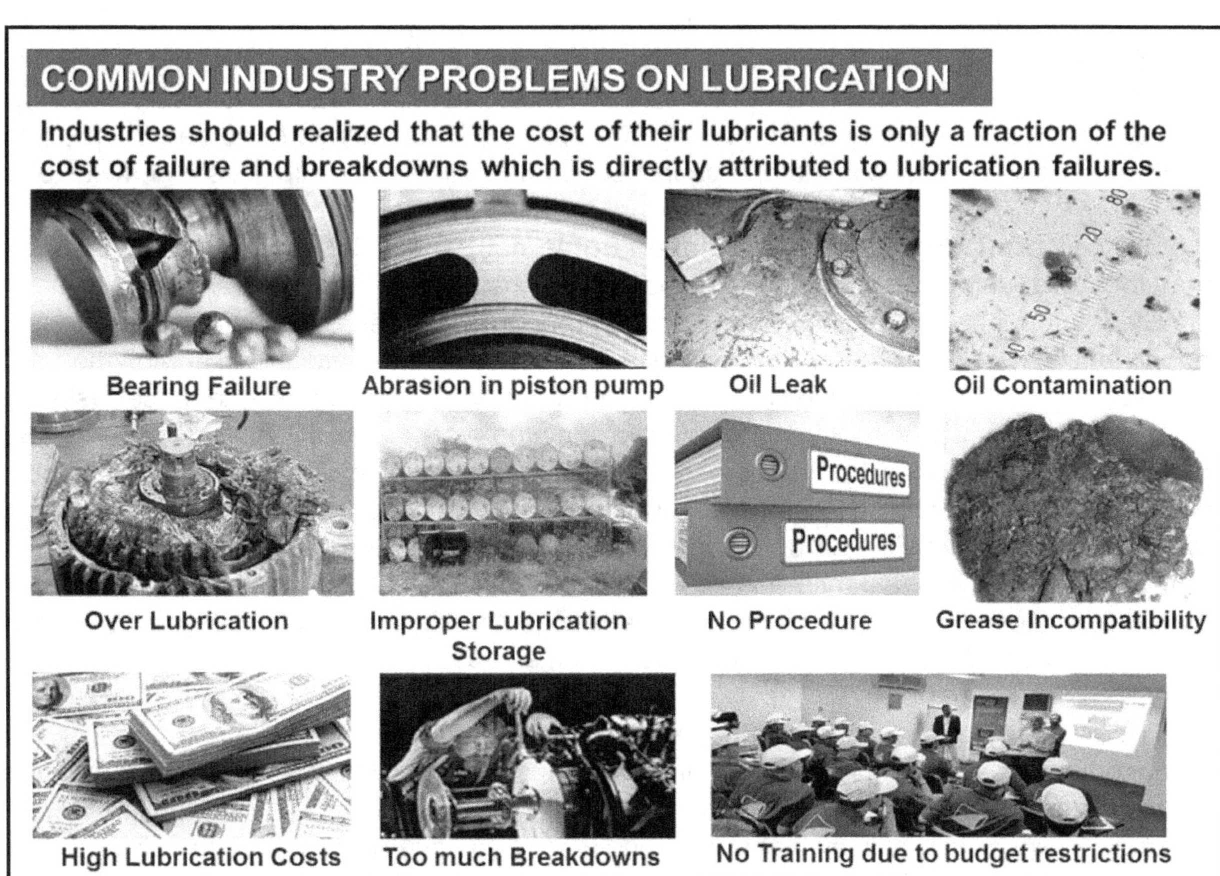

Figure 4.6: Common Industry Problems on Lubrication

6. Lubrication Attributed Failures: Industries may not realize this but, the majority of equipment failures can be attributed to lubrication. The abnormal operating condition often produces excessive heat and this lowers the fluid film thickness, which allows the metal-to-metal contact. Poor lubrication can lead to other problems such as overheating, contamination, fatigue, and corrosion. According to experts, lubricant failures constitute around 36% of bearing failures. Lack of lubrication can generate heat and the possibility of having metal-to-metal contact. Heat will cause the lubricant to decrease in viscosity, causing more heat as it loses its ability to support the load and form a varnish residue. Improper lubrication can generate problems such as corrosion, oxidation, increase in heat, and generation of more contamination, which can lead to excessive abrasion that can occur between tight metal clearances.

7. Hazardous Environment: The environment can also be a culprit to machine failure. This includes a dusty environment, extremely humid environment, and exposure to toxic chemicals, which can affect certain parts of the equipment including mechanical, electrical, and electronic parts as well. Metals can corrode easily depending on the environment the equipment or assets is exposed. Corrosion is the deterioration of metal due to chemical or electrochemical reactions with its environment. Although corrosion is not catastrophic from a safety standpoint it can be disastrous if it will result in the fracture of metal components. Not only will the equipment be at risk but also the people involved in operating and maintaining the equipment as the environmental conditions can stress and fatigue humans easily. Human exposure to these forms of stress can lead to reduced sensory capacity, fatigue, and reduced mental alertness. These are all manifestations of human fatigue and greatly increase the chances that the people concerned will make a slip, lapse, or mistake.

8. Lack of Knowledge on How to Maintain Equipment and Assets: Many industries think that they are doing the correct maintenance on their equipment and assets. Lack of knowledge on how to maintain may actually end up in over or under-maintaining the equipment as well as human errors, which can leave to premature and infant mortality failures. In most cases, an interval is provided to carry out specific maintenance tasks on the equipment, but the question is how did maintenance arrive at the interval? In the majority of cases, these intervals are a result of guestimates or OEM recommendations, which may be conservative. The question is are we doing this particular task too soon or too late? As I emphasize strongly and have written in my other books training will play a crucial role in the maintenance function as this is the starting point on where we acquire knowledge. The knowledge we learned will serve as a basis to develop our skills so we can perform the correct maintenance on our equipment and assets.

9. Top Management Cost Cutting Schemes: Although the intention is good, Top Management or the people involved in these cost-cutting schemes may be doing the industry more harm than good. A good example of this is purchasing going for the lowest bidder, perhaps in fairness, all that purchasing want is to save cost for their industry, but the strength of the material the vendor submits may compose of a weaker material compared with the original or existing part, which can turn out to have a shorter lifespan than expected. Cost cutting may be good for non-maintenance applications but can have repercussions and can backfire if implemented in the maintenance function. A classic example of this is the case of Bhopal India discussed in Chapter 12, The Conclusion of this book, which caused the lives of 16,000 people.

4.6: Creating an Equipment Sustenance

A piece of equipment may compose of different parts, which can include, mechanical, electrical, electronics, and electromechanical parts. When a piece of equipment is running, these stresses automatically happen inside our equipment. All parts inside the equipment will experience stress or perhaps a combination of them. It is just a question of which of these parts will happen to fail first.

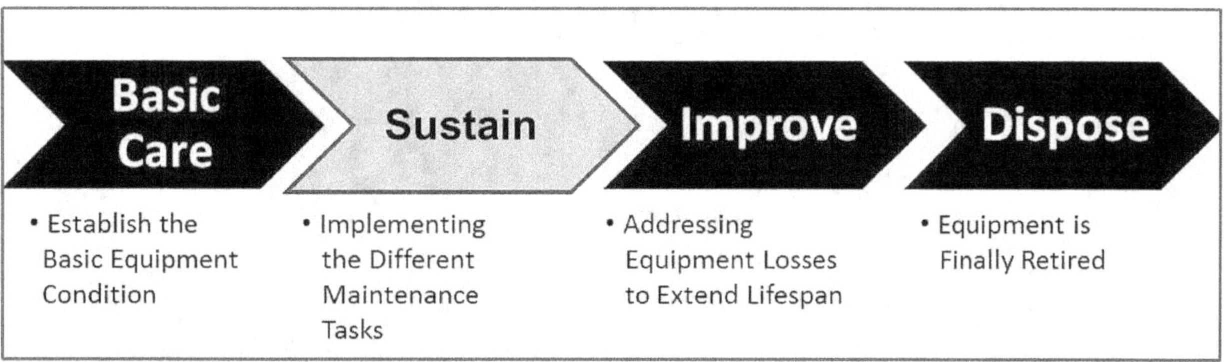

Figure 4.7: Creating an Equipment Sustenance 1

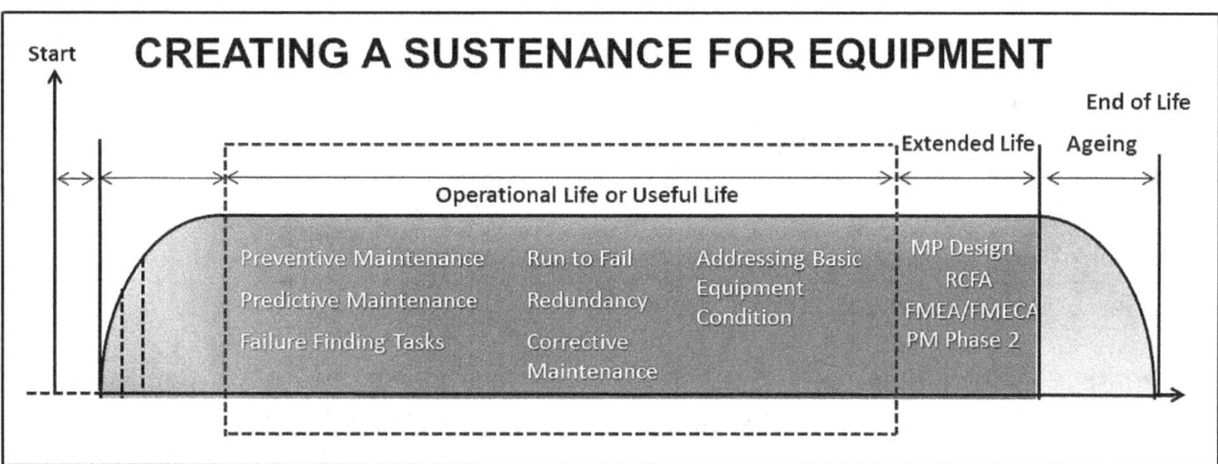

Figure 4.8: Creating an Equipment Sustenance 2

Now we understand that the different stress can provide problems that can end up in a functional failure. What is important is to counter these failures. A functional failure can be considered as a failed state while a failure mode is an event that can probably be caused by a failed state. A failure mode is defined as how the equipment could potentially fail to meet its intended function. It is stated as the most probable or likely cause of the failure. The goal of any maintenance in industries is to sustain and preserve their equipment against these failure modes. This means asking the question of what are we going to do to manage, prolong, sustain, prevent, predict or anticipate these failure modes in our equipment and assets. This is where maintenance will come into play. The main point of creating sustenance on the equipment is to do something before these failures occur. The main goal of sustaining the equipment is for the maintenance to plan ahead and be one step ahead of failure. It is always less expensive if maintenance is ahead of failure instead of the equipment failing and repairing the failure. Hence, to address the different possible failure modes that can occur on the equipment, maintenance tasks are deployed to counter these failure modes. What is important is that the degree and amount of maintenance tasks to be carried out should depend on the impact and consequences of the failure in the assumption that the failure is about to occur. Here are the different maintenance tasks that are performed on the equipment, machines, and assets.

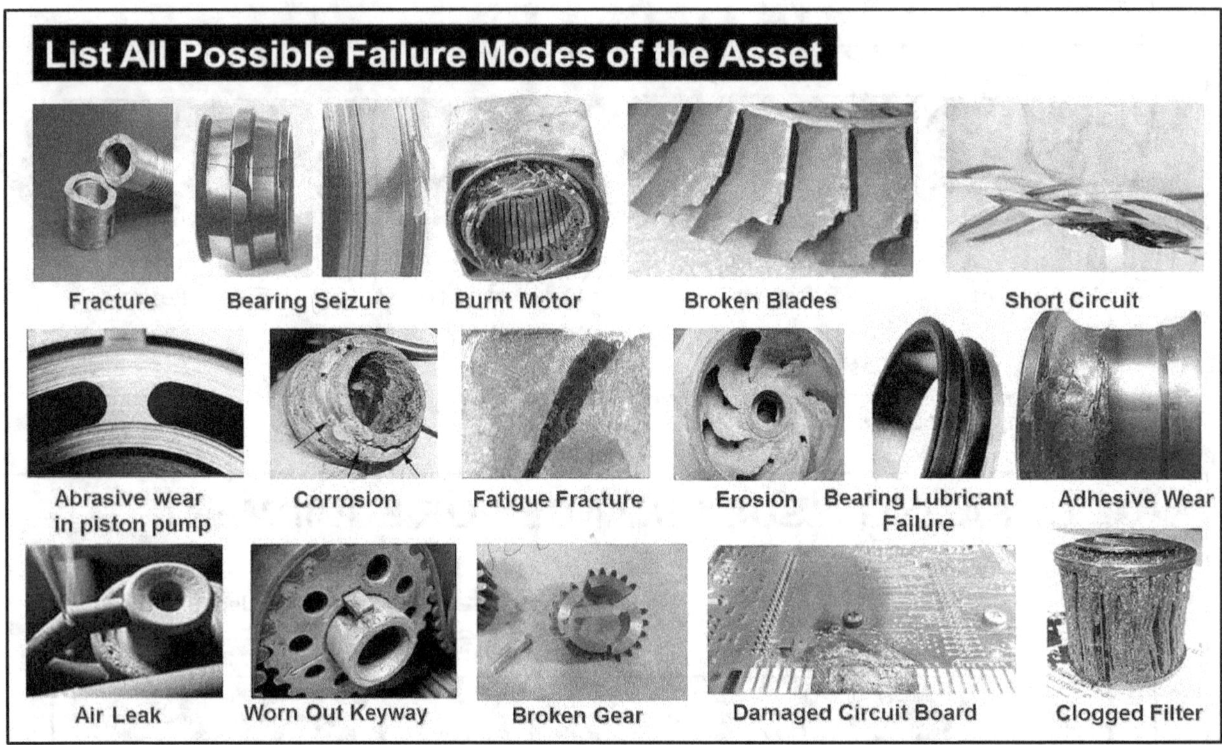

Figure 4.9: Different Failure Modes on Equipment

Preventive Maintenance: PM is a basic maintenance performed on the equipment and facilities. The main goal of performing a task on a scheduled basis is to extend the equipment's life and to assure its capacity in support of the plant's goals and targets. PM is also a series of tasks performed at a defined frequency dictated by the passage of time, the number of machine hours, or mileage that either extend the life of the asset or detect that an asset had a critical wear and is about to fail or break down in operation. The different stress inside the equipment will cause an asset to deteriorate by lowering its resistance, exposure to stress includes output, distance traveled, operating cycles, calendar time, and running time. The key to using Preventive Maintenance will be feasible if the failure is age-related. This means that the part or item has a dictated life and will wear out based upon its operating age. This means that there is a direct relationship between the part's operating age and its rate of wear or deterioration. The basic law to consider is that the cost of Preventive Maintenance must have to be lower than the cost of the consequences of the failure it is meant to prevent.

Predictive Maintenance: Unlike Preventive Maintenance, there will be parts that will fail randomly, but before they fail these parts will provide signs and symptoms that they are on the verge of failing. These failures can be captured through the use of these Predictive Maintenance instruments. This is a maintenance task that aids us in determining the potential failure or symptoms that a piece of equipment is in the process of failing. Changes or increases in the following conditions can denote a potential failure. A potential failure is defined as an identifiable physical condition that indicates that a functional failure is either about to occur or is in the process of occurring. Predictive Maintenance is a maintenance activity geared to indicating where a piece of equipment is on the critical wear curve and predicting its remaining useful life. Maintenance is based on the condition of the equipment, which differentiates it from Preventive Maintenance, which is based on a dictated interval or schedule. What is important for the user is to determine the P-F interval, which is the interval between the emergence of the potential failure and its decay into a functional failure. This is also the failure development period. Most of these Predictive Maintenance instruments have a threshold or limit to alert the user to make a recommendation on what to do before reaching the functional failure. The key to this maintenance task is to carry out tasks before reaching its functional failure.

Precision Maintenance: According to the dictionary, precision is the state or quality of being precise or exact. Precision Maintenance is the act of maintaining equipment in a consistently reproducible manner. This means that the maintenance performed is done exactly the same way and delivers the same results regardless of who performs the activity. It is a method of performing maintenance in a consistent, precise, and accurate way. If Precision Maintenance is properly implemented, this means that maintenance should yield exactly the same results no matter who is performing the tasks whether the crafts person is the most or least experienced in the craft. These tasks can be integrated into other maintenance tasks to minimize the chances of human error.

Failure Finding Tasks: Also known as functionality inspection which is usually performed on protective devices and redundant functions. Performing scheduled failure finding entails checking a hidden function at regular intervals to find out whether it has failed. Failure finding tasks applies only to hidden or unrevealed failures, which include protective devices and redundant functions. A hidden failure is one whose failure will not become evident to the operating crew under normal circumstances if it occurs on its own. The consequences of a hidden failure are not felt immediately unless the second failure will take its toll. If the second failure were to fail after an undetected failure of the first one, it will result in a critical failure. In other words, the consequence of any hidden functional failure is increased exposure to the consequence of multiple failures. The reason for conducting a failure finding tasks is to reduce the risks of multiple failures wherein both the protective and protected device is in a failed state

Run to Fail: Maintenance is done at a point when there is repair or actual breakdown. Usually, a repair or corrective maintenance takes place after a failure. It occurs when repair action is taken on a problem only when the problem results in the machine's failure. Unplanned downtime, in its simplest definition, simply means fixing it when it fails. The use of run to fail will only be valid if the consequences of failure are minimal which will only result in the cost of repair. The only question to raise is if allowing an item to fail will induce secondary failures on the equipment. Run to fail will not be valid if the failure consequences will result in safety or environmental consequences. Spare parts and component failures that will limit the failure to the component itself with no chances of secondary failures.

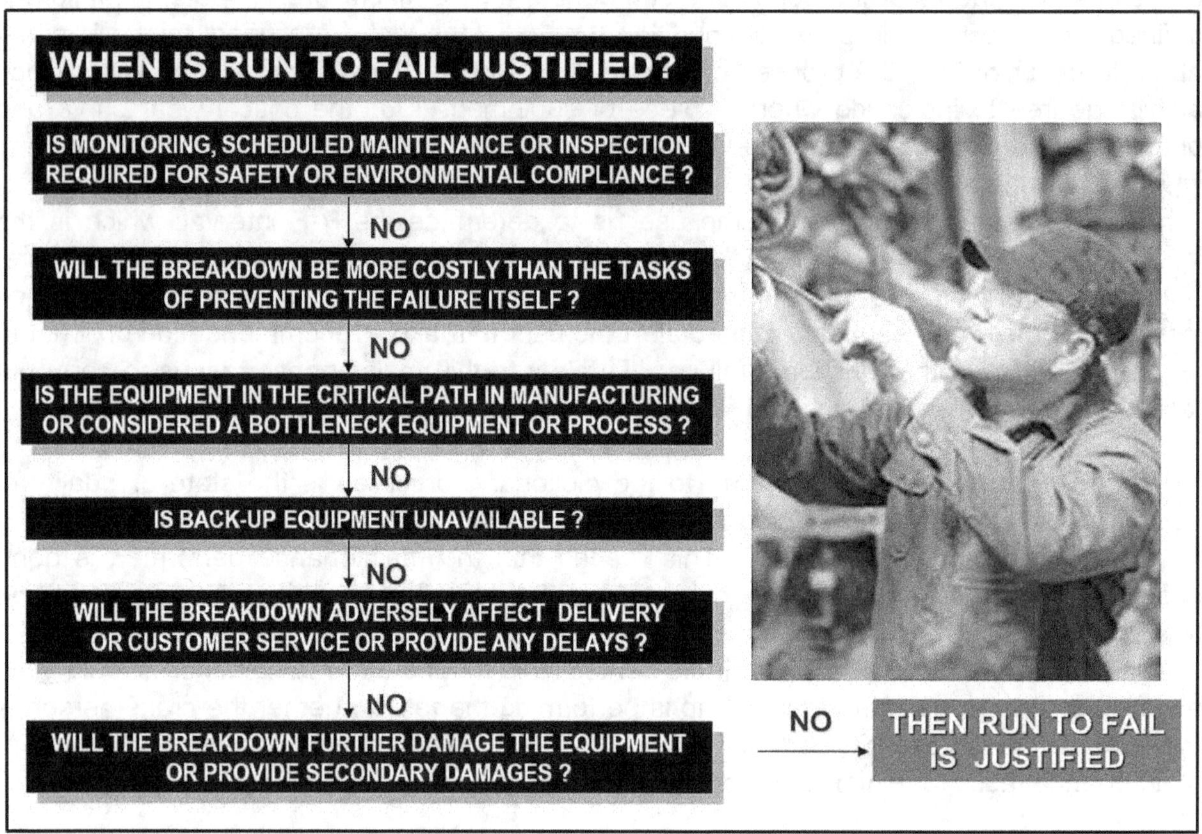

Figure 4.10: When is Run to Fail Justified?

Redundancy: Duplicating the system or component. Failures can be allowed or tolerated through redundant or duplicated functions making the system less critical. The presence of redundancy or alternative means of production is a feature of the operating context, which must be considered in detail when defining the functions of the asset in its present operating context. Oftentimes termed as standby-unit, even with the same equipment type, standby units have different degrees of maintenance requirements as the duty unit, and most failures for standby units are hidden.

On-Condition Tasks: This task refers to any maintenance tasks that entail checking the actual condition of the equipment whether using Predictive Tasks, monitoring gages, and the use of human senses to detect problems. Similar to Predictive Maintenance, which will use instruments to predict failure, based on the condition of the equipment. The main difference is this will be on a much wider scope which may also include, condition monitoring techniques that involved the use of specialized instruments to monitor the condition of the equipment, techniques based on variations in product quality such as

Statistical Process Control Charts, primary effects monitoring techniques which entails the intelligent use of existing gauges, process monitoring instruments, smart sensors, chart recorder, and inspection techniques based on the human senses. On-Condition tasks are any tasks that monitor the actual condition of the equipment by any means. As much as possible inspection using the human senses must be quantitative

Addressing Basic Equipment Condition: For those implementing TPM (Total Productive Maintenance), addressing equipment's basic condition simply means keeping the equipment clean, applying the correct lubrication, all bolts secure with the correct torque, and equipment with no form of any leaks. What TPM believes is addressing these small problems can prevent bigger problems from occurring. This activity is usually a joint effort from both Planned and Autonomous Maintenance respectively. Establishing the basic equipment conditions simply means eliminating the causes of accelerated deterioration. This is when a certain part of its component in our equipment does not reach its useful life. In hindsight, dirt, dust, foreign material, contamination, inadequate lubrication, and excessive vibration cause parts to fail prematurely. This is called accelerated deterioration. Contaminants and other foreign objects can cause abrasion and can come in direct contact with rotating parts, hydraulic systems, and other critical parts of the equipment. The goal of establishing these basic equipment conditions in our equipment is to expose these abnormalities and flaws and allow these parts to reach their natural deterioration and useful life. Many failure modes can be addressed if the basics have well been established in our equipment.

OPERATORS CHECKLISTS SHOULD BE QUANTITATIVE

Standards for cleaning, lubrication, inspection should be detailed and quantitative and not subjective. Operators should know what is acceptable and not

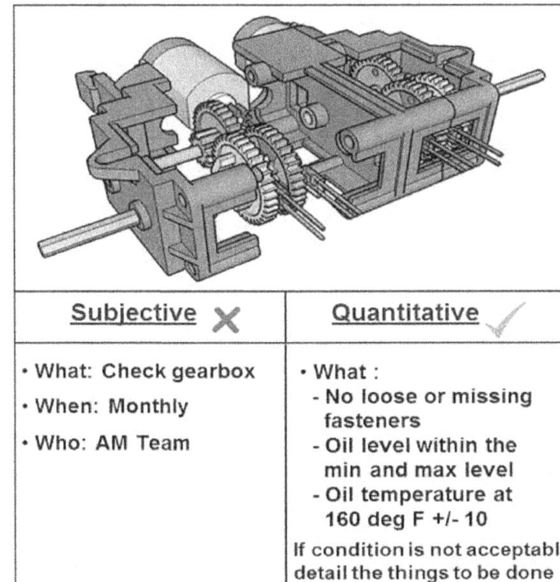

Subjective ✗	Quantitative ✓

• What: Check gearbox
• When: Monthly
• Who: AM Team
	• What :
	  - No loose or missing fasteners
	  - Oil level within the min and max level
	  - Oil temperature at 160 deg F +/- 10
	If condition is not acceptable detail the things to be done

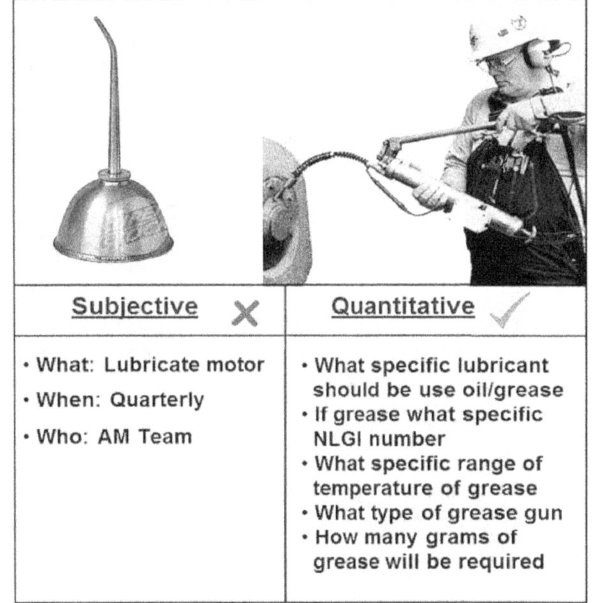

Subjective ✗	Quantitative ✓
• What: Lubricate motor • When: Quarterly • Who: AM Team	• What specific lubricant should be use oil/grease • If grease what specific NLGI number • What specific range of temperature of grease • What type of grease gun • How many grams of grease will be required

Figure 4.11: Human Inspection Should Be Quantitative

Maintenance leaders must understand that failure occurs in three patterns, infant mortality, random, and age-related failures. This means that developing and implementing the maintenance tasks on the equipment and assets should be done three folds. Those that will address infant, random and age-related failures. These maintenance tasks include all the activities needed to preserve, conserve and sustain our equipment and assets.

Precision Maintenance can be integrated into Preventive Maintenance tasks to minimize the chances of human error. The use of Failure Finding Tasks will be for protective devices to determine if they are still functioning or not. Preventive Maintenance tasks replacement and overhauls will be directly used for those parts that have an age-related pattern. For random failures, run to fail will be feasible for those failures with minimal consequences. Predictive Maintenance will be used for failures that provide signs and symptoms that the part or item is on the verge of failing or nearing its rupture. If both run fail and Predictive Maintenance is not feasible then there is the possibility of adopting a redesign or modification to change the pattern of the failure. Once these tasks had been identified, the next step is to determine the correct interval for the tasks.

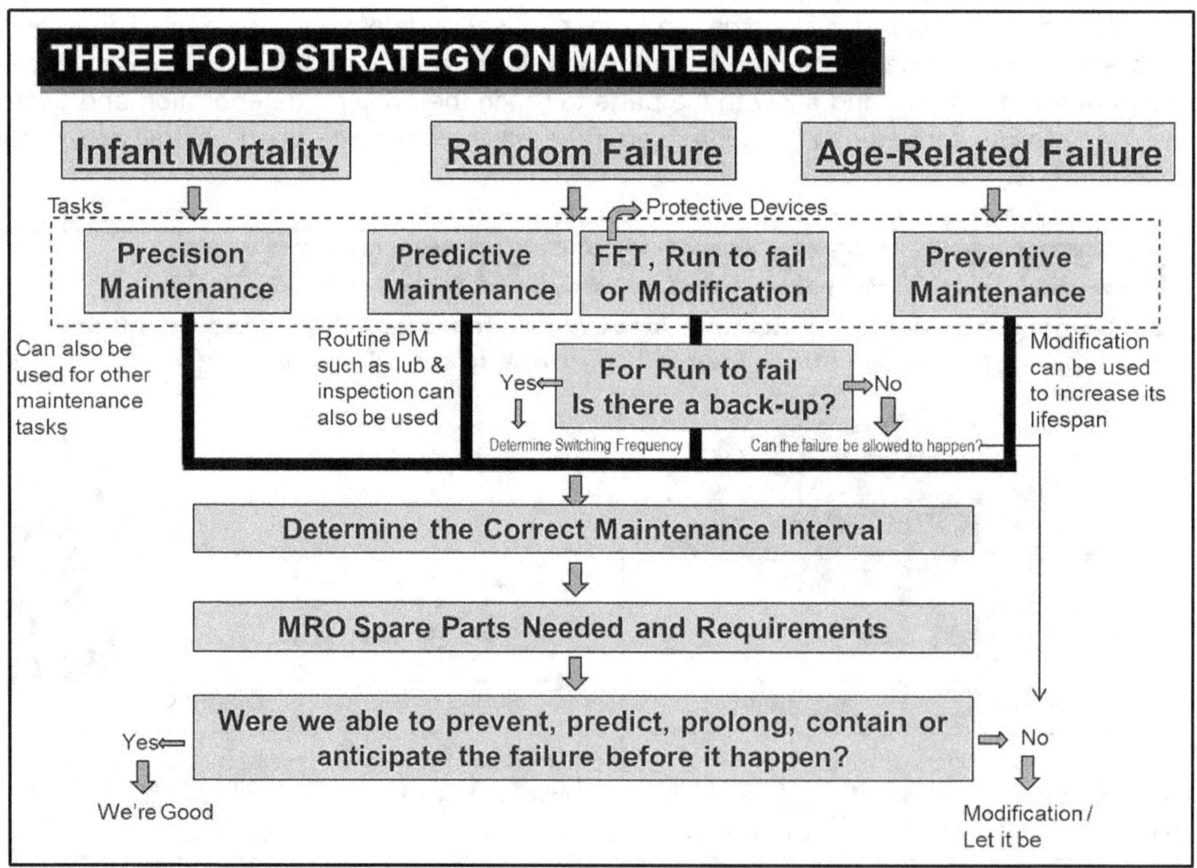

Figure 4.12: Three-Fold Strategy for Maintenance

The best strategy for MRO Spare parts and storeroom is to have a plan of action that will address both inside and outside the storeroom. Failures and losses on the equipment must be addressed so maintenance can maximize the lifecycle of their equipment and assets. If breakdown and downtime can be reduced, then we can lower the MRO spare parts inventory. This means that if these different maintenance tasks are utilized correctly, then breakdowns are not only reduced but are also controlled and managed. Likewise, important in any maintenance strategy is determining the correct maintenance interval for the different maintenance tasks performed on the equipment. If the interval is done too frequently, then this will be a complete waste of money, time, and resources, while if the interval is done too long, then there is the possibility that we have missed the failure. All these tasks are meant to preserve and sustain the equipment. The best that maintenance can do is to sustain reliability. The only way to improve reliability is to go beyond maintenance and walk beyond its boundaries, which is to redesign or modify the equipment, which will be covered in the later chapters of this book.

4.7: All Maintenance Tasks Should Address a Failure Mode

Whatever maintenance tasks we adopt, whether Preventive, Predictive, Failure Finding Tasks, inspection, or others, what is important is that each maintenance task deployed and specified should address a particular failure mode on the equipment. If there are maintenance tasks written that do not address a particular failure mode then this should be removed from the lists as this will just be a complete waste of resources and time.

TASKS	INTERVAL	YEARLY	RESPONSIBLE	ACTIVITIES	TOOLS	FAILURE MODE
1	Shift	365	Operator	- Inspect drop in Pressure Gage	Visual	Clogged Strainer
2	Shift	365	Operator	- Inspect Motor Temperature	Thermal Gun	Insulation Failure
3	Shift	365	Operator	- Inspect Shaft for Alignment	Visual	Misalignment
4	Shift	365	Operator	- Inspect Motor for Unusual Noise	Hearing	Bearing Failure
5	Shift	365	Operator	- Clean Motor Housing	Dry Rags / Solvent	Short Lifespan
6	Shift	365	Operator	- Monitor Drop in Flow Rate	Visual	Impeller Erosion
7	Weekly	52	PM Group	- Inspect base bolts for looseness	Torque Wrench	Excess Vibration
8	Weekly	52	PM Group	- Apply Lubrication	Grease Gun	Bearing Failure
9	Weekly	52	PM Group	- Inspect the coupling for looseness	Visual	Coupling Failure
10	Monthly	12	PdM Group	- Perform Vibration Analysis	Vibration Analysis	Bearing Failure
11	Monthly	12	Contractor	- Perform Monthly Greasing	Grease Gun	Bearing Failure
12	Monthly	12	PM Group	- Inspect Pressure Gage	Visual	Clogged Strainer
13	Quarterly	4	Contractor	- Replace motor bearing	------------	------------
14	Quarterly	4	PdM Group	- Thermography Scanning	Infrared	Overheating
15	Quarterly	4	Electrical	- Loose Rotor Bars	Vibration Analysis	High Vibration
16	Annual	1	Contractor	- Perform Shaft Alignment	Laser Alignment	Misalignment
17	Annual	1	Contractor	- Replace Impeller	New Impeller	Low Flow Rate
18	Annual	1	Contractor	- Check for Pipe Leaks	Ultrasonic	Leaks

Figure 4.13: All Maintenance Tasks Should Address a Failure Mode

Try to select a piece of equipment, machine, system, or sub-system and determine all the existing activities and maintenance tasks. List all the activities being performed by the different people in the equipment. Determine all recorded and unrecorded, formal, and informal activities and tasks being performed on the equipment or asset. It is believed that maintenance is being performed by a wide cross-selection of people including the operators, maintenance, and even 3rd party contractors. Compilation of Maintenance Tasks and activities is simply a matter of writing down all the activities different people are performing in their equipment. It is uncommon for organizations to have an informal PM system in operation and rare for an organization to have none. In this step, we list all the maintenance task activities performed by the different people involved and indicate what particular failure mode each of these maintenance tasks addresses.

• Operators of the equipment
• Preventive Maintenance group
• Electrical and Mechanical group
• Sustaining and Repair crew
• Third Party Contractor or OEM
• Calibration group
• Condition-Monitoring or PdM Group
• Facilities group

Once all the tasks had been listed together with the failure modes the tasks are meant to address, look for any duplicated tasks. Duplicated tasks are similar maintenance activities that are done by different people. This will create redundant work and a waste of manpower and other resources. By sorting the data by failure mode, task duplication and redundancy can be easily identified. Tasks duplication is when the same tasks are being done by more than one person. We ask ourselves do these maintenance tasks really need to be carried out by more than two trades or can be performed by a single group and who is the best trade to perform these tasks? In this step, the team reviews the different tasks generated. Once these are completed, try to list possible failure modes that can possibly occur that do not have any existing maintenance tasks and add them to the lists, and indicate the interval. The team must decide who is the best trade to perform the tasks on each failure mode identified. Determine if duplicated tasks really need to be performed by different people involved in the equipment. Thoroughly reviewing each task will reveal that there are maintenance activities that do not address any particular failure modes and should be removed from the lists. Again determine what are the possible failure modes that can occur which is not currently reflected in the tasks. Once these redundant or duplicated tasks are removed, the team will brainstorm if any other failure modes have been missed that should be included together with the task that needs to be done to finally address these failure modes. Once an additional failure mode has been added, the interval, responsible person, and maintenance tasks will be added to the list. Do this for the rest of the equipment and machines in the plant. This is how to check and assess if our existing and current maintenance activities and tasks on our equipment and assets are effective or not.

Chapter 5

Mitigating Human Errors in Maintenance

> *Not all human error is necessarily the fault of the person who committed the error. In many cases, human error is either forced by external circumstances or by obsolete rules and policies. Therefore, if blame is to be allocated for any error, care must be taken to identify the real cause of the problem. Remember, industries will never learn from the things that go wrong if we continue to blame people for their errors.*

5.1: Human Errors Explained

Before attempting to extend equipment's Life Cycle, we need to understand human errors. By definition, human errors can be defined as an action planned but not executed according to the plan. Research by Dr. James Reasons of the University of Manchester in England found that humans commit an average of 6 errors per week. As of 2022, the total world's population is 7.9 billion people, hence, at an average of 6 errors per week for every single person, this will be around 47.4 billion errors committed on our planet every single week. Some of these errors may not be critical but there will be errors that will be fatal or can even cause us harm or even death. A pilot error can cause the lives of all the passengers on board the plane. Errors committed by physicians, doctors, and nurses can compromise the lives of their patients. We watched it on the news regarding road accidents, which are being reported almost daily by the news anchors. Errors from management decisions, flaws in systems, obsolete procedures, and policies can have a detrimental effect on its product, operations, and services that can affect its clients, customers, and services, and even the existence of our business. But the most important factor in understanding human error is learning from the failure itself. What I believe is that whether from the lessons of life or from industries, if people created problems then people must also be capable of correcting the problem itself. Here are some definitions of human errors based on the people who have studied them.

- [5]An error is an out-of-tolerance action where the limits of tolerable performance are defined by the system. By Swai and Guttman 1983

---

[5] Whittingham R.B., *The Blame Machine, Why Human Errors Causes Accidents,* (Elsevier Butterworth – Heinemann Publications, 2004), Pages 2 to 5

- A generic term to encompass all those occasions in which a planned sequence of mental or physical activities fails to achieve its intended outcome and when these failures cannot be attributed to some chance agency. An error can be defined as an action planned but not executed according to the plan. By James Reasons 1990
- An erroneous action can be defined as an action that fails to produce the expected result and/or produces an unwanted consequence. Hollnagel 1993
- A failure of a common sequence of psychological functions that are basic to human behavior, stimulus, organism, and response. When any element of the chain is broken, a perfect execution cannot be achieved due to the failure of perceived stimulus, inability to discriminate among various stimuli, misinterpretation of the meaning of stimuli, not knowing what response to make to a particular stimulus, physical inability to make the required response and responding to the sequence, Meister 1996

Despite the difference in the meaning of human error, one thing common is that the action must be accompanied by an intention to achieve a desired outcome or result but failed to achieve it. Human errors can be measured by their severity and consequences. Human errors that have no or little consequences would not be important, but what matters is to identify human errors that can lead to major accidents and severe consequences.

[6]According to the book by James Reasons and Alan Hobbs on Managing Maintenance Errors, the largest number of maintenance errors in the aviation industry is associated with reassembly and installation activities during Preventive or Scheduled Maintenance. Analysis carried out on Boeing identified the top seven causes of In-Flight Engine Shutdown (IFDAS) as follows;

- Incomplete installation at 33%
- Damage on installation at 14.5%
- Improper installation at 11%
- Equipment not installed or missing at 11%
- Foreign object damage at 6.5%
- Improper fault isolation, inspection, and test at 6%
- Equipment not activated or deactivated at 4%

Human error occurs during the different stages of the life cycle such as the design, fabrication, installation, commissioning, and throughout its operating lifespan. The good thing is that we can identify human error in the early asset of the life cycle design. Human errors also exist throughout the running life of the equipment which can either occur during operations and maintenance, which results in downtime and can impact the revenue and performance of the asset. Traditional efforts to investigate errors are often aimed at identifying the person who committed the error. The traditional result will end up decertifying the operator ending up in retraining and recertifying the operator once again which often adds little value as the error may continue to recur once again. In addition, by the time the employee is identified, information about the factors that contributed to the error has been

---

[6] Reasons, James and Hobbs,Allan, *Managing Maintenance Errors, A Practical Guide*,(Ashgate Publishing Limited, 2003), Page 5

lost because the factors that contributed to the error remain unchanged, and the error is likely to recur, setting what is called the blame and train cycle in motion once again.

According to a dear friend of mine. C. Robert Nelms, quote, that the issues of blame, discipline, fear, punishment, and accountability are near the root of why industries do not learn from the things that go wrong. There is most certainly a need for rules, regulations, policies, and procedures in our businesses. Likewise, there is most certainly a need to enforce them by applying discipline to those who do not abide. Unfortunately, most organizations have it backward. They do not consistently apply their disciplinary policies ahead of time, instead, they wait for something awful to happen and then go after the person. This is why industries never learn from their problems. According to Dr. Normal Peal, a religious leader, quote, if we take a problem and we look at it, it becomes a situation. If we were to analyze this situation, it becomes a challenge, and when we think of our ability to solve and overcome the problem, then it becomes an opportunity.

If we speak about equipment-related problems, all breakdowns are man-made. From the way, the equipment was designed, fabricated, installed, commissioned, operated, or maintained. There is no such thing as a piece of perfect equipment. Every piece of equipment has a built-in design weakness that should be addressed by maintenance, reliability, and even the operators themselves. Humans play an important role during the design, fabrication, installation, commissioning, production, and maintenance phases of the equipment. In maintenance, human error may be defined as the failure to perform a specified task that could lead to disruption of scheduled operations or result in damage to property or equipment. While human errors had existed since the beginning of mankind, only in the last 50 years has it been the subject of scientific inquiry and studies. There were various reasons for the occurrence of human errors such as stress, human fatigue, working too long without adequate break time, incorrect tools, forgetfulness, complacency, distractions, insufficient knowledge, lack of knowledge and training, lack of communication, misinterpretation, ignorance, lack of standards, working under pressure, lack of awareness and the lists goes on. However, with every error that humans make, there is typically an associated change or something out of the ordinary occurring in the environment. The difference between humans and machines is that people can sense change, hear, smell, see, or feel something different and take the necessary actions to correct the anomaly. In industries, human errors exist either because we tolerate them or because we are just plain ignorant. Let me state some cases.

Try to loosen some bolts in your equipment and place a toolbox near it. Let us say that the toolbox will contain a hammer, pipe wrench, adjustable wrench, pliers, socket wrench, double-ended open-type wrench, and a torque wrench. Ask one of your technicians in the plant to tighten the loose bolts and observe how he tightens them. After he has tightened the bolts, ask him to loosen them again. Try to call 4 to 5 more technicians, one at a time, and give them the same instruction, which is to tighten the bolt and write down how each of them performed the tightening process. Perhaps you will observe that some have similar ways of tightening the bolt and have used the same tools, while others simply used other methods to tighten the bolts. Why? Because you just ask them to tighten the bolts, but if we are explicit in our instruction and procedure, we can instruct them to use a torque wrench with this amount of torque or perhaps a warning indicating not to use an adjustable wrench, hammer, or pipe wrench so that each one of them will do the same procedure on tightening the bolt.

5.2: Categories of Human Errors

When considering the interaction between people and equipment, human errors can be classified into four categories: Not all human errors are necessarily the fault of the person who made the error. In many cases, the error is either forced by external circumstances or other factors; therefore, if blame is to be allocated for any error, care must be taken into consideration to identify the real root cause of the problem. The Categories of Human Error include Anthropometric Factors, Human Sensory Perception, Physiological Factors, and Psychological Factors. Let us discuss each one of them.

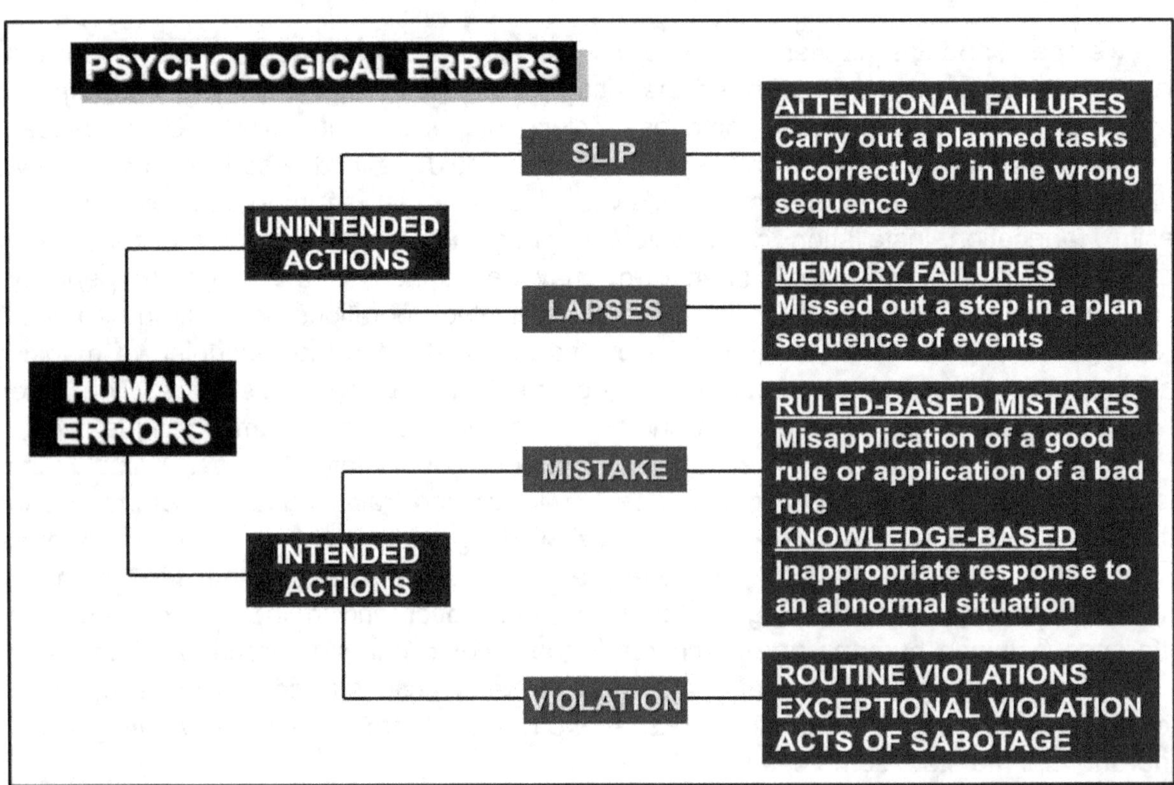

Figure 5.1: Categories of Human Error

A) Anthropometric Factors: During the hiring process, there are instances that human resources will include special requirements such as height, build, vision, and so on because they will be part of the requirements for the job. These are the factors that relay to the size, shape, and strength of the person. Errors occurred because the person simply cannot fit into the space provided, cannot reach something, or is not strong enough to move or lift something. Industries aiming for multiskilling should also consider anthropometric factors to avoid human errors. The human resource department hires people with a special built, height or perfect eyesight vision because it is mainly required in their job, hence, if a taller person is required to operate certain equipment, we simply cannot replace him with a shorter person because the height requirement is needed to perform the job. Many industries that practice multi-skilling should also take note of its limitations and one of them is anthropometric factors.

I recalled when I was working in the semiconductor industry here in the Philippines way back in 1994 when the operations manager called all his supervisors for a meeting and the story goes on like this. He said that he was not happy with the way the operation was being run. One supervisor said, boss, we are right on target this week. The boss said that is not what I meant. What I want is that the operators from Marking Station should know how to trim and the Trim Station operators should know how to mark. Since these were dedicated operators and before they can operate the equipment there will be a certification process for operators. I was handling the Marking Station. So after the meeting I talked to the supervisor of the Trim Station and discussed our plans. I send him five operators to learn how to trim and likewise, he send me five operators to learn how to Mark. We swap operators in short. After a month, again we send another 5 operators as there were around 30 operators per station per shift for both trim and mark. After three months of swapping operators, we reported to the boss and told him that all Marking operators know how to Trim and all Trim operators know how to operate the Marking equipment. There was still no smile or delight on the face of this boss as I had never seen him smile in my entire life. Instead of being satisfied with what we have accomplished, he said. I am not happy with your performance. What I want is for all Trim and Marking operators to know how to Mold and for all Molding operators to know how to Trim and Mark. So after that meeting, I and the supervisor on the Trim station talked to the supervisor of the Mold Station and we swap operators once more where the Molding operators were sent to the Marking and Trim Form station to learn how to operate the trim and marking equipment. Likewise, we send the operator from the Marking and Trim station to learn how to Mold. After doing this, the problem began. Both trim and marking operators were female while the Molding operators were male. The height of the equipment was taller compared to the trim and marking equipment. There were cases when the operator was short on height and would jump to push the button ending up pushing the wrong button. The effort of certifying operators on the Mold Station was discontinued. It was simply a case of anthropometric factors.

B) Human Sensory Perception: Humans are gifted with five senses which are the sense of smell, sense of touch, sense of hearing, feeling, and sense of taste. For inspecting the equipment, forget about the sense of taste. Some failure modes will give some sort of warning or symptom that they are on the verge of happening such as smell, excessive vibration, noise, and heat, which can be detected by our human senses. Therefore if errors are occurring or are thought to be likely to occur for any of these reasons, then human error again is not the root cause of the failure and we need to dig deeper into the real cause of the problem. When maintenance cannot smell a burnt motor because he has a terrible cold or running nose and was not allowed to go home by his boss then he or she should not be blamed for the failure. In this case, human error is not the root cause of the problem.

C) Physiological Factors: The term physiological factor refers to environmental stress, which can affect human performance. This stress can include high or low temperatures, loud or irritating noise, excessive humidity, high vibration, exposure to toxic chemicals, radiation, or working too long without adequate break time, not to mention the day-to-day pressures from the boss. Exposure to these kinds of stress can lead to reduced sensory capacity, fatigue, and reduced mental alertness. These are all manifestations of human fatigue and will greatly increase the chances that the people concerned will either make a slip, lapse or mistake. Therefore if errors occur are thought to occur for any of these reasons then again human error is not the root cause of the problem. What we should be addressing is how we can improve the environmental factors and surroundings in which both operators and maintenance are working. For example, reducing temperature or providing hearing protection should minimize the person's exposure to this noise and stress.

D) Psychological Factors: These factors can be grouped according to those, which are unintended, and those errors, which are intended. Unintended errors can be grouped into slips and lapses while intended errors can be grouped according to mistakes and violations.

Slip occurs when somebody does something incorrectly or does something in the wrong sequence. If a person is going to perform something in sequences such as step 1, step 2, step 3, step 4, and step 5, and the person does step 1, step 2, step 4, step 3, and step 5, then the person committed a slip. These are human actions that take place that were not intended. The occurrence of a lack in one train of thought, which is derived from unconscious behavior

Lapses occur when someone misses out on a key step in a sequence of events or activities. For example when a mechanic leaves a tool behind after working on a machine or simply forgot to fit a key component while reassembling it. If a person is going to do steps 1, step 2, step 3, step 4, and step 5, and the person missed out on step 4 in the process, then he committed a lapse. As humans age, our memory will not be as sharp compared to when we were young just like in our high school or college days. People tend to forget many things as they age.

Mistake is a planning failure, where actions go as planned but the plan was simply a bad one. These are errors in judgment. Mistakes are the real challenges in the analysis of human errors. They are influenced by external factors and can be grouped as knowledge-based, ruled-based mistakes, or violations.

Knowledge-Based Mistakes are types of mistakes that occur when someone is confronted with a situation that has not yet occurred before and which had not been anticipated. In other words, there are no rules or procedures to follow. In situations like this, the person has to make a decision quickly about an appropriate course of action, and a mistake occurs due to the wrong decision. This suggests that the first and foremost way to avoid knowledge-based mistakes is to improve the knowledge of the people who have to make these kinds of decisions.

Ruled-Based Mistakes are mistakes that usually occur when people believe that they are following the correct course of action when doing a task based on a specific rule or procedure but the course of action was not appropriate. An example will be replacing parts in the equipment to comply with their PM procedure even if the part to be replaced is still in working or in good running condition.

Violation occurs when someone knowingly and deliberately commits an error. Violations fall under three categories, which are routine violations, exceptional violations, and acts of sabotage. If this is a case of sabotage, this will be a different matter and will be out of the context of this book. But I recommend that people that commit sabotage in your industry should be terminated immediately as they are no different from terrorists.

5.3: When Maintenance Errors Cost Lives

It can be said that traveling by plane is among the safest way to travel. However, on the contrary, it may also be the most fatal; when aviation accidents happen resulting not only in the loss of the aircraft but also the loss of lives of the passengers and crew. Although aviation accidents may have been reduced for commercial aircraft, but not with private owners and smaller planes. The Convention on International Civil Aviation (CICA), defines Air Traffic Fatalities as incidents where a person is fatally injured due to an occurrence associated with the operation of the aircraft. This definition starts at the time from when the first person boards until the last person disembark the plane. Usually, personal own corporate jet and military transport accidents are generally excluded from this.

One of the TV Series I loved to watch is about Air Crash Investigations. Whenever there was an aircraft incident, a group of investigators will go to the crash site to investigate the incident. The National Transportation Safety Board (NTSB) is an independent US Government body whose responsibility is to investigate every civil and commercial aviation incident in the United States. Other countries also have their own independent investigation team. The NTSB will be responsible for incidents involving US-based airlines, and planes scheduled to arrive and depart from the United States. The FAA or Federal Aviation Administration is an agency of the Department of Transportation that regulates all aspects of aviation in the United States. This group is responsible for all Air Traffic Control and set standards for maintenance, manufacturing, and certification of new planes as well as pilot training. They may also be involved in aviation accidents to determine if their protocols and regulations have been violated, and consider necessary legal action. Although there was a decline in the Accidents per Million Take-Offs due to the lockdown caused by the **Covid 19** Pandemic in 2020. Here are some major causes of why aircraft accidents happen.

Pilot Error: Pilots are involved at every stage of the flight, and pilot errors can occur at any one of them. The pilot error refers to an action, decision, or simply a failure to make a decision during a situation on the part of the pilot that eventually led to the accident. Inadequate training, lack of experience, fatigue, lack of sleep, decision-making flaws, and intoxication are all factors that can contribute to pilot error.

Maintenance Errors: Equipment failures account for about 20 percent of aviation accidents. This can be anything from engine failure to poor repair or short-cutting the maintenance process. For example, a government investigation body analyzed the crash of Japan Airlines Flight 123 in 1985 to be the result of improper repairs by Boeing Company on the plane's pressure bulkhead. Improvements in design and manufacturing have made planes much more reliable in this day and age. Nonetheless, with the hundreds of thousands of parts failures still happen during flights.

Weather: Flying becomes more dangerous in bad weather. In fact, the National Transportation Safety Board (NTSB) studies show that more than two-thirds of all weather-related aviation crashes have been fatal. Snow, fog, lightning, thunderstorms, heavy rainstorms, and other natural elements can all make flying more difficult, which is one factor that delays a plane's flight. For countries with snow, deicing is done on the wings of the plane. Deicing fluid is a mixture of a chemical called glycol and water, which is heated and sprayed under pressure to remove ice and snow on the aircraft's wings and body. While it removes ice and snow, deicing fluid has a limited ability to prevent further ice from forming.

Bird Strikes: One of the classic cases of bird strikes happened on January 15, 2009, on US Airways Flight 1549 where Captain Chesley Sullenberger and his co-pilot decided to land the plane on the river. All 155 passengers and crew survived this incident and were rescued by nearby boats. A few minutes after the plane touched down the river. This incident today is known as the Miracle in Hudson River and a movie was even made called Sully, which stars veteran actor Tom Hanks. In May 2010, the NTSB released a report, which confirmed that Captain Sullenberger made the right decision. The plane is now on display in the Carolinas Aviation Museum in Charlotte.

Hijacked Situation: After the 9-11 incident, security measures for passengers boarding aircraft have been upgraded and changed worldwide and no incidents of hijacked have been reported ever since. Hence, this will no longer be an issue or threat.

Missile Strike: Malaysia Airlines Flight MH-17 was a scheduled passenger flight from Amsterdam to Kuala Lumpur that was shot down on July 17, 2014, while flying over Eastern Ukraine. All 283 passengers and 15 flight crew were killed. Flight MH-17 aircraft was a Boeing 777-200. It was hit by a missile when it was about 50 kilometers from the Ukraine–Russian border, and wreckage from the aircraft fell near Donetsk Oblast, Ukraine, and 40 kilometers from the border.

Other Human Errors: 80% of aviation accidents are caused by human errors. Pilots are not the only ones responsible for ensuring the safe arrival of the plane. Air Traffic Controllers (ATC) has also several cases of human errors committed where two planes can collide on the runway or during taxing.

When an accident happens, what is important for the investigator is to locate the plane's Black Box. There are actually two black boxes for a commercial plane that records all the conversation of the pilot in the cockpit and a Flight Recorder for the investigators to study the flight path. All flight information is recorded into the black box with a specific algorithm. This makes the recorded flight data accessible to the investigative body when needed. Black boxes can resist damage in extreme situations. An aircraft black box helps to clarify the causes of accidents, the decisions made by the pilot, and what eventually happened during the last seconds before the accident. The first use of the device dates back to 1947. After 1958, it was mandatory to have them installed for all commercial planes according to the rules set by the Civil Aeronautics Board. The black box stores all kinds of information and conversations about the plane. Here are three classic cases of fatal plane accidents caused by human errors.

5.3.1: Case 1: Alaska Airline Flight 261

Alaska Airlines Flight 261 was an Alaska Airlines flight of a McDonnell Douglas MD-83 plane that crashed into the Pacific Ocean on January 31, 2000, killing all 88 people on board including 5 crews and 83 passengers. The flight route started from Licenciado Gustavo Díaz Ordaz International Airport in Puerto Vallarta, Jalisco, Mexico, to Seattle–Tacoma International Airport in Seattle, Washington, United States.

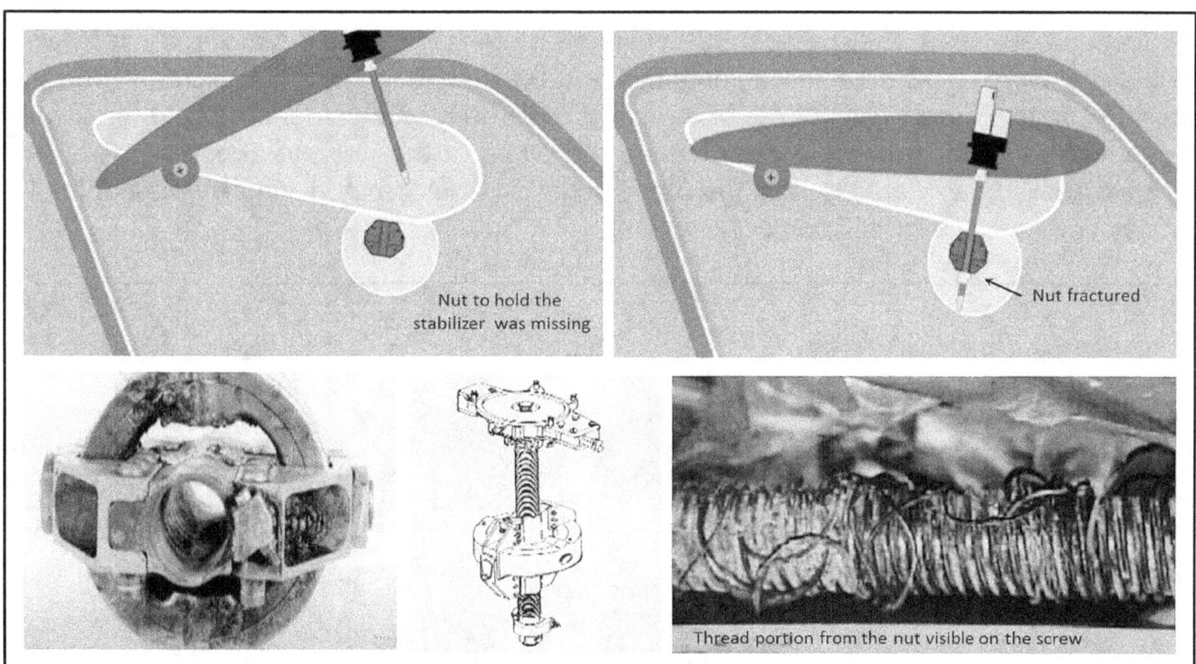

Figure 5.2: Failed horizontal Stabilizer on Alaskan Airline Flight 261

Root Cause: Lack of Lubrication and Management Cost Cutting Schemes: In the case of Alaska Airliner 261, the horizontal stabilizer was jammed keeping the nose of the plane down. The plane plunges from 18,000 feet upside down. The horizontal stabilizer controls the plane's pitch where it can tilt the nose up and down. If the stabilizer moves up, the nose of the plane tilt down while if the stabilizer moves down, the plane's nose moves up. A motorized jack screw moves the stabilizer up and down when engaged by the pilot. The NTSB investigators found out that the jack screw was not bolted in with the nut. This caused the plane to move beyond its aero-dynamical limit. Evidence showed that the screw and nut were separated. The reason for this is that there was no lubrication on the screw. The FAA ordered an inspection for all 34 units of all Alaskan Airliner ND3 and found out that 6 out of 34 stabilizers possess the same situation of having no lubricant.

Further investigation shows that the problems started two years before the Alaska Airline 261 crash. The maintenance crew was under pressure and scrutiny from management to put the plane in service ASAP. Maintenance was forced to falsify records indicating the scheduled activity had been done, but the reality is that it was skipped. Records showed that the stabilizer of this aircraft was scheduled to be replaced however, it was overruled and was not actually done. FAA suspended two maintenance supervisors for falsifying records. Maintenance at the Alaskan Airlines records indicates that management cut costs on corners so that their plane continues flying by shortening the time for doing maintenance, parts needed for replacement were not delivered on time for their scheduled maintenance which forced the maintenance to allow the plane to be put in service so the plane can earn money.

5.3.2: Case 2: British Flight 5390

On June 10, 1990, British Flight 5390 was en route from Birmingham Airport in England to Malaga Spain. After a few minutes after takeoff, the plane's windscreen collapsed sucking Captain Tim Lancaster out of the plane. His body was pressed against the window frame. The first officer successfully landed the plane at the nearest airport. First Officer Alastair Atchison was able to land the plane safely at Southampton Airport in London. The Captain survived this horrible ordeal.

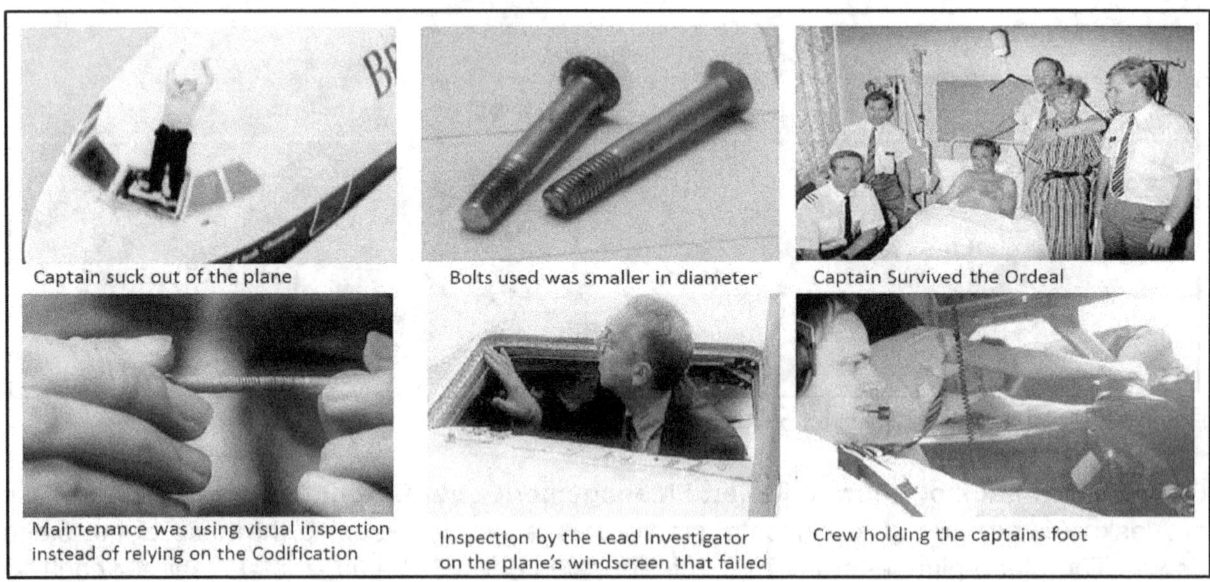

Figure 5.3: Wrong Bolt Used on the Plane's Windscreen

Physical Cause of Failure: Maintenance Used the Wrong Bolt on the Windscreen: Although there were no fatalities on this flight. The captain survived this horrific ordeal. Lead Investigator Stuart Culling from the Air Accident Investigation Branch (AAIB) found out that scheduled maintenance was performed on the windscreen the day before the flight. The correct bolts were A211-8D, although A211-7D were on the old windshield. He matched the old bolts to new ones using his eyesight in a store's bin but chose A211-8C, which was the correct length but the wrong thread and diameter. The diameter of the bolt used on the windscreen was much smaller in diameter. Although the error was unintentional on the part of the maintenance, it almost cost the life of the Captain. The problem was that the storekeeper advised him that the bolt was out of stock. The maintenance decided to went to the other side of the hangar to locate the bolts himself. He found some similar bolts thinking that it was the same bolt for the windscreen which actually is smaller by 200 of an inch. In this case, the maintenance did not follow the procedures. The question was why was the bolt not ordered by the storeroom or purchasing if this will be needed for a scheduled Preventive Maintenance shutdown.

5.3.3: Case 3: Boeing MCAS – A Fatal Design Flaw

Every time a commercial plane crash, there will always be an investigation. However, if two planes crashed for the same reason, what is important to the victims of the families is the true reason why it did happen so these people can have closure. This was the story of the Boeing 737 Max, the fastest-selling plane in Boeing's history.

Figure 5.4: Families of Boeing's MCAS Victims

On October 29, 2018, Lion Air Flight 610 was on a scheduled route from Soekarno–Hatta International Airport in Jakarta, Indonesia to Depati Amir Airport, Pangkal Pinang in Indonesia when it crashed into the Java Sea 13 minutes after takeoff. It killed all 189 passengers and crew. On March 10, 2019, five months after the crash of Lion Air Flight 610, Ethiopian Airlines Flight 302 another Boeing 737 Max plane was on the route on a scheduled international passenger flight from Addis Ababa Bole International Airport in Ethiopia to Jomo Kenyatta International Airport in Nairobi, Kenya. Six minutes after the flight, it crashed and killed 157 people on board the plane. Both fatal accidents have zero survivors. This was the second plane from Boeings' 737 Max that crashed and killed a total of 346 people. This time, Boeings' name and reputation was in question. The question is, were these two accidents from Boeings' 737 Max connected?

After this incident, China Airlines with Boeing 737 Max models grounded their planes, and other international regulators grounded their planes followed by the European Union group, but FAA the Federal Aviation Administration still allows all Airlines with this model to continue flying since their reason was that they were still waiting for the reports from the investigation.

The problem with the new Boeing 737 Max begins with an upgraded software called Maneuvering Characteristics Augmentation System (MCAS), which was totally hidden and unknown to the pilots. The fatal design flaw started as a result of stiff rival competition between Airbus and Boeing 737 Max. In 2010, Airbus introduced the A320 NEO, which is a more fuel-efficient aircraft that is a direct competitor to the Boeing 737. Boeing's 737 models were first introduced on January 17, 1967, and were the fastest sold plane by Boeing, however, with their competition, Boeing had to come up with their own fuel-efficient plane to cope with their Airbus competition. The problem with the redesign of the Boeing 737 Max was their plane was lower than their competitor, which is the Airbus 320, which means that putting a new turbine that is fuel-efficient means that the turbine will be of a larger size. Since Boeing's 737 is lower compared to the A320, and the landing gear cannot be built any higher, they mounted the turbine engines further forward the wings, which protrude beyond the wing of the aircraft. This will cause the aircraft to pull upward when subjected to a high thrust. This was the reason that the MCAS software was installed to correct the problem. In this case, when the nose is too high, the MCAS software will automatically adjust the elevator by pressing the nose downward. However, in the case of both planes that crashed, the Angle of Attack, (AOA) malfunctioned, causing the MCAS system to push the nose further down too far making it impossible for the pilot and first officer to correct the problem and control the plane. The problem is that every time the nose of the plane tilts down, it will be corrected by the pilot but after a few seconds, the MCAS will activate once more taking control of the plane.

Figure 5.5: How the MCAS Should Reacted

Root Cause: Management Cost Cutting Scheme: Boeing's decision-makers and management emphasize coming up with a design quickly with as much as little cost as possible. Boeing Management made it clear that any design changes should not require any new additional simulator training for the pilots as this would be Boeing's additional investment. An agreement with one of Boeing's clients Southwest indicates that any new airplane delivered by Boeing that has design changes or upgrades in the system would cause Boeing to pay $ 1,000,000.00 for every plane that was purchased from them. The end in mind for Boeing is to produce a plane as fuel-efficient as possible with as little or no pilot simulation training as this will cause a rebate to Boing's clients. This means that if they emphasize that MCAS was a new function, then simulator training for the pilot is needed and Boeing will be scrutinized much more as new design planes need to be certified by the FAA before they can be airborne. That is why Boeing deleted MCAS on their flight manuals, which makes the pilots unaware it actually exists. Boeing Top Management insists that this is still the same plane as before but just upgraded.

The next question is should it be that new design planes should be certified before they can become airborne by the regulatory board? Hence, why did the Federal Aviation Administration (FAA) certify and allowed Boeing's 737 Max to be airborne so easily? The answer is that FAA delegated a handful of Boeing people capable of certifying new design model planes. This means that it was no longer FAA that certifies the planes but Boeing itself. This means that the FAA delegated some authority to several Boeing employees. The US Congress even passed laws that pushed the FAA to hand down more responsibility to the Airline companies even to the point of certifying their own planes. Captain Mark Forkner, a 737 Chief Technical pilot recommended removing MCAS from the pilot's manual. He was responsible for concealing this software from the pilots. However, Boeing even further modified the MCAS system to control the Angle of Attack (AOA). The problem is that when the Angle of Attack (AOA) malfunctions, the MCAS system will continue to push the nose down unaware that the AOA is not actually working, providing the wrong data to MCAS. Flight data recorder from both Lion Air and Ethiopia Airlines indicates that their AOA had malfunctioned. However, even before the crash itself, Boeing engineers raised this concern what about if we have a faulty AOA sensor that happens all the time? However, this concern was dismissed since advance sales from Boeing 737 Max was estimated at 370 billion US Dollars. American Airlines have ordered 100 units and Southwest Airlines 200 units not to mention sales from the Asian Countries where Lion Air was one of them. Both Lion Air and Ethiopian airlines crashed have their AOA uncalibrated sending the wrong signal to the MCAS allowing the software to push the nose of the plane down.

In the case of the two accidents, the plane continually pushed the nose down where the pilots were trying to gain control of the plane. The problem with Boeing 737 Max is that they placed a software MCAS which also takes control of the plane's AOA or Angle of Attack. Boeing never informed the pilots of the new design software and instead blames the pilots for human error. The MCAS function is to correct the Angle of Attack of the plane and to avoid it from stalling. In this case, the MCAS will take over from the pilot and MCAS is in control of the plane if the pilots take over automatically the MCAS software will overrule the pilots until the software was again in charge of the plane. The MCAS system is like a plane having a mind of its own and ignoring the human pilots flying the plane.

There was a lot of stake in this case for Boeing, leaving 346 passengers dead from the two plane fatalities all because of a flawed software that was made unknown to the pilots. What is on the mind of Boeing's Top Management is that this is the same plane that requires no pilot simulation training. Boeing Management clearly sacrifices the safety of its passengers over their profits, which is the same reason why the NASA Space Shuttle Challenger exploded killing all astronauts including the teacher Christa Mc Aullife. However, what Boeing also should realize are the individual families at that time that were taken away from them without any chance of survival and the resentment of these families towards the Boeing Industry. Boeing can upgrade and replace its planes, and software but the damage and agony they have inflicted on the victims was something Boeing has to reckon with.

5.4: Maintenance Induced Errors

Maintenance Induced Errors are failures that are directly caused by maintenance themselves. Examples of Maintenance Induced Failures include infant mortality failures or doing intrusive maintenance, intrusive overhauling, and inducing early failures right after endorsing the equipment back to operators. Perhaps the maintenance reassembled the equipment incorrectly. Non-Maintenance Induced Failures are failures on the equipment, which are not caused by maintenance but by some other factors such as how the equipment was operated, design errors, commissioning errors, or cutting costs schemes by Top Management on maintenance, just like the case of MCAS on Boeing 737 Max.

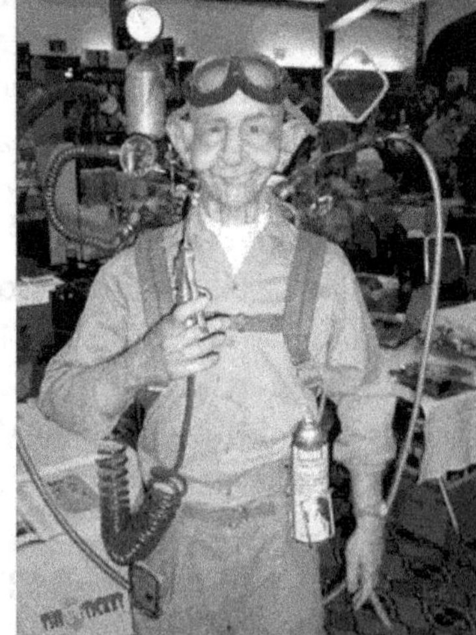

Figure 5.6: Common Causes of Human Error

[7]In the book of R. Keith Mobley, An Introduction to Predictive Maintenance, he quotes, that from studies of equipment reliability problems conducted over the past 30 years, maintenance is responsible for about 17 % of production interruptions and quality problems.

---

[7] Keith Mobley, **An Introduction to Predictive**, (Elsevier Butterworth - Heinemann, 2002), Pages 10 -11

This means that the remaining 83 % are totally outside of the traditional maintenance function's responsibility. Inappropriate operating practices, poor design, non-specification of parts, and a myriad of other non-maintenance reasons are the primary contributors to production and product quality problems and not by maintenance themselves. Although the percentage of human errors induced by maintenance is not that big, still the irony lies behind that the worst industrial disasters that occurred globally were a result of maintenance and of course human errors. Human errors will have a wide range of consequences and the problem is that industries only react if the consequences of the failure already occurred. Even if we already knew the problem for as long as nothing happens, then industries will always result in the same old ways of doing things. Here are some possible situations where the failure is caused by maintenance itself or Maintenance Induced Errors:

1. Infant Mortality Failures: These are early failures that can occur on the equipment right after a major Preventive or Scheduled Maintenance is done on the equipment. Operators often say sarcastically that if maintenance did not perform their regular overhauling and replacement on this piece of equipment, I bet it will be running without any problems. But isn't it that the main purpose of Scheduled Preventive Maintenance is to ensure that the equipment performs as it is intended, but the opposite seems to happen whenever maintenance does something on the equipment. Something goes wrong. Is it profound? No. Honestly, we are just a victim of what we called Infant Mortality Failures. In almost all cases, Infant Mortality Failures are caused by human errors and intrusive or forced maintenance. Many things can go wrong when we try to dismantle the equipment especially for overhauling purposes. Human errors such as slips and lapses can occur on the part of the maintenance performing the overhauls. Infant Mortality Failures are failures that occur at the beginning of life, others refer to them as commissioning failures, start-up failures, or debugging failures. Many factors affect Infant Mortality Failures, which include poor equipment design, poor quality of parts manufactured, incorrect installation, incorrect commissioning, incorrect operation, unnecessary or intrusive maintenance, human errors, or simply bad workmanship.

The case of infant mortality failures, which is pattern F of the six failure patterns, starts with a high incidence of early failures which eventually drops to a constant or very slow increasing conditional probability of failure and ends up in a no wear-out zone. If Infant Mortality Failures exist frequently, the question raised, is can we eliminate them? The answer is no, we cannot eliminate infant mortality failure completely but it can be reduced. Reducing unnecessary PM activities performed on the equipment can reduce infant mortality failures. In layman's terms, the more we touch, the more it fails, the less we touch, the less it fails. If we review the lists of every single activity that is being performed on the equipment, we can find out that there are maintenance tasks that duplicate themselves which it is already performed by regular maintenance yet are still being performed by other people such as third-party contractors. There may also be maintenance tasks that are deemed unnecessary and do not address any failure modes. These should be removed from the PM tasks lists of activities. There might also be activities that should be done but in reality, do not exists at all in the Preventive Maintenance lists of tasks.

2. Installation and Reassembly: Preventive Maintenance includes risks. There is always an assumption that everything will be put back in place well together after a PM initiative. Likewise is the possibility of human errors, which leaves to infant Mortality Failures of newly installed components, which eventually leads to additional failures after endorsing back the equipment to production.   The process of overhauling includes disassembly and reassembly. Human errors are minimal during the disassembly process but are more prone to happen during the reassembly process, which can definitely lead to infant mortality failures.   Human intervention is involved during a major overhauling activity on the equipment.   Although there is little or no human error can occur when dismantling equipment. Human error happens mostly when reassembling or putting the equipment back in one piece. What is important for maintenance is that the people involved in these tasks should be fully focused on completing the tasks correctly. A detailed and guided procedure on the correct way to overhaul can minimize the chances of human error. There should be no shortcuts on how to dismantle and reassemble the equipment as well as using the right tools to dismantle the equipment. Maintenance should also know that there are ***do not disturb*** portions on the equipment that should be left untouched. Maintenance must have the correct skills and the right tools in doing the installation and reassembly of the equipment. Remember that when dismantling the equipment, human error is negligible. Human errors usually happen during re-assembling the equipment.

3. Faking Maintenance Documents to Comply with PM Specs: If we can refer to the Top Ten Problems on Preventive Maintenance, this problem ranks third among the lists. The maintenance thinking is that it is better to fake the documents than to be incomplete and besides this is just a minor thing that no one will know or nothing happens after all if I place a check in this activity. Besides, I will be audited if I did not fake this document or my shift is ending and I will be late, as the company shuttle bus will be leaving in 5 minutes. This is one minor problem that can lead to a dangerous situation. This is one of the major issues if maintenance will be done by maintenance 100 % of the time. Operators are never involved because the operations manager just wants these operators to operate and produce output. You see, if Autonomous Maintenance is established in an industry, there will be checks that will be done not only for maintenance but also for operators themselves. Another reason why maintenance people do this is they are given limited time by management to execute their Preventive Maintenance activities so that the equipment or machine can be back to operating mode once again.

4. Not Performing a Functionality Check on Protective Devices: A protective device is placed in the equipment to serve a particular function and that is to protect something. That something is called the protected function. Protective devices should be inspected at the correct frequency because the failure is hidden. The only time that we know that these protective devices had failed is if the protected function also fails. When both protective and protected function fails then, we experienced multiple failures. This is what we are totally avoiding. Try to think of your protective device as a parachute. If you jump off the plane with your parachute and it malfunctions or it did not open when you press the button, do not panic as there is always a reserve parachute in place. When it is time to open the reserve parachute and still it did not open then all I can say is, that will be the end of your life.

118

Protective devices are like a parachute. These protective devices are in the form of led, alarms, sensors, beacons, and lighting arresters, that are strategically placed in the equipment as a defense to alert operators and maintenance of upcoming problems, deviations, and abnormal conditions.

Here is a classic case problem. On April 20, 2010, Deepwater Horizon experienced an oil spill in the Gulf of Mexico. Vital warning systems on the Deepwater Horizon oil rig were switched off at the time of the explosion to spare workers from being awakened by false alarms, according to a federal investigation officer. The revelation that the alarm systems on the rig at the center of the disaster were disabled and that key safety mechanisms had also intentionally been switched off came in a testimony by a chief technician working for Transocean which is the drilling company that owned the rig. Mike Williams, who was in charge of maintaining the rig's electronic systems, was giving evidence to the federal panel in New Orleans that is investigating the cause of the disaster, which killed 11 people. Mike Williams also told the hearing that no alarms went off on the day of the explosion because they had been inhibited or switched off. Sensors monitoring conditions on the rig and in the Macondo oil well beneath it were still working, but the computer had been instructed not to trigger any alarms in case of any adverse readings since it always gives a false alarm waking up the people. Both visual and sound alarms should have gone off in the case of sensors detecting fire or dangerous levels of combustible or toxic gases.

5.5: Non-Maintenance Induced Errors

As said previously, the majority of errors, which account for 83 % are way outside of the maintenance function, in which maintenance has very little or no control whatsoever. These errors are caused by the different functions of the entire organization, which should be addressed directly by the Top-Management and C-Level people. While Non-Maintenance-Induced errors are those that were not manifested by the maintenance function, the causes are beyond maintenance. This is a much bigger problem because maintenance may have little or no control over the situation. Let me give some examples of Non-Maintenance Induced Errors

1. PM is waived by Operations: When the equipment was subjected to maintenance intervention or Scheduled Preventive Maintenance, operations or management frequently deferred or waived the equipment because the output is always the top priority and supersedes everything. When the demand for the product is high, or there were months when production is at its peak, maintenance will find it very difficult or impossible to get the equipment for a scheduled Preventive Maintenance activity. As a result, more unexpected failures occur, and the feud between operations and maintenance gets worst and pretty much alive over time. Maintenance ends up always on the losing end because the operation's decision prevails. Operations will give the equipment to maintenance only when the demand is low.

2. Making Maintenance a Jack-Of-All-Trades: For dynamic industries that keep on reorganizing their people, today you are the mechanical technician, next week you will be transferred and become an electrician. When new Top Management was hired and after a week of observing the plant he decided to reorganize his staff and unluckily, you were a part of it. I have been employed in a dynamic industry where almost every month there are changes in the organization. Is it good? Honestly speaking, I just cannot tell. Nevertheless, I see a lot of wasted projects and improvements that were abandoned as a result of the reorganization. All I can say is that if you plan to make maintenance a jack-of-all-trades, then try also to consider the anthropometric factors.

3. MRO Spare Parts are Not Managed by Maintenance: In many organizations today, MRO Spare Parts are not controlled nor managed by maintenance themselves. This is a very big problem as the best people to manage the storeroom are the maintenance people because they are the users of the parts and they know these parts better than any other people in the organization. While different consultants have different views regarding this, my point is that how can we manage something that we do not control. We just cannot. While in some industries they can be managed by the Supply Chain, Warehouse, Purchasing, Finance, Admin, or even other functions. The role of the storekeeper is not only to issue and receive parts from the vendor. The role of the storekeeper is also to maintain the parts inside the storeroom. For example, belts should not be placed in a cold area, as the property of resiliency will be lost. Bearings that are in boxes should not be placed vertically but horizontally. Large motors and rotating components inside the storeroom should be rotated once or twice a week to avoid any chances of false brinelling, especially if there is some form of vibration being felt inside the storeroom. Rotating components should be greased every 6 months if kept idle inside the storeroom. Lubricating oil should be properly stored inside the warehouse. With a limited amount of storage space inside your plant, it is common to have an outside storage area for newly delivered oil drums. This practice creates a high risk for water and dust ingression into the new oil if the drums are not stored properly. This is especially true during the summer months when the temperature of the drums can reach 150 ° F or 65.5 ° C. According to Svante Arrhenius a Nobel Price physicist, the life of the oil is cut in half for every 10 ° C (18 ° F) increase or rise in operating temperature. This means that lubricants have exceeded their base activation temperature, and will degrade or oxidize twice as fast for every 10 ° C rise in temperature, which can oxidize the oil. There will be a wide array of problems that will happen if maintenance does not own the storeroom. I would strongly recommend that maintenance should be the ones to own and manage the storeroom, as they are the users and likewise to gain more control of the spare parts.

4. Purchasing Going for the Lowest Bidder: Although in most industries, purchasing will source for at least three different bidders for a particular part where the end result is that they will award the part to the lowest bidder. So the Purchasing Manager together with their people are celebrating because they have saved a few coins on the part or item but what they do not realize is that they make the life of maintenance much more difficult and miserable because the part breaks easily and costs thousands or even hundreds of thousands in US dollar a year not to mention the cost of downtime, repair, and revenue lost due to the downtime. There were three vendors for this type of oil and the purchasing awarded the oil to the lowest bidder. What purchasing do not know is that the Flash Point of the oil of this vendor is at 180 ° C which caused a lot of evaporation and tapping up of oil on the equipment by maintenance, and now they question maintenance why this equipment consumes too much oil. As a rule of thumb, the minimum Flash Point should not be lower than 204 ° Centigrade. This means that the lower the Flash Point of the oil, the oil will be more prone to evaporate easily.

For many industries, there is always the temptation to purchase equipment or spares based on the lowest or cheapest source. This might not be such a good idea since purchasing based on the initial costs only tells us one side of the coin. The true costs can be seen based on its performance and the total life cycle costs, not on the initial costs. In my experience, this is the problem with most procurement and purchasing departments since they make the decision to purchase parts based on the lowest bidder or lowest possible costs, without the advice or recommendation of the technical people in the plant. The savings that these departments claim are insignificant since both operations and maintenance can encounter many failures as a result of this. They always look at the initial cost of the part and not the cost of problems the part may provide the user in the long run. This is the problem with buying cheaper parts since the equipment can fail easily and will likely yield a much larger amount of cost in a period compared to a slightly higher cost of equipment or spare. The true cost of the equipment can be felt from the time the equipment was commissioned to the plant until the time it will be decommissioned or disposed of.

5. OEM Design Errors: Since God did not create the equipment, then no equipment is perfect by design. It will always contain flaws and inherent design weaknesses. This can be addressed by maintenance by deploying the Planned Maintenance pillar of TPM, which will be covered in the later chapters of this book. Planned Maintenance Phase 2 will address this issue. What is important for the maintenance is to observe the equipment and list all the parts that fail or break easily and the challenge is how to extend the lifespan of these parts through modification or redesign.

6. Operator Errors: Despite the training, certification, and recertification, operators will still continue to commit errors especially if their work is routine since humans commit errors which is part of our human nature. These errors committed by operators can lead to failures as well as product defects. Operators' retraining and recertification may not always be the solution in this case but adopting Visual Controls and Poka-Yoke can reduce operator errors in this case.,

5.6: Implementing Precision Maintenance to Reduce Infant Mortality Failure

In the book of James Reasons and Alan Hobbs on Managing Maintenance Errors, the largest number of maintenance errors in the aviation industry is associated with reassembly and installation. There are two ways people can go wrong. First, they can either do something they should not have done, which is called a commissioning error and second, maintenance fails to do something that they should have done, which is called omission errors. Omission errors during the installation process are the largest single category of maintenance error. The US Nuclear Power Plant reports that 64.5% of errors associated with maintenance activities involved omission of necessary steps. If the impact and consequences bought about by human error are not acceptable, then we need to do something not only to mitigate but if possible eliminate the risks involved in this situation completely.

Precision Maintenance involves performing maintenance work in a consistent, precise, and industry-accepted way. If properly implemented, this means that maintenance should yield the exact same results, no matter who is performing the tasks. Precision Maintenance is much more than merely having procedures on PM. This must be the correct culture of the organization. The good thing about doing Precision Maintenance is that it can reduce the chances of Infant Mortality Failures seen during the start-up of assets right after a Preventive Maintenance shutdown or when scheduled overhauls are performed. It involves performing maintenance work in an accurate and precise manner, which means that the maintenance should provide the exact same results no matter who is performing the work, whether the work is done by the most or the least experienced maintenance craft in the plant.

When different maintenance people perform the same PM on the same equipment, variations occur. These variations cause problems since they do not meet the requirements for the process. Precision Maintenance help to ensure that equipment will be maintained to the highest possible standard so that the variations that can cause defects and failures can be reduced, or eliminated enabling equipment and assets to run at their optimized and peak level of reliability. Precision Maintenance rebuilds machines and equipment to the highest standards so that fewer problems can occur during operation. It is a matter of ensuring the important things for equipment and machinery health are done correctly and accurately.

The key to Precision Maintenance lies in the knowledge and skill of the maintenance craftspeople. Precision Maintenance will take time to achieve, but the benefits that can be reaped are worth doing. Doing Precision Maintenance will not only reduce downtime, but it will reduce the cost of performing maintenance. Training maintenance to perform, the correct tasks on the equipment will yield outstanding benefits for the plant. Organizations that can build a culture of Precision Maintenance will reap the rewards of a better bottom line, a safer workplace, and more productivity throughout their facility.

To begin with, a skills gap is necessary to conduct, and assess the difference between the skills required to perform a specified task and the skills maintenance currently possesses. Precision Maintenance describes the maintenance culture in a facility. A scoring system will help determine the knowledge and skills maintenance currently has. With these scores, a sensible skills training plan can be developed to transform the maintenance workforce into a highly skilled team that is fully trained and qualified to perform these tasks in a precise state every time maintenance tasks will be performed on the equipment.

Installation and Commissioning of New Equipment: Equipment installation refers to setting the equipment in the plant while commissioning equipment refers to conducting all required tests and procedures required as per industry standards to show that the equipment can perform for the purpose for which it was intended. For smaller plants, both installation and commissioning can be conducted on the same day, but for bigger plants, the installation will be done first before commissioning. If industries purchase additional equipment in the future in support of their production needs and targets, they should ask for some data regarding the vertical start-up time from the previous equipment installed by the OEM, or simply ask how fast they can commission the equipment in the plant. Their previous data can be used as a way of having some benchmark on the process for both installation and commissioning. Determine the activities where time had been spent most. The important part of this is to evaluate their previous activities and provide an improvement in reducing the vertical set-up time or commissioning time needed for the equipment. The goal of equipment manufacturers should be to shorten the vertical set-up time of the equipment being commissioned. If the previous commissioning time had been reduced from seven to three days, this must be documented and highlighted regarding what they have done to reduce the commissioning time. This will be an added value for the organization since the equipment is now capable of operating and adding revenue to the organization. This new record time will now be their benchmark when trying to commission the same type of equipment in the future. Remember that the longer the equipment is commissioned in the plant, the less industry is not generating revenue from it. Profit only starts when the equipment has been handed over to operations, and production starts to generate products. Both equipment manufacturers and industries should consider the vertical set-up time as a record. This is like running a race and improving the previous record time in completing that race.

Preventive Maintenance Overhauls: As previously discussed, overhauls contain risks since it is assumed that when the equipment is disassembled, it will be put back again together in the same original condition as it was before. In most cases, this is often not the case. The number one cause of infant mortality failures is human error. During overhauls, error mostly happens when the equipment is now in the process of being reassembled or put back together once again. When the equipment cannot run smoothly right after a Preventive Maintenance overhaul, then there is something wrong with how it was overhauled. Maybe if you belong to maintenance, you might have experienced that whenever this person is the one who performs the overhaul, the equipment is running smoothly after it was endorsed back to operations. If another person was the one that performed the overhaul, then operations will find a difficult time running the equipment, and it needs to be debugged and corrected. One thing that is important in performing a PM overhaul is the sequence of how the overhauled was done. What did the PM did first, then second, then third, and fourth, until it was finally completed? This is very important, as the correct sequence should be very crystal clear and in detail for everyone. There is only one way to perform the maintenance task. For example, there is a correct way to remove the cylinder head from a car engine. The unbolting pattern is always fixed to assure no damage is done to other parts of the engine. What is the first bolt to be removed, and then followed by this and so on? The tools used should also be consistent with whoever is executing the tasks. This should also be indicated in the task lists. Also important likewise is to provide a thorough list of post-mortem or providing a checklist of all the activities done for a final inspection to ensure that everything has been covered before finally endorsing the equipment back to operators.

5.7: Reducing Human Errors in Maintenance

Although human errors cannot be eliminated as they are part of being human. The good news is that they can be well managed. According to James Reasons and Alan Hobbs, errors are not intrinsically bad. Success and failure spring from the same roots. We are error-guided creatures. Errors marked the boundaries of the path to successful action. What is important is that we learn from the failure as well as the error itself. Human errors will happen, and they cannot be eliminated 100%, but we can manage them more intelligently by learning from the things that go wrong in our industry. Failure is our best teacher. The most successful and well-known people in the world have failed many times themselves before becoming successful. Here are lists on how to reduce human errors in industries.

1. Ergonomics: Improve the Working Condition of the People: Ergonomics is the design of equipment, operations, procedures, and work environments that are compatible with the capabilities, limitations, and needs of the workers. It is a vital complement to other engineering disciplines that seek to optimize hardware performance and minimize capital costs with little or no consideration of how the equipment will actually be operated and maintained. For example, a salvaged dial thermometer attached to a nozzle just below the third level platform might be a very inexpensive way to accurately measure the temperature in a tower. But from a human point of view, the design is flawed itself as both operators and maintenance need to climb a ladder, which is around three floors, just to get the reading and log it on their records, which need to be done during their shift. While we always blame people for committing mistakes and errors, care should be given to improving the working environment and safety of both operators and maintenance themselves.

2. Develop a Functionality Inspection for Protective Devices: Protective devices are defenses we placed on the equipment to alert us of an upcoming problem or error. They may be in the form of sensors, led, alarms, lighting arresters, overload protection, ground fault protection, overcurrent protection devices, emergency stops, over-speed devices, and so on. They should be functional 99% of the time, depending on the criticality of these protective devices. Functionality inspection or Reliability-Centered Maintenance (RCM) called these Failure Finding Tasks. These will be the only maintenance tasks that can be provided for these devices to ensure that they are still functioning in the equipment, asset, or system that which they are placed in. Failure of most of these devices will be hidden, and the only time that maintenance or operators know they have failed is when the ones they are protecting also failed, which is called the protected function. Try to assess the risks of damage. If the risks are high enough and are not acceptable, then these functionality inspections should be carried out consistently and regularly. Everyone should be informed of the consequences and the risks involved in both the protective device and protected function. Remember that they are placed in the equipment for a reason, and that is to protect something that is why they are called protective devices. The questions to ask will be what could happen in the event under consideration did occur. Is anyone likely to be hurt or killed as a result? How likely is the failure to occur, and is the risk acceptable?

3. Tidiness and Housekeeping: It is less stressful to work and operate in an environment where good housekeeping is in place rather than working in a workplace where everything is a mess. Provide adequate lighting in the workplace. Implementation of 5's will be a good starting point to address housekeeping issues and reduce the chances of a slip, and trip hazards. People fatigue easily if the workplace itself is a mess. Slip and trip accidents in the workplace account for a large percentage of injuries, which can be attributed to poor housekeeping. This is less likely to happen if housekeeping is in place in the organization. It also creates discipline in us, and that good housekeeping starts in our home.

4. Simplify Maintenance Procedures: If we take a look at all the procedures we used in our organization, there will be many outdated or even obsolete procedures that are still being used and implemented up to this very point in time. Perhaps these procedures were effective at the time it was written but would no longer hold true for now. Today, there are pieces of equipment that had already been disposed of, and some procedures still exist and are still being followed up to this point in time. Procedures should be simple, precise, and should be easily understood by the user. Avoid complicated words. What is important is that if there is a way of simplifying the things being done, procedures should likewise be revisited, reviewed, revised, and simplified whenever necessary. Provide pictures or drawings whenever necessary. When industries invited me to their plant for in-house training, the moment I entered the plant, I can already have a sense of feeling about how things are being done in that plant. How long will security take for me to be allowed inside? Sometimes due to very strict procedures, it can take me 30 minutes, or sometimes more just to enter the plant. Although we understand their point, sometimes I am beginning to think that the bigger the industry, the more problems exist. I told my sponsors, especially if this will be an in-house training if they have coordinated my entry to their security ahead of time? I remembered in 2006, I was scheduled to go to this semiconductor plant for two days of training on Maintenance Indices and KPI; at their security, you need to enter a metal detector where you will be scanned just like in an airport. Very nice indeed but not impressive. When I entered, it alarmed, and the security told me to remove my belt, so I did; then, after removing my belt, I passed the detector again, then it alarmed me for the second time. Therefore, the security told me to remove everything in my pocket, which I complied with. Once again, I entered the metal detector scanner, and it alarmed me again. I told the security that I have some metal braces on my leg and told them if I should call my doctor and remove them so that I can enter? The security called his office, and the officer still cannot make a decision and called the human resources, and that was the only time I was able to get inside and started to conduct my training. The time I lost from security alone was around 45 minutes. If you ask me, what was the name of this plant? Do not bother; it is already closed as they moved to China. Another case I have was when I was in India for in-house training in their plants in 2010. When I entered the security, each one of the visitors will have a number and one security person will scan my face, print it, and hand me an ID with your face printed. If there are few visitors, then you are lucky if you can be inside the plant in 30 minutes, if not then you need to fall in line and wait for your number to be called.

I asked the security if this process would be done for the following days as I will be teaching here for five consecutive days and the security said it was their protocol. All in all, I wasted around 4.5 hours just sitting in the security where that 4.5 hours can be spent planting seeds on the maintenance brains. What a total waste of time. When you come, I'll instruct my dog Mr. Stones that you will be coming. When the dog barks, then you will be inside my home in a matter of seconds. Why can't industries be like that?

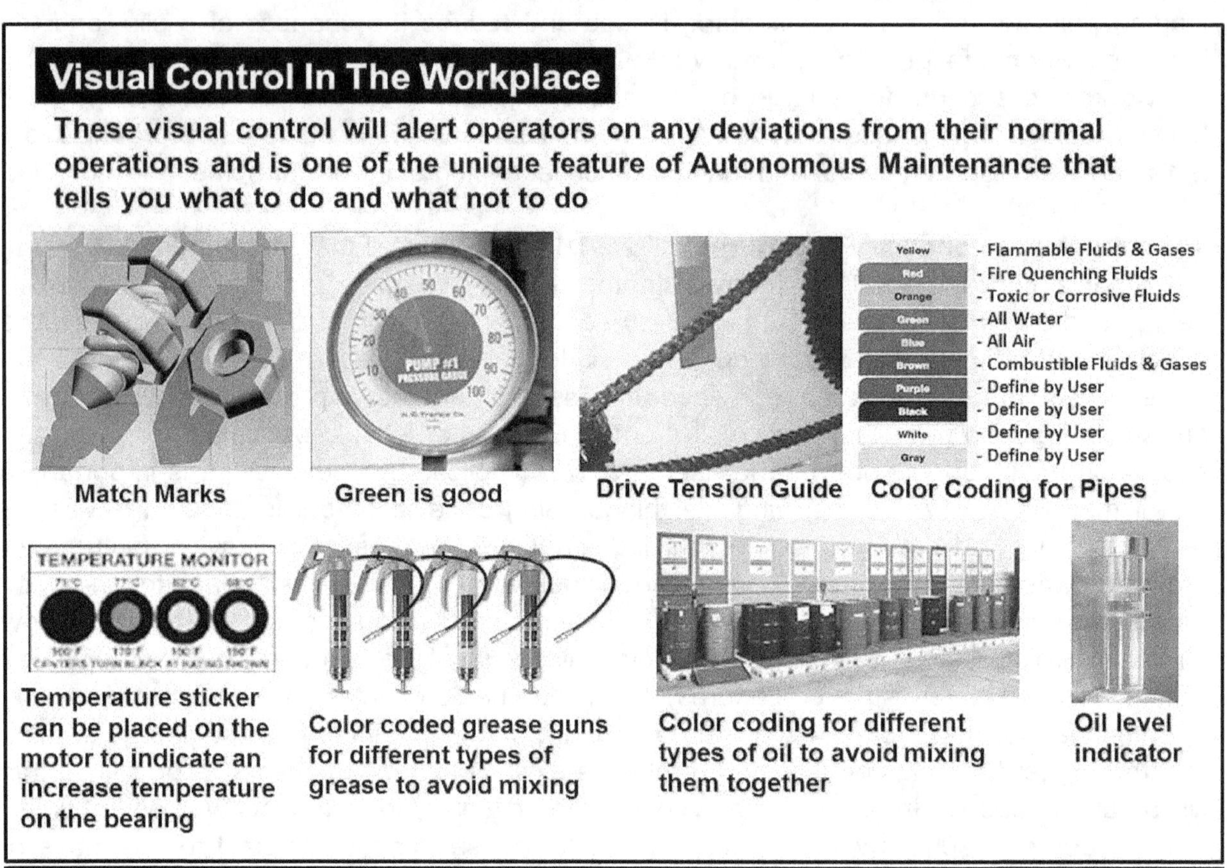

Figure 5.7: Samples of Visual Control Management

5. Use of Visual Controls in the Workplace: Visual controls are placed in the workplace to expose problems and deviations easily by just looking at them. It makes deviations easily seen by operators by alerting them that something is not right. They are designed to make the control and management of the plant as simple as possible. Visual Controls originated from the Japanese. They are not meant to make us dumb or look stupid. It simply entails making the problems, abnormalities, or deviations from the standards much more visible to everyone in the plant. When these deviations are clearly visible and apparent to all, corrective actions can be taken to immediately correct these problems and anomalies. Visual Controls allow us to see problems more easily. They are also meant to provide instructions and convey information. These are devices or mechanisms that were designed to manage or control our operations process to meet the following purposes. They make the problems, abnormalities, and deviations from standards easily known not only to the operator but also to everyone in the entire organization. It will display the operating or progress status in an easy-to-see format, provide instructions, convey information, and provide immediate feedback to people. Remember that the simplest solution that works will probably be the best. Good visual control management should tell the operator, immediately, if a process is going beyond the specified parameters. Ideally, the process would be stopped automatically. Like in an airport, signs can guide the passengers in a way that saves them a great amount of time instead of asking for directions most of the time on knowing where you want to go when you are inside an airport. Visual Control is also sometimes called Visual Communication. In the early 1970s, U.S. executives went on tours to different Japanese industries to see why they were being beaten so badly in terms of quality by their Japanese competitors. They noted one specific thing in which the workplaces were filled with pictures and diagrams of what to do and what not to do. With the way things were displayed in the plant, anyone passing by, whether a U.S. executive or a new employee, could actually sit down and do the assembly themselves without even knowing the Japanese language. The most important benefit of visual control is that it shows when something is out of place or missing.

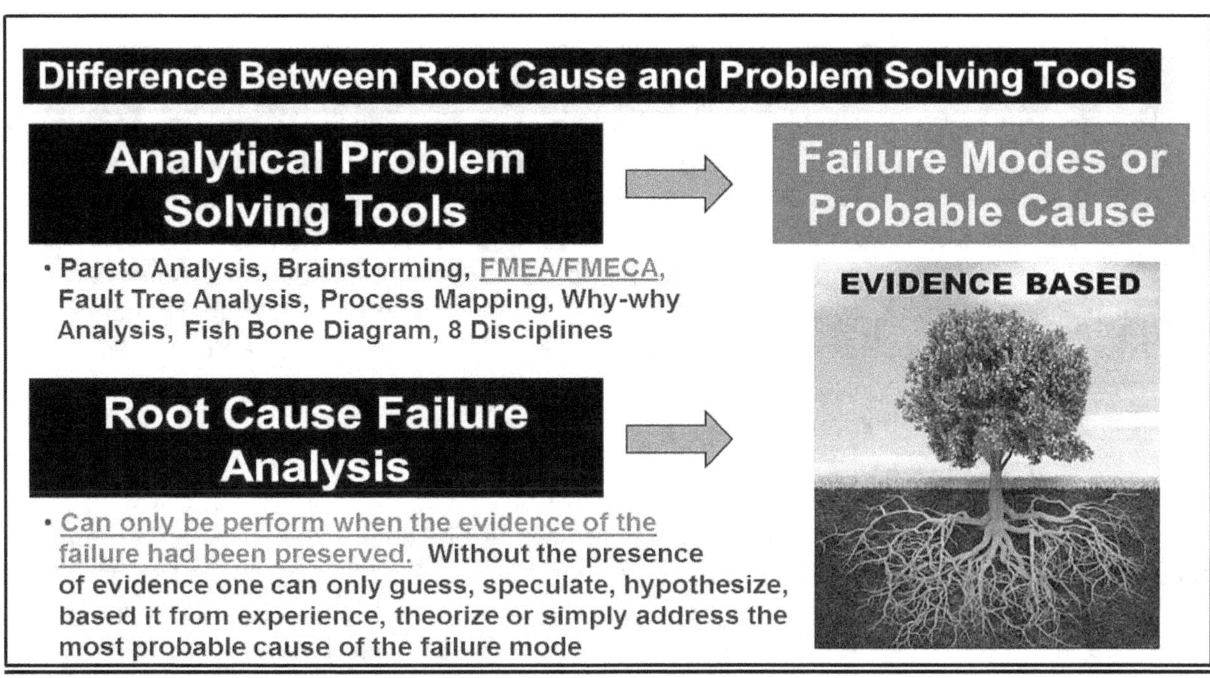

Figure 5.8: Difference Between Root Cause and Problem Solving Tools

6. Conduct Root Cause Failure Analysis on Recurring Failures: Although I would not recommend that for every single breakdown or failure, a Root Cause Failure Analysis investigation should be performed, as this will be a very tedious process indeed. However, if the failure keeps on repeating itself, then I think it is time to perform a Root Cause Failure Analysis investigation and take things slowly. Please take note that a Root Cause Failure Analysis is different from implementing the problem-solving or analytical problem-solving tools such as Fishbone, Pareto Analysis, FMEA/FMECA, 8-Disciplines, and the like. To conduct a Root Cause Failure Analysis, the failure should be fresh, which means it just recently happened. This will require a Principal Investigator and a group of people to gather pieces of evidence to shed light and see the bigger picture of why and how the equipment failed. Root Cause is evidence-driven. The goal of conducting a Root Cause Failure

Analysis investigation is to understand why and how the failure occurred so that industries can finally learn from the things that go wrong in their plant. Conducting a Root Cause Failure Analysis will allow the investigating body to determine the Physical cause of the failure, the human cause, the system cause, and finally, end up their probe and investigation on the latent cause of the problem. While there are many techniques for analyzing a problem that provides a quick answer, it does not mean that the answer will be correct every time. A true and meaningful Root Cause Failure Analysis will take the time to prove that what we say is fact and supports our hypothesis with evidence before we spend money to improve the design of the equipment. When the facts are finally backed up by evidence and science, and these facts are separated from the fiction, then we now have a better and clearer understanding of the real Root cause of the problem. Root Cause is 100% evidence-based. Without evidence, the root cause is not possible, and we can only get the most likely or probable cause of the problem.

7. Apply Poka-Yoke Solution: Poka-Yoke is a Japanese term that means mistake-proofing. Examples of Poka-Yoke include elevator alarms, 110/220 volt or auto volt appliances, low battery indicators, fuel indicators in the car, child lock doors, and so on. Poka-yoke is any mechanism mostly used by industries, especially in the manufacturing process that helps an equipment operator avoid errors and mistakes. Its purpose is to eliminate product defects by preventing, correcting, or drawing attention to human errors as they occur. It was developed by Shigeo Shingo as part of the Toyota Production System. The term Poka-yoke was applied by Shigeo Shingo in the 1960s to industrial processes in which its primary purpose is to prevent human errors. The aim of Poka-Yoke is to design the process so that mistakes can be detected and corrected immediately, thereby eliminating defects and human errors at their origin.

8. Enhance Maintenance and Operator's Training: As the saying goes, that people are the company's greatest asset is not always true. What I believe is that the right people are the company's greatest asset, and the wrong people are called liabilities. We can only have the right people if our people are equipped with the right knowledge to do their jobs right the first time around. However, nowadays, the only training maintenance got is from the University of the Hard Knocks. As the great leader Nelson Mandela said, Education is the most powerful weapon that you can use to change the world. In my first book on World Class Maintenance Management - The 12 Disciplines, I indicated that training is the foundation and backbone of any cultural change. This will be the foundation for setting up the basic maintenance strategy in the plant. Maintenance, as well as operators, should be trained and educated first. However, the sad part of it is that in most industries, training is one of the primary targets for cost-cutting schemes by Top Management. While industries invest in adding hardware resources to expand their business, they forgot to invest in their human resources, which is a huge mistake. If industries deprived their people of attending any training, my advice is to educate yourself, read, surf the internet and join maintenance forums. Training and knowledge can come both formally and informally.

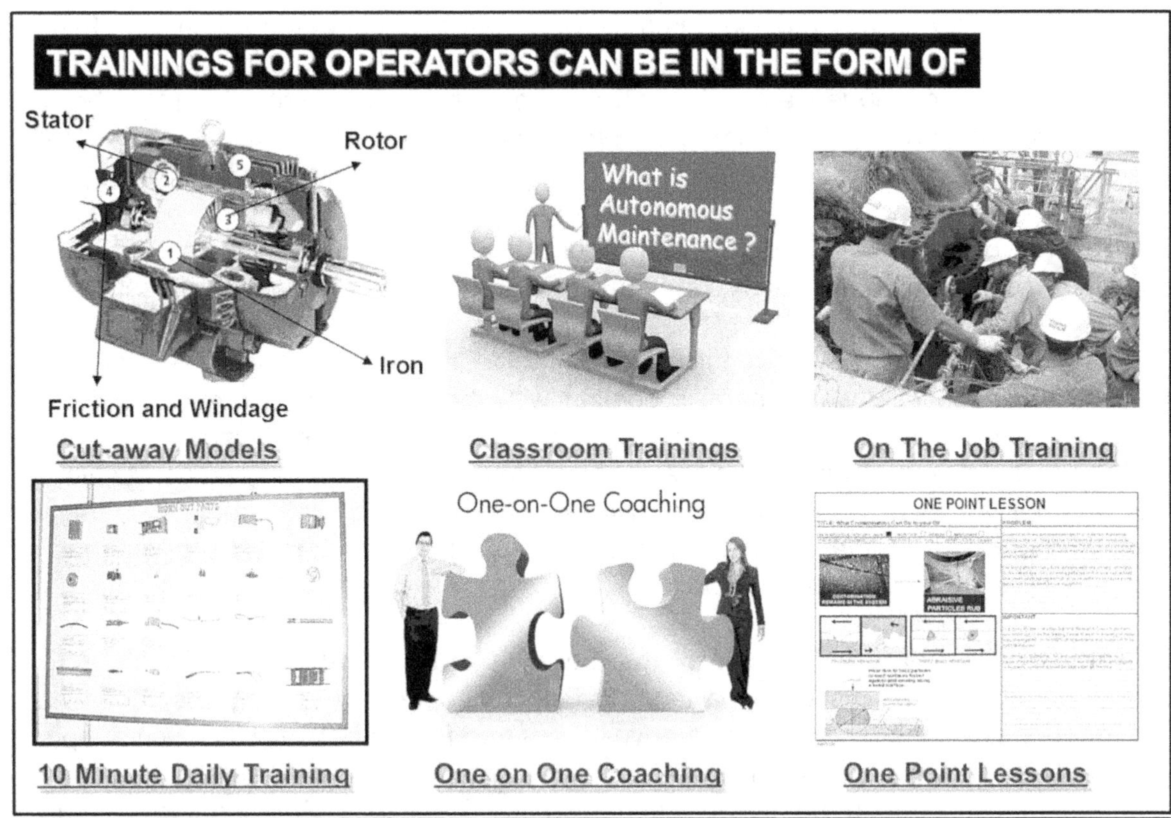

Figure 5.9: Different Ways of Training Operators

For the operator's training, their coach will be the maintenance people if they are to implement the seven Steps of Autonomous Maintenance, where the ultimate goal is to empower operators. My last work was in the mining industry in the Philippines. It was located in Mankayan, Benguet, which is 16 hours from Manila by bus or 45 minutes by plane. Our main office was located in Makati, but I used to teach in the plant in Benguet. When I teach, everyone was happy because of an abundance of food. They even have some takeaways at the end of each training day. Then one day, the finance manager called me and questioned me on why I had a big budget for food. My answer was simple. I do not want to teach a person with a hungry stomach. After a month, I went back to the plant for another batch of scheduled training, and I noticed that there was no food during break time, which I learned was the directive from the finance manager. When I went back to the plant, I confronted him and told him that you are saving on small pebbles while the big savings will come from the people I am teaching. That day on, I filed my resignation and never returned to that plant the following day.

9. Develop a Risk Criticality Assessment for Failures: I believe that this is one area where both Quality, Health, Safety, Environment, Reliability, Maintenance, and Operations people should sit down, relax and communicate as one team and discuss this matter thoroughly and develop a comprehensive list of all possible critical failures that can occur in their industry and assess the risks involved. Not all failures are critical; the problem sometimes or many times (I just can't tell) on maintenance is that all failures are considered critical. That is why their storeroom is loaded with excessive stocks, which were not even used and ended up as non-moving items. When the equipment failed, and the stock was not around, the boss yells at you, making the spare part critical. The criticality of the part will depend upon the consequences of the failure. Remember, for a critical failure, the risk is high, and the aftermath is unfathomable. What I am saying is that critical failures seldom happen, but it is possible to happen, and when they do, the impact and consequences of the aftermath will definitely hurt your industry's reputation, image, brand, and business. The worst failure you can ever imagine will be those failures that can affect the environment. Failures that can cause harm, injury, or even death are likewise considered critical.

10. Integrate Precision Maintenance on PM: Performing Preventive Maintenance includes risks especially when there is contact with the equipment. Integrating Precision Maintenance, especially during replacement and overhauls will minimize the chances of infant mortality failures. This means that whoever is executing the Preventive Maintenance tasks whether the person is the most or least experienced in the craft should yield the same results and outcome so that when the equipment will be endorsed back to operations after PM, there will be no problems whatsoever.

11. Remove Departmental Barriers for Better Communication: [8]According to Dr. Edward Deming, 14 Points of Management, telling people to work together does not do much good if the management system drives them to different behaviors. Such support for teamwork is merely a slogan without the necessary management commitment. Management can support but never commit to the initiative. We need to change the management system and the behavior of those in leadership positions in the organization. When we create incentives to optimize parts of the system, the overall system is sub-optimized. To achieve the best overall results, individual parts of the system may have to suffer to achieve the best overall result. When we evaluate people and provide bonuses and promotions based on optimizing a portion of the system, this creates pressure that works against cooperation across departments. Departments have to compete for their budget and staff, which can cause barriers between departments. When departments have their budgets to protect and spend, it often creates barriers among departments. Often this even progresses to the point where employees are considered more a part of one department than employees of the company are, and if they transfer to another department, that is seen as a sign of disloyalty. The supposition is prevalent the world over that there would be no problems in production or services if only the production workers would do their jobs in the way that they were taught. In reality, the workers are handicapped by the system, and the system belongs to the management. As I said, there are no proponents for more barriers between departments as a management strategy. Moreover, there is often talk about the importance of cooperation and that we are all one team, but when the management system is structured to undermine that notion, it is not much use to tell people to work as if optimizing the overall system is what is being desired. The management system needs to encourage the behavior the organization wants to see. Too many organizations still have difficulty breaking down barriers between departments due to the management systems they have already in place that work against their goals. Likewise,

---

[8] Dr. Edward Deming, **Deming's 14 Points for Management**

with Safety, Quality, and Reliability, people cannot eliminate human errors 100%. Stricter rules will only make matters worse. What we need is a time where these people set aside their priorities, pride, and ego and understand each other's strengths and weaknesses. There is no such thing as first. Learning from one another would make a difference instead of working in isolation.

Chapter 6

Reducing Equipment's Running Costs

Maintenance should focus on improving reliability and not on reducing cost. Why? Because if reliability starts to improve, then the cost will definitely go down. It cannot be the other way around. Remember that there will be times that focusing on reducing costs will hurt reliability, a lesson we should reflect upon. Having a low maintenance cost is always a consequence of good maintenance practice.

6.1: Maintenance Costs Explained

Although there is no distinct and universal standard on what maintenance cost includes as it varies from one industry to another. Maintenance cost is a universal and common measurement and indicator for all types of industries, unlike other KPIs, which can be exclusively for a particular industry only. By definition, maintenance costs are any expenses incurred on the equipment, machines, or asset by the maintenance function to sustain and preserve it. It can be said as a major part of the total operating costs for all manufacturing and non-manufacturing industries. Depending on the type of industry, the Total Maintenance Costs can represent around 3 to 50% or more of the costs of products and services produced. This means that if the ratio of the direct maintenance cost to the total value-added costs for the mining industry is at 50 % and that particular mining industry can produce 20 Kilogram of a gold bar at 1,165,721.8 USD, this means that 50 % of that amount is spent on doing maintenance.

Although industries may differ on what to include or exclude in their maintenance cost, still this is one indicator that almost all maintenance managers tracked consistently. An example of this will be the training for the maintenance function. While everyone agrees that this is an expense to industries, the question is should it be included in the maintenance cost especially if a third-party contractor will be providing the services? In other industries, the training for the maintenance people will be under the budget of the training department. For other industries, this will be divided into the different maintenance function cost centers, while in other industries, this will be under the budget of the maintenance department or even the human resources. It really depends on who will shoulder this cost.

Figure 6.1: What Maintenance Costs Can Include

In the previous chapters, we discussed the vertical setup, which is included as one of the main objectives of the IFCA or EEM pillar of TPM. All the activities covered in this field will be those expenses while setting up the equipment for a start-up production. Once the equipment has completed its vertical set-up, then the equipment will start its operations and the running cost of the equipment begins. The running cost of the equipment will include all the expenses incurred in operating and maintaining the equipment throughout its entire life cycle. While the majority of the expenses will involve maintenance, other expenses are non-maintenance related such as training of operators, operator's certification, expenses on raw materials, together with Quality and Safety expenses as well. Comparing these expenses included in the running cost with the initial or acquisition cost is simply just like imagining what is beneath an iceberg, which is always larger than what is above the surface of the sea that the human eyes can see. What we do not see is what is beneath the iceberg, which is always bigger than what is above the sea.

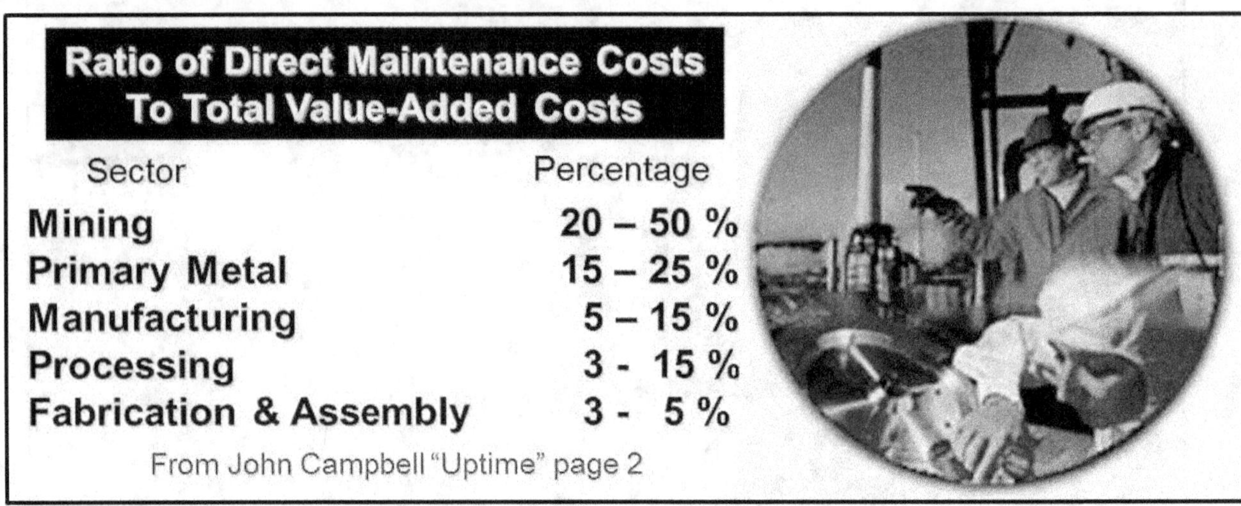

Figure 6.2: Ratio of Maintenance Costs to Total Value Added Costs

6.2: Using the CMMS to Track the Maintenance Costs

In a manufacturing plant with around 2000 pieces of equipment, it will be a very daunting and cumbersome process to track the maintenance cost for every single piece of equipment manually. What is important is to automate the process. As I have mentioned several times in my previous books what is important in any automation is that the software should adjust with the user's needs and not the user adjusting to the software. That is the reason why I staged this maintenance discipline into the last part. What is important in the initial stages is for the maintenance to know what they want to automate and how the software will benefit them from using it. Whether the maintenance function is using a CMMS or EAM software, this software is only as good as, what we populate in the system. Before deciding on tracking the maintenance costs, the key people in maintenance and engineering need to sit down and discuss what it will and will not include in the maintenance cost. How we can sort the maintenance cost by equipment type or even provide the top equipment and machines that consume the highest maintenance cost. Likewise, it is also important to provide a code for every single maintenance cost. Here are my recommended steps in automating the overall maintenance cost. Having this information can provide us a great detail on which machines and equipment to prioritize.

Step 1: Determine What will be included in the Maintenance Costs: As said previously, the maintenance cost will include all expenses incurred on the equipment during its entire life cycle. The majority of the running cost will include the maintenance costs. For example, decide if obsolete and non-moving parts currently stored in the storeroom should be included in the maintenance costs or not? If the industry invests in the acquisition of

Predictive Maintenance instruments, will it be included in the maintenance costs? How about the certification of the users of the instruments? If the answer is yes, then the overall cost of the Predictive Maintenance instrument and certification should be amortized by all equipment that will undergo the Predictive Instrument that was purchased. If there are around 500 rotating equipment in the plant and the cost of the Vibration Instrument and certification is PHP 1,300,000.00, dividing PHP1, 300,000.00 by 500 will be PHP 2,600.00. This will be the amortized cost for these 500 rotating equipment.

CODE	EXPENSES	CODE	EXPENSES
SPC	Spare Parts Costs	PM BOM	PM Bill of Materials
MDT	Machine Down Time Costs	MRO CC	MRO Consumable Costs
LUB	Lubricant Costs	MOD	Modification or Redesign Costs
TNE	Training Costs	WAR	Warranty Costs
CC	Contractor's Costs	DC	Decommissioning Costs
RNM	Repair and Maintenance Costs	OBS	MRO Obsolete Parts
PdM	Predictive Maintenance Instrument	NON	MRO Non-Moving Parts
PdM CC	Predictive Maintenance Certification	FC	Fuel Costs
EC	Energy Costs	MOT	Motor Pool Costs
OT	Overtime Costs	SMART	Online Smart Sensors Application
CAL	Instrument Calibration Costs	REP	Replacement Costs
MAN	Manpower and Labor Costs	OTHERS	Other Maintenance Costs

Figure 6.3: Coding Maintenance Costs

Step 2: Provide a Code for Each of the Maintenance Costs Identified: To clearly visualize the overall representation of what the Overall Maintenance Costs include, it is important to provide a code for each of these costs individually such as in figure 6.3. Categorizing every cost incurred on maintenance can provide us a clear understanding of these maintenance costs individually so we can target our efforts to reduce what seems to consume where most of the money is spent on maintenance. This is very important and maintenance should be consistent on how they enter the cost in the system and the codes that generate with them.

Step 3: Educate All Maintenance and Operators on the Codes: As said previously, the software is only as good as what we populate. In this case, it is important for those who generate the code to educate all maintenance people on the codes and when encoding these codes on the system, to ensure that they indicate the correct code on the maintenance costs. What is important is to be consistent. All users of the software should be informed and must familiarize themselves with the codes. Once the codes are entered into the system and the people are familiarized with the codes, the software must be capable of generating the maintenance report according to the codes, which can be sorted depending on the report we want to see. These codes can be sorted from the highest to the lowest per equipment, per equipment model, and by period, say weekly, monthly, or quarterly. A bar or line graph can also be generated indicating what codes generate the highest cost of maintenance.

Step 4: CMMS to Generate the Report Based on What We Want: The system should generate a report indicating the highest maintenance costs, and what machines seemed to generate the highest costs based on a given period. A second-level report can also be visible to indicate all the costs incurred on a particular machine. For example, if the top three costs of maintenance include lubrication, repair or corrective maintenance, and non-moving parts, then we can have our people train on lubrication and MRO Spare parts, and deploy strategies based on the learnings from the training such as establishing an Oil Contamination Control Awareness and application of High-Efficiency Filtration System.

Step 5: Provide the Necessary Training Needed before Initiating Improvement: Before initiating any improvement plans and implementation, it is important to train the people on the different tools and maintenance strategies needed to reduce the maintenance costs. This will be an industry investment that will yield profits in the long run. We just cannot assign anyone to carry out an improvement project if the team is not knowledgeable on how to approach the problem.

Step 6: Start Deploying Improvement Teams to Reduce Costs: Once the report can now be provided by the system. We can now have clear data on which machines to focus. What seems to be the main contributor as to why the overall maintenance costs of particular equipment are high compared to other equipment? Who will be the team to be deployed to address the problem, what particular strategies to be deployed, and so on? Having this kind of report and data can allow us to focus clearly and prioritize improvements on the equipment that needs to be improved or modified.

Code	Description	Maintenance Strategy to Adopt
SPC	Spare Parts Cost	MRO Spare Parts Management Strategy
MRO CC	MRO Consumables	MRO Spare Parts Management Strategy
OBS	MRO Obsolete Parts	MRO Spare Parts Management Strategy
NON	MRO Non-Moving Parts	MRO Spare Parts Management Strategy
LUB	Lubricant Costs	Lubrication Strategy and Oil Contamination Control
RNM	Repair and Maintenance Cost	RCFA, Planned Maintenance 4 Phases
PM BOM	PM Bill of Materials	Reliability-Centered Maintenance
MDT	Machine Downtime Costs	Planned Maintenance 4 Phases, RCM
MOD	Modification Costs	Planned Maintenance Phase 2
REP	Replacement Costs	Study of Increasing Replacement Interval
PdM	Predictive Maintenance	Compute for the Payback or ROI
PdM CC	PdM Certification	Compute for the Payback or ROI
EC	Facility Energy Costs	Use of Predictive Maintenance
MOT	Motor Pool Costs	Lubrication Strategy Program
CC	3rd Party Contractor Costs	Review for duplicated tasks, or can be done in-house
OT	Overtime Costs	Adopt Autonomous Maintenance so other tasks can be delegated to operators

Figure 6.4: Different Maintenance Strategies to Adopt

Step 7: Compare Before and After Deploying the Strategy: Once the strategy has been deployed, monitor the cost on a weekly or monthly basis and compare these with the previous cost if it was reduced. If the cost was reduced then the team can move to the next equipment with the highest cost.

6.3: Strategies to Reduce Equipment's Maintenance Costs

Once we have determined the different maintenance costs and expenses involved in the equipment, we can deploy different strategies to reduce the maintenance costs. Listed are some of the cost that eats up on maintenance and how we can reduce them.

6.3.1: Reducing MRO Spare Parts Costs

There are several ways to reduce the cost of MRO Spare Parts in industries such as removing obsolete parts and providing an algorithm or decision diagram for making decisions on which parts and items to store and those that need not be stored inside the storeroom. The 5th book I wrote on Problems and Solutions on MRO Spare Part and Storeroom discusses the most common problems in the industry's storeroom as well as the solutions to these problems. One of the main highlights of this book is that it provides an MRO Decision Diagram or algorithm to allow the users to make a decision on whether to stock or not to stock parts inside the storeroom.

Maintaining a high level of inventory accuracy is important to any storeroom industry since this is usually the beginning of a domino of problems in the storeroom. What is also important in managing and controlling the industry's spares and storeroom is that the strategy that needs to be applied should come not only from inside but also from outside the storeroom. This means that if we have lesser breakdowns as a result of conducting root cause, then we just need to store fewer parts inside the storeroom. Remember that the storeroom is simply a contingency plan since, in reality, we do not exactly know what machines will fail when they will fail and what particular part will be affected. However, on the other end, some failures can be prevented or predicted. What is important is instead of keeping these parts in the storeroom, they can be ordered on a need basis depending on the lead time to deliver these parts. If an impeller will be replaced after a couple of years, and the lead time to acquire the impeller will take 6 months, then we can order the impeller 7 to 8 months before the PM schedule. If vibration monitoring can predict the failure of the bearing 3 months before it fails and the lead time for the bearing to arrive is one month, then we do not need to keep a stock of the bearing in the storeroom.

6.3.2: Reducing Preventive Maintenance Costs

One of the important points to consider in performing Preventive Maintenance is that all tasks indicated should address a particular failure mode. If there are tasks that are written that do not address a particular failure mode, then they should be removed from the Preventive Maintenance task lists. Another important consideration is to review the interval of conducting Preventive Maintenance. Does this PM tasks of replacing the item be done monthly? Can this failure mode be predicted using any PdM instrument? If the answer is yes, then we remove them from the PM tasks and this will be under the watch of the Predictive Maintenance group. According to John Moubrey, a common problem with mature maintenance programs is that if not correctly designed, then between 40 to 60% of the PM tasks will serve very little purpose, therefore, evaluating our current Preventive Maintenance System should lead us to the following:

• Many tasks duplicate other tasks where the same tasks are done by different people
• Some tasks are done too often while others are not enough
• There are failure modes that do not actually have any maintenance tasks
• Some tasks serve no purpose whatsoever
• Many tasks will be intrusive or forced and overhaul based where they should be Condition-Based
• Some tasks cost more to do than the failure it is meant to prevent
• Maintenance is costly by replacing perfectly good parts since we are basing replacement on Time-Based Maintenance

For the PM Interval, what is important is we are not doing Preventive Maintenance too soon or too late. Performing PM too soon will be a complete waste of time and resources, while doing the PM too late means the failure may already have occurred sending us back to the reactive mode. Review the PM tasks if there is a possibility of extending the interval of performing them. If performing overhauling is done every year and no signs or traces of wear exist, would it be possible to base the overhaul and replacement based on Oil Analysis Wear Metal Debris Test? If a monthly task on PM can be done bi-monthly or quarterly, this means that instead of doing these tasks 12 times in a year, then this will just be done 6 times or 4 times a year and still obtain the same results. Try to review all the maintenance tasks identified on the PM lists and the interval of doing it and check the possibility of extending the interval. Below is an example of how to extend the interval of performing the tasks.

For Revision 1 on the PM Interval
• 24 tasks performed every shift were reduced to 12, increasing the daily tasks from 20 to 32 tasks.
• 15 weekly tasks were reduced to 10, increasing the monthly tasks from 25 to 30 tasks.
• 25 bi-monthly tasks were reduced by 15, increasing the quarterly activities from 30 to 45 activities with additional 5 new tasks.
• 35 semi-annual tasks were reduced to 25, increasing the annual activities from 40 to 50 tasks.
• Performing this revision decreased the number of PM tasks from 35,040 to 26,070 PM tasks, which eliminated around 8970 tasks.

A second revision was made once again by the PM Planner together with the PM crew and the final revision includes the following changes from Revision 1 to Revision 2:

• 12 tasks done every shift were reduced to 9 tasks, increasing the daily tasks from 32 to 35.
• 30 monthly tasks were reduced to 20 tasks, increasing the monthly tasks from 15 to 25.
• 45 quarterly tasks were reduced to 35 tasks, increasing the bi-monthly tasks from 25 to 35.
• Again, performing this second revision reduces their annual tasks from 26,070 to 23,800 tasks, which reduced around 2,270 tasks in a year.

While most industries trust to use the maintenance interval dictated in the equipment manual prepared by the Original Equipment Manufacturer or OEM. The problem, in this case, is that in most cases, the interval may be too conservative, or the conditions in which the equipment is currently operating may not be exactly the same as dictated in the PM operating manual. In many instances, PM replacements are done too soon, especially if the spare part will be coming from them. I think the reason here is obvious. Another case is that if the equipment operating environment is hot and humid like in the Philippines, then the frequency of performing greasing and the grease to be used might not be exactly and precisely the same grease that is written in the OEM manual. While it will be easy to derive the maintenance tasks for a failure that has a symptom, or what we called the Failure Development Period where we can perform the maintenance tasks less than the Failure Development Period. This means that if the Failure development period is 1 month, then we can perform the task every week or every two weeks for inspection purposes until the limit has been reached. For PM replacement, we want to maximize its lifespan; hence, the PM replacement will be done when the useful life reaches around 80 to 90%. If the failure development period will be 30 days, then the actual replacement will be done on either the 24th or the 27th day period. In this case, we have more or less maximized the useful life, and the failure has not yet occurred, but the thing is, there will also be failures that will just happen without any form of signs or symptoms. What is important is to identify the consequences of failure and if the consequences are unacceptable, then that item cannot be allowed to fail by any means.

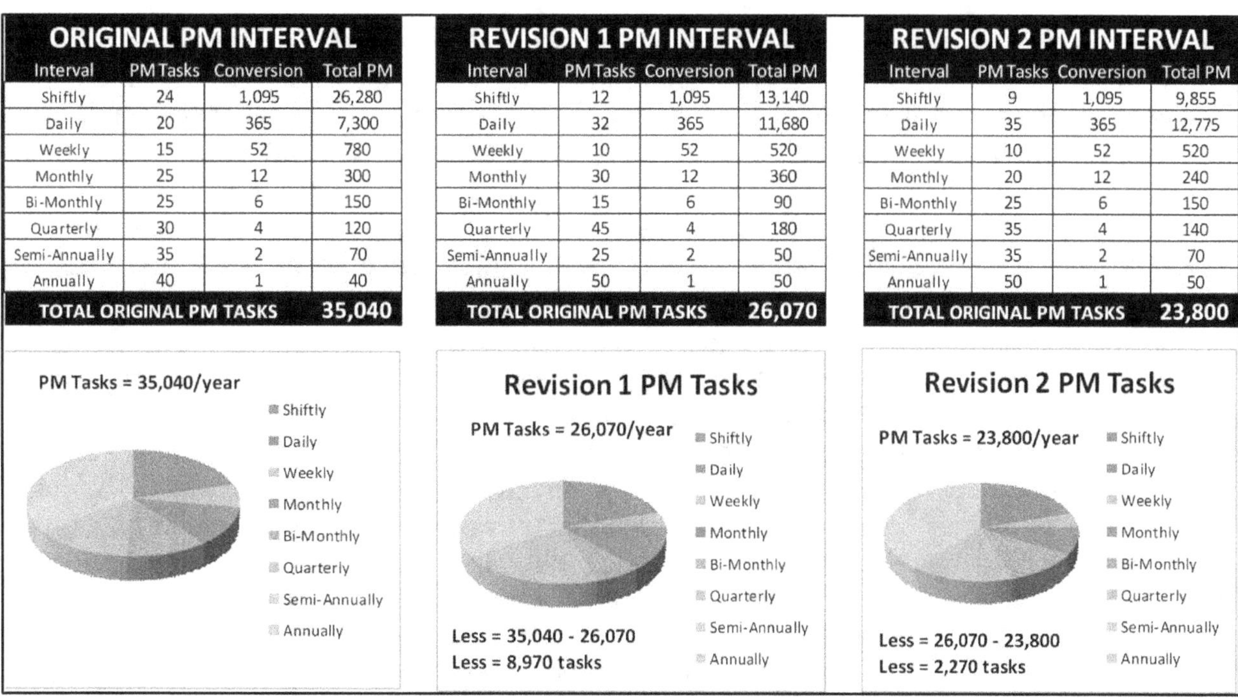

ORIGINAL PM INTERVAL

Interval	PM Tasks	Conversion	Total PM
Shiftly	24	1,095	26,280
Daily	20	365	7,300
Weekly	15	52	780
Monthly	25	12	300
Bi-Monthly	25	6	150
Quarterly	30	4	120
Semi-Annually	35	2	70
Annually	40	1	40
TOTAL ORIGINAL PM TASKS			**35,040**

REVISION 1 PM INTERVAL

Interval	PM Tasks	Conversion	Total PM
Shiftly	12	1,095	13,140
Daily	32	365	11,680
Weekly	10	52	520
Monthly	30	12	360
Bi-Monthly	15	6	90
Quarterly	45	4	180
Semi-Annually	25	2	50
Annually	50	1	50
TOTAL ORIGINAL PM TASKS			**26,070**

REVISION 2 PM INTERVAL

Interval	PM Tasks	Conversion	Total PM
Shiftly	9	1,095	9,855
Daily	35	365	12,775
Weekly	10	52	520
Monthly	20	12	240
Bi-Monthly	25	6	150
Quarterly	35	4	140
Semi-Annually	35	2	70
Annually	50	1	50
TOTAL ORIGINAL PM TASKS			**23,800**

Figure 6.5: Reviewing the PM Interval

Once the PM tasks have been used for a year, we need to have a review of whether both the PM tasks and the interval we indicate are effective. Our main goal is to provide the necessary PM tasks at the right interval, which means that the failure has not yet occurred.

There are two ways to refine the PM Tasks. The first is to review all the existing tasks if it really needs to be done or not on the dictated interval. The second is to review the frequency or interval if we are doing the PM tasks too soon. Ask the following questions.

• Are there any PM tasks done every shift that can be done daily?
• Are there any PM tasks done daily that can be done weekly?
• Are there any PM tasks done weekly that can be done monthly?
• Are there any PM tasks done monthly that can be done every bi-monthly?
• Are there any PM tasks done bi-monthly that can be done every quarterly?
• Are there any PM tasks done quarterly that can be done semi-annually
• Are there any PM tasks done semi-annually that can be done annually?
• Are there any PM tasks done annually that can be done every couple of years or more?

In figure 6.5, adjustments have been made by the maintenance planner together with the PM crew who are responsible for executing the PM tasks. The original total PM tasks done on one piece of equipment totaled 35,040 activities yearly, which was reduced to 23,800 tasks yearly. This means a reduction not only in maintenance costs but also in manpower resources. The following changes have been made in the interval, which is designated as Revision 1 of the PM Tasks Interval.

6.3.3: Reducing Lubrication Costs

Having a Lubrication Strategy for industries not only saves on the lubricant costs but way beyond that such as the costs of failures attributed to lubrication, the cost of manpower resources on repair, the costs of spare parts, wasted energy costs, fuel costs, and others. In fact, experience tells me that if the cost of your lubricant is PHP 1,000,000.00 per month, the cost of failures attributed to lubrication is a minimum of two times the overall costs of the lubricants themselves. In this case, the overall cost of failures attributed to lubrication failure will be around PHP 2,000,000.00 per month and in most cases even more. The biggest barrier and mistake industries make on lubrication is depriving their people of education.

Although lubrication is an important factor in reducing friction, it is the lubricant that we need to focus on and provide our attention. Other factors include the type of motion, speed, temperature, load, and operating environment of the equipment to better understand the correct type of lubrication for our equipment and machinery types. Again, with limited knowledge of lubrication, everything else depends on what the OEM tells us to do. I have one client, where OEM recommended that they have their diesel engines sampled through Oil Analysis, which their OEM will cater for free, which I readily agree with them. When I asked them at what interval they change their oil, they told me that it is still the same at 500 running hours. When I told them that isn't one of the main reasons for conducting oil analysis is to precisely determine when to change the oil based on its condition and no longer running hours. They told me that it is what the OEM wants, and so I told them that your oil analysis report is just plain useless and your OEM is just mocking you. The reason why we need to educate our maintenance and operations people on lubrication is for them to perform the correct practices on lubrication. Nothing less and nothing more. If every single decision we do is based on the OEM or oil manufacturer, then this will just benefit them and not the industry itself.

Leakage Rate	Est. Monthly Cost	Estimate Yearly Cost
1 drop/5 sec	6.6 gal = $ 26.40	80 gal = $ 320.00
1 drop/sec	34 gal = $ 136.00	408 gal = $ 1632.00
3 drop/sec	113 gal = $ 452.00	1356 gal = $ 5424.00
Steady Flow	720 gal = $2880.00	8640 gal = $34,560.00

Figure 6.6: Cost of Oil leak in Equipment

Figure 6.7: Industry's Wrong Practices on Lubrication

Essential to any industry are competent people who understand their equipment intimately. Training will be the very basic foundation of any cultural change. It is mainly the missing link ingredient in any change or continuous improvement effort. Training is where we acquire knowledge and also where we build our culture. The knowledge gained in training is used to build our skills so maintenance can do their jobs correctly the first time around. Skill is the ability to do one's job, and apply knowledge and experience correctly in all kinds of events over an extended period. Training should not be the focus of any cost-cutting measures. Top Management people must understand that it will play a vital role in improving the skills of their people. Here are some ways to reduce the cost of lubrication attributed to failures in the plant, extracted from my book on Lubrication Tactics for Industries Made Simple.

1. Address Oil Leaks: One of the most common forms of contamination is process leaks. The simplest method of identifying and correcting leaks is to tag them. Locate the oil leak on the equipment and identify its source, type of leak, and severity of the leak. Then tag the leak; the tag should only be removed once the leak had been addressed. Implementation of a simple, easy-to-use system such as this can save hundreds of thousands of dollars. In figure 6.6, the minimum cost of $4.00 per gallon is used for the oil, which does not include the significant cost of freight, handling, storage, drum charges, labor, and waste disposal. Oil leaks cause money in the long run and safety issues as well. It makes the floor slippery, which can cause injury to a person. In TPM or Total Productive Maintenance, one of the reasons why the machines are kept clean is to easily spot the oil leak, underneath the equipment. These leaks can also be a source of ingression of moisture and dust in the equipment. If an oil analysis report indicates the presence of silicon (Si), then this is dirt or dust.

2. Use Desiccant Breathers on Oil Reservoirs: Desiccant breather offers better filtration to protect the equipment from even the smallest contaminants that can destroy the machinery's effectiveness and cause downtime as well as costly repairs. As air is drawn into the equipment through the breather, the desiccant filter elements remove particulate while the desiccant prevents moisture and dust from ingressing inside the reservoir. Moisture contamination can accelerate oil oxidation and there is only a certain amount of moisture that the lubricating oil can handle. Desiccant breathers will be the first line of defense against contamination. It uses a filter media to remove particles greater than three microns and a silica-gel desiccant to remove moisture from the air entering the reservoir. Studies have shown that 50-70% of lubricant contamination comes from outside the equipment. Most reservoirs are designed to breathe, so protecting the source of airflow from moisture and dust is extremely important. This is true for most sizes and types of machines.

3. Use a Dedicated Pump Dispenser for Each Type of Oil: If you have different oil types being used in your industry, such as engine oil, hydraulic oil, gear oil, compressor oil, etc., each of these types of oil must have its own pump dispenser. Never borrow the oil pump dispenser of engine oil to be used on different types of oil. These oil pump dispensers should be dedicated per oil type. What we are avoiding here is trying to mix the different additives for the different types of oil, which can result in incompatibility issues where both the life of the oil and the part it lubricates can be shortened.

4. Using Offline Filtration for Transferring Oil: Never assume that new oil is also clean oil. New oil also contains contaminants. In tapping or adding oil during change oil. It is best to use an off-line filtration to remove contaminants from the new oil. This portable offline filtration can be used for various purposes such as cleaning stored oil, scheduling hydraulic equipment for re-filtering to ensure that the hydraulic oil complies with ISO 4406 cleanliness standards, flushing after machine repair, and for adding or tapping new oil during change oil. It is best to use a 3 to 6 micron with a beta rating of at least 200 with an efficiency of 99.5%. It is also recommended that offline filtration be dedicated per type of oil used in your industry. Do not use the same offline filtration for transferring engine oil, hydraulic, turbine, and other oil types to avoid mixing both the additives and their base.

Figure 6.8: Use Dedicated Oil Dispenser for Every Type of Oil

5. Apply Color Coding for Different Grease: When using several types of grease, it is a good practice to provide a color coding for every type of grease, grease gun as well as grease nipple used to avoid the chances of human error in mixing different types and brands of grease as their thickeners, additives or base may or may not be compatible with each other. Grease incompatibility creates many issues, including oil base separation, thermal breakdown, thinning of the grease, and formation of hard deposits, all of which can accelerate the early failure of the part or subject it lubricates. The purpose of color-coding is to minimize human errors where either operator or maintenance may select the wrong grease for a piece of particular equipment and avoid the chances of mixing lubricants of different types. The use of color-coding is no rocket science; our main goal is simply to reduce human error. The problem with Grease Compatibility Charts is they only based the compatibility on the thickeners, which excludes the base and the additives.

6. Consolidate All your Lubricant Vendors: Every oil supplier or vendor have their own recommendation regarding the brand of oil they want you to use on the equipment. The key is for maintenance to have a full understanding of selecting the correct lubrication for their equipment. Several factors are considered, such as the new oil specification, the equipment's operating temperature, room temperature, equipment load, and whatnot. Having too many vendors on lubricants can create incompatibility issues where different brands are mixed. This occurs when changing over to a brand new type of lubricant that is incompatible with the previous lubricant used. The mistake, in this case, is failing to check the compatibility of the lubricant and failing to fully drain the oil.

7. Oil Analysis Sampling Extraction should be Consistent: Oil sampling must be consistent in that it should be extracted on the same reference point and on the same interval. The preferred method is to collect a sample as it flows through the system with as much turbulence as possible. The best place to do this is at the elbow for good mixing. The ideal port will be located downstream of the part or components to be monitored, such as pumps and bearings, but it must also be placed before the filter unless the filter is monitored. A minimess valve can be installed directly in a port on a pressurized system. There are applications where a drain line or a return line can't be accessed, or no such line exists, such as diesel engines, circulating gearboxes, and circulating compressors. In these applications, sampling should be done from the pressurized supply line leading to the gears and the bearings before the oil enters the oil filter.

8. Feasibility of Using By-pass Filtration: Unlike offline filtration, these by-pass filters can be installed permanently on the equipment, which is separated from the main system where both the pressure line filter and return filter are installed. The main purpose of installing a secondary or by-pass filtration is to remove solid contaminants that cannot be captured by the existing full-flow filter. The by-pass filter will have its own suction and return line hose, while offline filtration activities can be included during weekly, monthly Preventive Maintenance routine activities, or when the particle counter register that the cleanliness code is above the desired standard. It is highly recommended that a pressure gauge be installed to dictate when to change the filter cartridge. When using bypass filtration on mobile equipment, a pump is not needed, since the flow of the oil will be based on gravity.

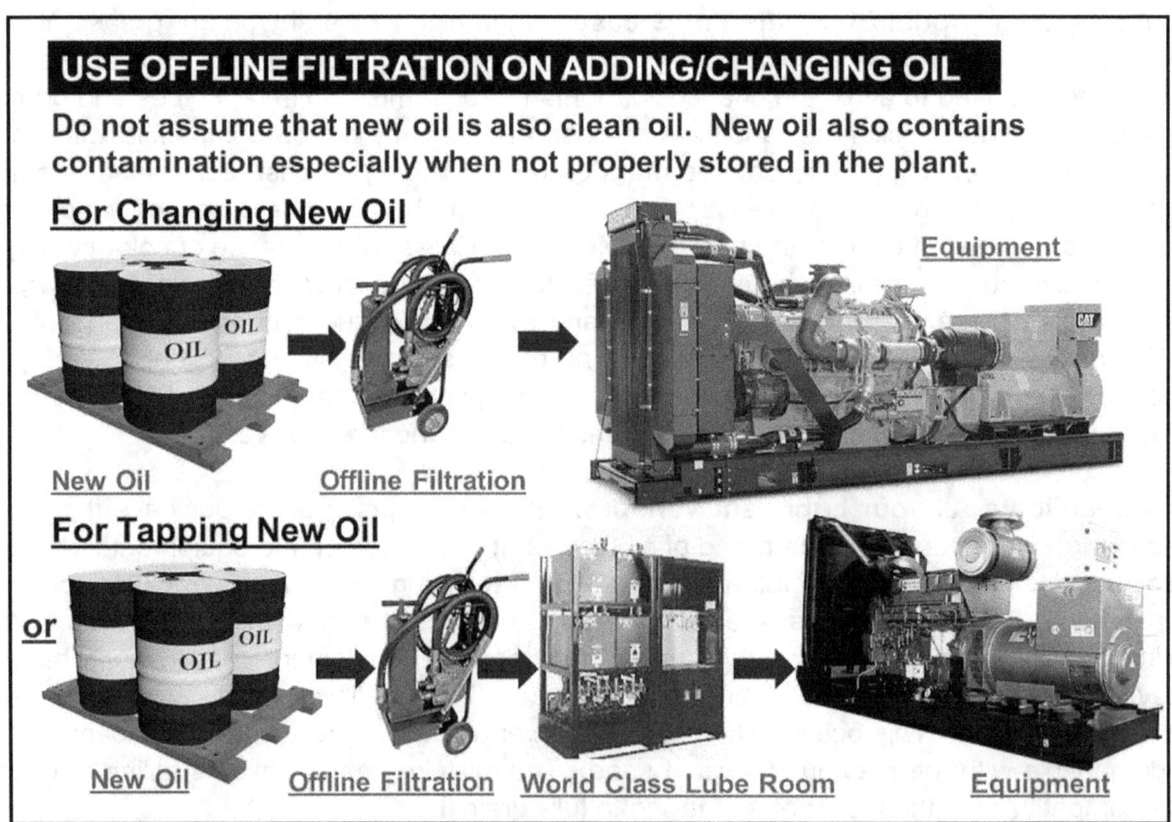

Figure 6.9: The Use of Offline Filtration When Adding or Tapping Oil

9. Cover the drums When Not Used: It is highly recommended that new oil from the drum be covered when not in use. They should be placed in a warehouse and should not be exposed to the outside environment. Suppose there is not enough storage for these lubricants in the plant; in that case, it is important to discuss this matter with the purchasing people to control its purchase. Oxidation does not only happen when the oil is used in the equipment; this can also happen to new oil when exposed to a humid environment. In the Philippines alone during the summer months, the temperature can reach around 38 °C and sometimes higher. For lubricants stored outside in an open environment, it is best to extract a sample from the new oil and test it in an independent oil analysis laboratory to check the oil for any degradation signs. Remember that improper storage for lubricants can accelerate oil oxidation and other forms of problems in the lubricant.

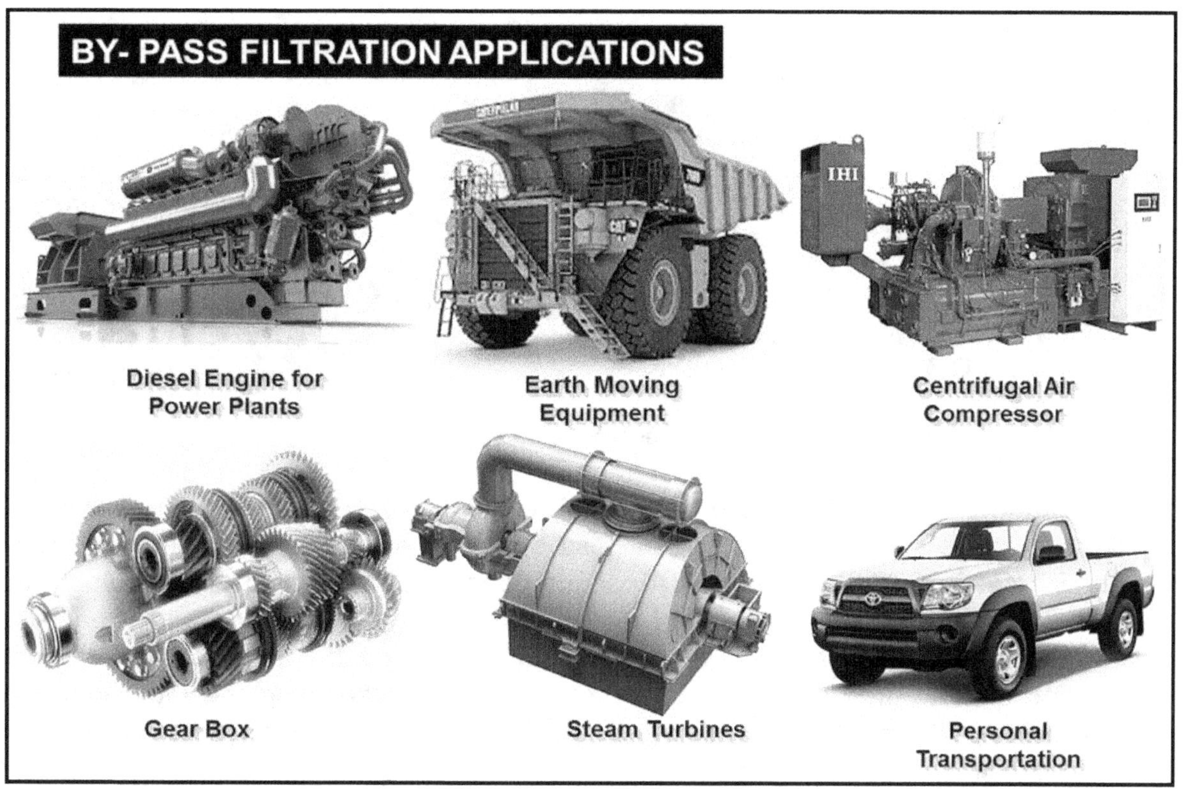

Figure 6.10: Feasibility of Using By-Pass Filtration for Better Protection

10. For Large Lubricant Consumers, Adopt an Oil Analysis: If you are a mining industry or any large consumer of lubricating oil. It is highly recommended that you need to have your own oil analysis laboratory. Although there will be an initial investment, in this case, it is important to conduct a feasibility study on the benefits that can be derived from having your own. Calculate the ROI or return on investment. The number one benefit of having an in-house oil analysis will be the speed of the report. Another benefit is analyzing the actual condition, not only of the oil but also of the parts and components of the equipment itself, which will result in savings on the cost of lubricants, waste, lubrication failures, downtime, spare parts, and manpower as well.

11. Generate an Oil Contamination Control Awareness to Everyone: Generate an Oil Contamination Control Awareness Program, especially for both operation and maintenance on how oil is being contaminated. Contamination can be in any form such as solid, liquid, or gas. These are those that should not be present in the oil. Create awareness of what these contaminants can do inside the equipment. Example of lubrication practices that can generate contamination includes, using an open type container when tapping or adding oil, no dedicated pump dispenser for the different types of lubricant, and wrong storage of lubricants where lubricants are exposed to humidity, temperature, rain, sunlight, and dust. According to lubrication experts, there can exist around 400 metric tons of suspended dirt in one square mile of space and some of them can ingress into our equipment and assets. Contamination is the major cause of why oil degrades, and why equipment fails. This means that the more contaminants present in the oil contained, the more prone the equipment is to failure.

Figure 6.11: Creating Ann Oil Contamination Awareness Program

12: Application of ISO 4406 Cleanliness Code for All Lubricating Systems: All lubricating systems have a form of cleanliness standard which can be determined either through ISO 4406 or NAS Standards. A particle counter is an oil analysis instrument that provides the procedure recommended by the ISO for establishing fluid cleanliness standards. The objective of having a particle counter is to control fluid cleanliness levels. This method aims to check the cleanliness of the oil by determining the total number of particles in several different size ranges. This quantification is often referred to as the particle size distribution. This simply means that the lesser the number of contaminants present in the oil, the lesser the failures and the cost of maintaining the system. Having an ISO 4406 Cleanliness Standard for all lubrication systems in the equipment and assets provides us a measurement of how clean the oil is in the system.

• Roller Bearings	16/14/12
• Journal Bearings	17/15/12
• Industrial Gearboxes	17/15/12
• Mobile Gearboxes	17/16/13
• Diesel Engines	17/16/13
• Steam Turbine	18/15/12
• Paper Machine	19/16/13
• Servo-Valve	13/12/10

- Proportional Valve   14/13/11
- Variable Volume Pump   15/14/12
- Fixed Piston Pump   16/15/12
- Vane Pump   16/15/12
- Gear Pump   16/15/12
- Ball Bearing   14/13/11
- Turbine   17/15/12
- Hydraulics   16/15/12

6.3.4: Reducing Repair Costs through Root Cause Failure Analysis

There are several ways to reduce repair and maintenance costs. First, a repair happens when a failure or breakdown is experienced on the equipment. Unless the system is duplicated, it will result in downtime on the equipment or system. The TPM pillar on Planned Maintenance will definitely reduce breakdowns on the equipment, which is discussed further in the later chapters of this book. Another approach to reducing repair and maintenance costs is understanding the failure itself, and this is done by conducting an actual Root Cause Failure Analysis Investigation. However, here's the thing, many think that conducting an analytical problem-solving tool and conducting a root cause investigation are one and the same which they are not. These analytical problem-solving tools are not meant to address the Root Cause of the problem but only to address the most likely which are the failure modes. Root Cause is 100% Evidence-Based. Without evidence, Root Cause is impossible to perform. This simply means that the first step in conducting any root cause failure investigation is preserving the physical evidence.

Root Cause Failure Analysis investigation can only be conducted when the physical evidence has been well preserved. This means that when a failure is experienced on the equipment, people in charge of collecting evidence should collect all physical evidence on the machine such as extracting the part that was affected, collecting debris, taking photos of the equipment at different angles, noting all charts, and gauges at the time of the failure. Note that if the equipment is repaired after it failed, then it is unlikely to conduct a root cause investigation as this will only end up in witch-hunting, but an analytical problem-solving tool can be conducted which can lead to the probable or most likely cause. The thing is, to determine the root cause is to base our conclusion on the evidence that was unfolded. Therefore, if a failure warrants a root cause investigation, the physical evidence needs to be well preserved. Depending on the magnitude of the problem, a Principal Investigator and Evidence Gathering team will be needed. Here is a detailed step-by-step on how to conduct a Root Cause Failure Analysis Investigation. More detailed coverage of root cause discussion is covered in my book on *Investigating Equipment Failures through Root Cause Failure Analysis*.

Step 1: Determine if the failure is a Mini, Midi, or Maxi event: It is strongly recommended that the industry should categorize which problems and failures should be considered mini (small-scale failures), midi (medium-scale failures), or maxi (large-scale failures) event. The basis for consideration can include the cost of the consequences or impact of the failure in the plant. Even if the impact of the failure is small, the industry still needs to learn from the things that go wrong in their industries.

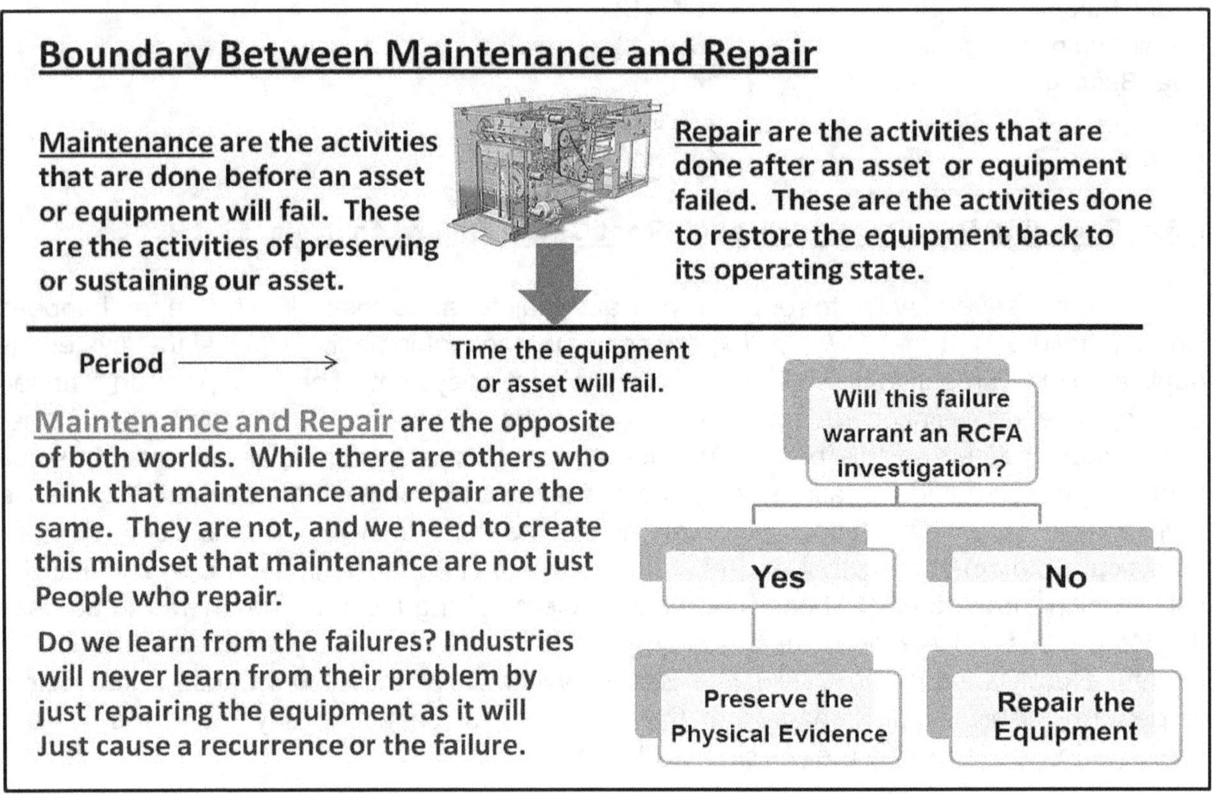

Figure 6.12: Boundary Between Maintenance and Repair

- **Mini-or Small Scale Event:** RCFA is performed on Mini or minor small-scale events. The impact of the failure is small but we still need to understand the cause of the failure. Mini-event can be performed by a single person or whoever comes in direct contact with the problem. The principal investigator will also be responsible for gathering the three types of evidence. (physical evidence, people evidence, and paper evidence) This will not require a stakeholder meeting. The analysis and investigation should include the physical, human, system, and latent causes of the problem.

- **Midi or Medium Scale Event:** RCFA is performed on Midi-or medium-scale events. They are led by an insider or someone from the affected site but as detached as possible from the specific event. This will involve a Principal Investigator and one to three evidence-gathering people. This will require a stakeholder meeting. The analysis and investigation should include the physical, human, system, and latent causes of the problem.

- **Maxi, or Large Scale Event:** RCFA is performed on maxi or large-scale events or failures with a large impact on the industry. It is recommended that this event should be done by an outsider or independent person to act as the principal investigator. Three evidence-gathering teams will be required. Maxi events will usually require a stakeholder meeting and the analysis/investigation should include the physical, human, system, and latent causes of the problem.

Step 2: Set up the RCFA Team: For midi and maxi events, this will require a Principal Investigator and three people to compose the Evidence Gathering team. Note that the team to composes the RCFA should have been trained in the RCFA process as a preliminary requirement.

Step 3: Freeze the Evidence: RCFA can only be performed on fresh failures and when the parts that failed and debris have been well preserved. This means that the failure just recently happened. This will be part of the physical evidence. The three evidence-gathering teams should be independent of one another. One will be assigned to interview all people involved in the event, one evidence-gathering person will be assigned to collect all physical evidence, and the other evidence-gathering person will be assigned to collect all paper documents relevant during the time of the failure itself.

Step 4: Proceed with the Evidence Gathering Event: Sufficient time will be allotted to the three people in charge of gathering the different pieces of evidence. A timeframe will be provided by the Principal Investigator. All pieces of evidence should be summarized independently by the three evidence-gathering teams.

Step 5: Principal Investigator and the Three Evidence Gathering Team to Meet: Once the evidence gathering team had completed summarizing all the pieces of evidence generated, the principal investigator and the three evidence-gathering teams will meet and each of them will summarize their evidence. The Principal Investigator will start performing the RCFA Logic Tree Diagram based on the evidence unfolded by the evidence-gathering teams.

Step 6: PI and Evidence Gathering Team to Conduct RCFA Logic Tree: Both the evidence gathering team and the principal investigator will perform an RCFA logic tree. If there are causes that cannot be answered by the group, they proceed to Step 4 once again. After completing the RCFA logic tree diagram, the Principal Investigator and Evidence Gathering Team will determine the physical cause, human cause, and the system cause of the problem.

Step 7: PI and Evidence Gathering Team to Plan for the Stakeholder Meeting: After completing the physical, and human cause of the problem, the Principal Investigator and the Evidence Gathering team to name the people who will be involved in the stakeholder meeting, and together they will set a meeting with them. Note that the stakeholder meeting will only be applicable for both the midi and maxi events. Mini events do not require a stakeholder meeting.

Step 8: Conduct the Stakeholder Meeting: The Principal Investigator will explain the reason for the meeting and the evidence-gathering team will present the summary of their investigation and why the stakeholder was involved. Stakeholder meetings would be required to determine the physical, human, system, and latent cause of the problem.

Step 9: Stakeholder to Determine a Smart Countermeasure: A **SMART** countermeasure should be generated by the stakeholder. Note that countermeasures will be required for both the physical and human causes. Change in oneself will be required for the latent cause.

Step 10: Translate the Findings: The Stakeholder to create a plan to translate the findings to other people, departments, and areas in the plant that they think should also learn from the problem. The translation may be in the form of posters, comic storybooks, memos, or creating an awareness that can be readily read by people in the plant.

Step 11: Principal Investigator and Evidence Gathering Team Report: The Principal Investigator and the evidence gathering team finally conclude and generate a report on the incident and submit it to the head of the plant and to all concerned people.

6.3.5: Reducing Maintenance Costs Through Predictive Maintenance

The use of Predictive Maintenance provides maintenance an edge against breakdown and failures since there are failures that provide some symptoms or warning that they are on the verge of failing which can be captured by these PdM instruments. This gives maintenance the advantage to be one step ahead of the failure. However, this does not happen easily as we think. Many industries have these instruments but do not benefit from them. Industries must understand that before they can benefit from these instruments, they must understand there are three investments in Predictive Maintenance. First, we have the PdM instruments, second, the instruments need to be regularly calibrated by the manufacturer. The most important investment is in the training and certification of the users. Although, there may be some vendor training on the use of these instruments which may not be sufficient. Before the user can use and benefit from these instruments, the users must be well versed in the principles of the instrument itself. In using infrared thermography, the user must be well versed in the principles of heat transfer. This means that the instrument will always have an error that the user needs to correct as well as the concept of emissivity as each material will emit a different form of heat depending on the material used.

In using the infrared thermography instrument, the user is interested in only the heat that is being emitted by the equipment. If we have three pieces of equipment that are loaded and the user is scanning the middle equipment, the other equipment can also transfer heat to the equipment being scanned and vice-versa. The user must also be familiar with the materials being scanned as different materials have different emissivity. A user that is not well versed in heat transfer may actually be measuring the overall heat induced by the equipment.   Remember that heat can be transmitted, absorbed, reflected, refracted, scattered, or emitted. What the user is interested in is only the heat that is being emitted or released by the equipment. Reflected and transmitted energy must be filtered out to provide an accurate measurement of data.

Case 1: One of the reasons why Predictive Maintenance is important is that it will not only reduce the amount of reactive maintenance but will also reduce the number of activities for Scheduled or Preventive Maintenance. With a PM shutdown scheduled, a major mining industry had to determine ways to reduce their manpower requirements on their PM Schedule. Normally, they would change out all of the valves on their 2000 HP compressor, but they decided to use Ultrasonic Monitoring as a method of determining which valves needs to be changed. After checking all the valves, they found that only a small fraction of the valves needed to be changed since most of them are still working.

Case 2: Compressed air is one of the most costly utilities in plants today. The Department of Energy has estimated that around 30% of all compressed air produced in the US is lost due to leaks. An ultrasonic compressed-air leak survey can offer fast payback. One plant averaged an electric cost of $1,760,000 to generate compressed air. The survey using Ultrasonic Monitoring identified 3,561 leaks totaling 6,340 CFM at an annualized cost of $597,000 of energy waste. A program of leak detection and repair can have a dramatic effect on the profitability of a company such as this.

Case 3: The following is an extract from his paper on "Proven Benefits of Precision Maintenance," presented by Ralph Buscarello the father of Precision Maintenance as his keynote address to the Vibration Association in New Zealand in May 1998. He quotes that in over 46 years of teaching vibration reduction and conducting seminars in over 34 countries, he could produce hundreds of outstanding financial case histories on the benefits of doing Precision Maintenance, many involving millions of dollars. However, for those who remain skeptical, he suggests you accept none of the figures. Instead, start with the first idea of measuring the vibration levels taken at the worst point on about 50 ordinary but common machines, such as motor and pump assemblies. Plot the trends of their previous year's maintenance costs on the same graph as their vibration amplitudes. Then make up your mind as to whether reducing vibration to precision level pays off or not.

Machine Type	Highest Velocity mm/s	Dollars Spent in USD	Lowest Velocity mm/s	Doing Precision Maintenancee	Percent Savings
Single Stage Pumps	5.6	$3,200	2	$650	80%
Multi Stage Pumps	4.8	$6,100	1.5	$1,100	82%
Major Fans & Blowers	9	$900	2.8	0	100%
Single Stage Turbines	3.8	$8,200	1	$2,000	76%
Other Machines	7.8	$11,850	3	$3,700	69%

Figure 6.13: Relationship between Vibration and Cost on Breakdowns

Chapter 7

Built for Reliability

> *While others say that maintenance cannot improve the reliability of the equipment but it can only sustain it. My take on this is two-fold. It depends on how you define maintenance. While I agree that considering maintenance as a task will only sustain reliability, considering maintenance as people and human beings can definitely improve the reliability of the equipment without a doubt.*

7.1: Why the Maintenance Tasks Alone Cannot Improve Reliability

I used to believe that maintenance is the driver of reliability. My belief has changed these past few years, and I would like to thank a very dear and good friend, R. Keith Mobley author of 22+ books on maintenance and reliability, for enlightening me. While there are a wide variety and types of industries worldwide, each of these industries has something in common, and that is, they have their equipment and assets to maintain whether an industry is producing quality goods or providing services. The main difference lies in how industries approach reliability as their focal point of strategy. While there are industries that focused on quality such as manufacturing, semiconductor, food industries, pharmaceutical, electronic industries, and others, there are also those industries that focused heavily on safety, such as shipping, construction, mining, aviation, milling, cement, oil and gas, and so on, yet very few or none of these industries I know focused on reliability. I suppose that power plants focus on reliability, but not all.

Does it mean that if an industry has good and solid safety measures and procedures in place, can we conclude that this industry is also reliable? I do not think so. What I believe is that if reliability is in place, then industries will be safe to operate. I know of many plants with very good safety measures, procedures, and very strict protocols. They orient everyone in their plant about the importance of safety, which is good. There were times when I conduct in-house training, where I experienced several times that as part of their plant's requirement, all visitors need to attend a safety orientation from a safety officer for an hour or so to calibrate all people in the plant including their visitors that they are dead serious about safety. If I may ask, if there is a plant in which their reliability and maintenance people do the same and orient all their people about reliability, just like safety?

Perhaps there might be a few, but honestly, I have never heard or experienced any one of them in my lifetime. Still, I hope that one day people involved in reliability and maintenance will start educating their people, including top management about reliability and the role they have to play with it.

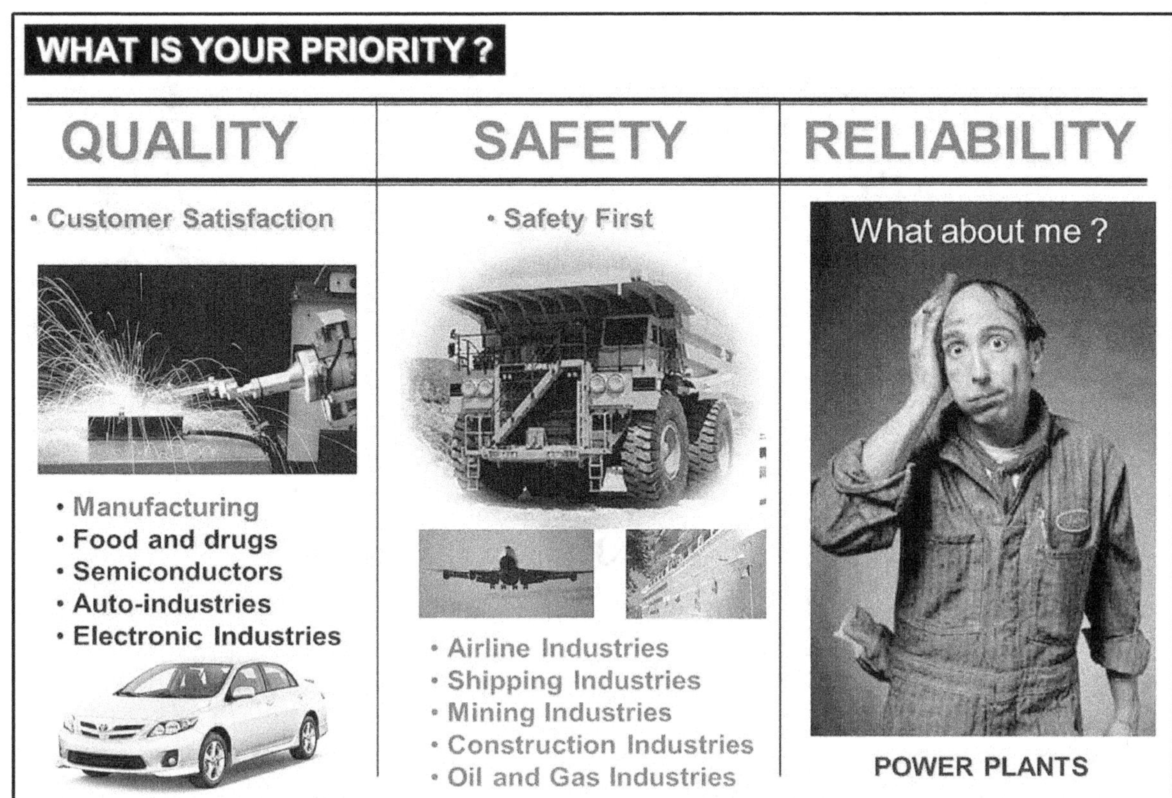

Figure 7.1: What's your Industry's Priority

Here lies the problem; doing maintenance tasks cannot improve the equipment's reliability. What maintenance can do is just sustain it. That is on the assumption that maintenance people have everything they need, which includes knowledge, tools, skills, instruments, and software. Assuming that even if maintenance has all the tools and knowledge they need, they still cannot improve reliability. Why, because reliability is everyone's responsibility in the plant, starting from the C-Level people down to the shop floor in the plant. C-level and executive decision-makers should be the drivers of reliability and not maintenance. My question is, do the C-Level people in industries currently understand what reliability is, or even what their role is supposed to be? If not, then there lies the problem.

[9]In the book Ivor Bazovsky on Reliability Theory and Practice on page 11, he quote, that reliability is the probability of a device performing its purpose adequately for the time intended under the operating conditions encountered. This definition implies that reliability is the probability that a device will not fail to perform as required for a certain length of time.

---

[9] Bazovsky, Ivor, *Reliability Theory and Practice,* Dover Publications Incorporated, Page 11

Such probability is also referred to as the probability of survival. Reliability is the probability that no failure will occur throughout a prescribed operating period. The modern concept of reliability is simply the capability of equipment not to fail or breakdown in operation. When equipment works well and performs to do its job for which it was designed to do, then such equipment is said to be reliable.

Figure 7.2: Formula for Reliability

The formula for reliability is shown in figure 7.2 and is expressed in percentage. Many will be asking isn't it that if the reliability of the equipment goes up from 90 to 95%, then we are increasing the reliability of the equipment? The answer is no. Every piece of equipment or machine has its inherent design reliability. This means that this is what the equipment or machine can actually do based on its design. We cannot surpass the design of the equipment since this is how it was designed. Therefore, if the design of the equipment is given 100 % of its reliability. Hence, if the machine has a design speed of 1000 units per hour at 100 % reliability, the machine cannot produce 1001 units or more since the machine is not designed to operate beyond its limit. The only way to go beyond 1000 units per hour is to alter the design of the equipment.

Still, Bazovsky's definition of reliability lacks something. If we speak about TPM or Total Productive Maintenance, failures, and breakdown are not the only losses the equipment may suffer that will eventually lead to downtime. There are other losses in the equipment that can stall or slow down the process. An example of this is the design speed loss. This loss refers to the difference between the actual operating speed, compared to the design or ideal speed of the equipment. For example, if we have a piece of equipment, which is designed to produce 1000 units per hour, but currently, it is producing 920 units per hour with no breakdown or downtime. A pump has a rated capacity of 1000 gallons per minute, and it is discharging fluid from point A to point B, after a year, the flow rate drops from 1000 to 950 gallons per minute. What happens is that the impeller is eroding, which creates an internal leakage since the distance from the tip of the impeller to the volute increases, thereby creating an internal leakage. This means that the pump can no longer discharge the fluid at its rated capacity since some of the fluid is returned back to the system. This internal leakage is the result of excessive clearance in the pump, creating a reduction in the volume it can pump. In fact, availability, utilization, and MTBF will be very high or near perfect. However, still, the equipment's efficiency and output are way below what was expected. Another loss experience mostly in manufacturing is set-up, changeover, or conversion in which the machine will be stopped since a new product will run on the same equipment but with different parameters and jigs. This is not a failure but an equipment loss. Hence, the question is if the equipment suffers from these losses beyond equipment failure, can we declare that the equipment is reliable? I do not think so. You see these equipment breakdowns and failures are just a subset of the bigger problems, which are the losses.

We also have errors, or minor stoppages, which mostly exist if the machine is fully automated and contains many electronic parts in it. A minor stoppage is different from a breakdown since nothing actually failed, and the equipment just stopped. Idling and Minor Stoppages are caused by temporary problems with the equipment, which can easily be corrected just by restarting the equipment. For example, a work-piece jam in the chute or a quality sensor shuts down the machine; as soon as someone removes the jammed workpiece and resets the machine, it operates once again. Minor stoppages are errors in automated processes where the workpiece flow stops, the operator resets, and the machine runs again. This means that failures and breakdowns are not the only losses equipment can suffer that can decrease its design reliability. What I am pointing out here is that failures are not the only factor that can affect reliability; we also need to consider the other losses the equipment can experience.

Now let us get the definition of reliability from the dictionary. According to Cambridge Dictionary, reliability is the quality of being able to be trusted or believe because of working or behaving well. Reliability is also defined as how well a machine, a piece of equipment, or a system works. It is the ability to be relied on or depended on for accuracy, honesty, or achievement. When we say that our equipment must be reliable, it means:

• Consistently good in quality and performance
• Able to be trusted or being trustworthy by its customers
• Dependable, well-founded, authentic, valid
• Genuine, committed, unfailing
• Dependability or dependableness
• The ability to be relied on or trusted on

• Can be applied to anything that can be counted upon to do what is expected or required

This means that to consider the equipment reliable, we not only consider the failure the equipment experience, but we also need to consider the bigger picture if the equipment can be depended on to deliver as expected so that it can generate revenue for the industry. According to R. Keith Mobley, reliability is an enterprise-level issue. It can only be addressed as an enterprise program. That means commitment, leadership, and enforcement from the executive level, including the boardroom and the CEO. Reliability has everything to do with everything. It means having stability, consistency, trustworthiness, and repeatability in everything we do in our organization. Reliability does not only refer to our physical equipment and assets, but it has more to do with the business and its reputation. Meaning what good is reliability to a company if it will close or run out of business permanently in the next few months to come. Remember that maintenance can create availability, but it cannot force production to use or utilize it. Production cannot use the equipment if there is no demand or sales for the product manufactured.

Now let us answer the question of why maintenance tasks alone cannot improve reliability. Most, if not all, of this non-maintenance, induced failures are things over which maintenance has little or no control whatsoever. Nevertheless, maintenance will always be blamed for the failure even if they did not cause the equipment to fail. Decision-makers must understand that reliability is not just a maintenance or reliability engineer's responsibility. Still, it is each and everyone's responsibility in the plant. Industries should focus on improving reliability and not on cutting costs on corners. Why, what I believe is that if we can sustain and preserve the reliability of the equipment, then cost will definitely go down. It cannot be the other way around. Remember that there will be times that cutting costs will affect the reliability of the equipment, a lesson we should reflect upon. Here are some of the many things that maintenance has no current control over that can affect reliability.

• Cost-cutting scheme by top management
• No training budget for maintenance
• MRO and storeroom not managed by the maintenance
• Government laws and regulations on shutdown
• Physiological factors and environmental stress
• Poor quality of raw materials supplied
• No budget for Predictive Maintenance Instruments
• No budget for Predictive Maintenance User's Certification
• Not sufficient time is given to maintenance to conduct their scheduled PM
• Operations deferring equipment for a PM
• Poor quality of spare parts
• Purchasing going for the lowest bidder
• Third Party Contractor Errors

These are just a few of the things that maintenance cannot control since it is beyond their responsibility, however, these C-Level people can do something about this but on the initial condition that these people know, and understand what reliability is all about.

7.2: A Deeper Meaning of Reliability from R. Keith Mobley

When I wrote my 4th book on Cutting-Edge Maintenance Management Strategies, I asked a dear friend of mine R. Keith Mobley to write the foreword message of my book. He told me that he could only write a foreword message if he thinks that the book has some value. In short, my feeling was that he was a bit hesitant but nevertheless I provided him a softcopy of my book. He wrote back to me after reading and sent his forward message. His foreword message was an eye-opener on what reliability is all about and it deepens the horizon of my understanding of the subject. Let me share his foreword message just in case the reader has no copy of the book.

Rolly is totally unique, and perhaps the most passionate person that I have met, especially when it concerns Reliability and Maintenance. We have been sharing philosophies for the past year, and while I sometimes have trouble following his logic, it is clear we share far more than we disagree on. I admire his passion, his energy, and his absolute commitment to helping others in their journey towards reliability. At times, our philosophical discussions had become heated. Still, in the end, we have always come to a consensus that both could accept. This book can be difficult to follow at times, but there is knowledge to be gained when you take the time to really read and understand what he is trying to help you understand.

As you read his book, remember that reliability is more than Mean-Time-Between-Failure. Reliability is universal, dimensionless, and immeasurable; it either is or is not. It, or the absence of it, surrounds us every minute of every day and is ingrained into every facet of our personal, family, social, and professional or business life. Reliability manifests itself as the consistency of purpose, and dependability by always meeting commitments as well as the expectations of others; being steadfast and focused on goals, objectives and creating a positive future; and being unfailing in meeting responsibilities. At first glance, these appear to be human characteristics, and they are. Still, they also apply to physical assets as well as the business and work processes that all enterprises depend upon each day.

Those that continue to believe that reliability only pertains to physical assets and is measured by Mean-Time-Between-Failures (MTBF) cannot see that the reliability of the processes, procedures, and practices that govern asset design, procurement, operation, and maintenance are the variable that determinate that result in the measurable MTBF, which is really the symptom of unreliability. Without reliable business, work processes, and employees who execute them, physical asset reliability is impossible. Instead, the volatility and instability of the unreliability of the infrastructure, not the assets, determine their reliability.

Understanding the true magnitude of reliability defines a reliability leader. It is also what differentiates reliability leaders from their peers. It enables them to lead their organizations into a brighter future, one that would not have happened without their leadership. Reliability is an essential, foundational part of a viable enterprise's Life Cycle Asset Management process. Still, it must be more than just reliable assets as defined by MTBF.

Capital or tangible assets are the heart of any enterprise that relies on its ability to produce or manufacture products or, as in the case of facilities, to maintain conditions that occupants require. They are the engine that drives the revenue stream but can quickly become a demanding drain on the enterprise's cash flow and net operating profit when not effectively managed. Reliable processes, procedures, and practices are the framework, the infrastructure that binds the entire enterprise operations, including asset management together. It provides stability, consistency, and dependability that are crucial to all aspects of the organization.

R. Keith Mobley
Thought Leader and Executive Advisor
Renowned Reliability and Maintenance Book Author

7.3: Is It Possible to Improve the Reliability of the Equipment

In Section 7.1 of this Chapter, we discussed that the maintenance tasks cannot improve the reliability of the equipment. It can only sustain it. Now, the question is posed, is it possible to increase the reliability of the equipment. The answer is an astonishing yes but not by doing maintenance tasks but by moving beyond maintenance which is about redesigning or modifying the design speed and capacity of the equipment. Although this will be tough since most redesign and modifications aim to increase the lifespan of a part or spare with inherent design weakness. This means that the part has a short lifespan. If a modification or redesign is designed to increase the speed or capacity of the equipment, then we can therefore conclude that the reliability of the equipment has been increased and improved. We also need to take into account the capacity and efficiency of the equipment. Every piece of equipment has its inherent or given reliability. Improving reliability simply means making the equipment perform beyond its intended function. If a particular machine can produce 1000 units per hour according to its design speed, then it cannot surpass that speed since that is the given inherent reliability of the equipment.

I have a friend, his name is John, and our friendship goes a long way ever since we were in high school in Don Bosco Makati. His father owns a shipping business and has several cargo ships. They were rich. One of those ships was called MV Sea Raider, which was my first taste of work experience way back in 1986. One day, John called me and asked for some favor and if I can help him. I said ok and asked him what he wanted me to do. John said that he bought a tag boat and wanted to convert it into a fishing boat. He said that he extracted the old anchor winch of his father's ship, which was the MV Sea Raider, where I used to work, which was already decommissioned and scrapped. John's idea was to use the anchor winch and attached it to the net to catch some fish. Since this was where I used to work with, it has some form of sentimental attachment to me. I accepted his offer to help him. We went to the shipyard, salvage the old anchor winch, and we started to open it. I started to measure the different gears inside the old winch. Hence, I went to their shipyard and opened up the anchor winch, and take measurements. Once I have taken the measurements, I went home and started doing the calculations for the unknown for the rpm (N) on N2, N3, N4, N5, and N6.

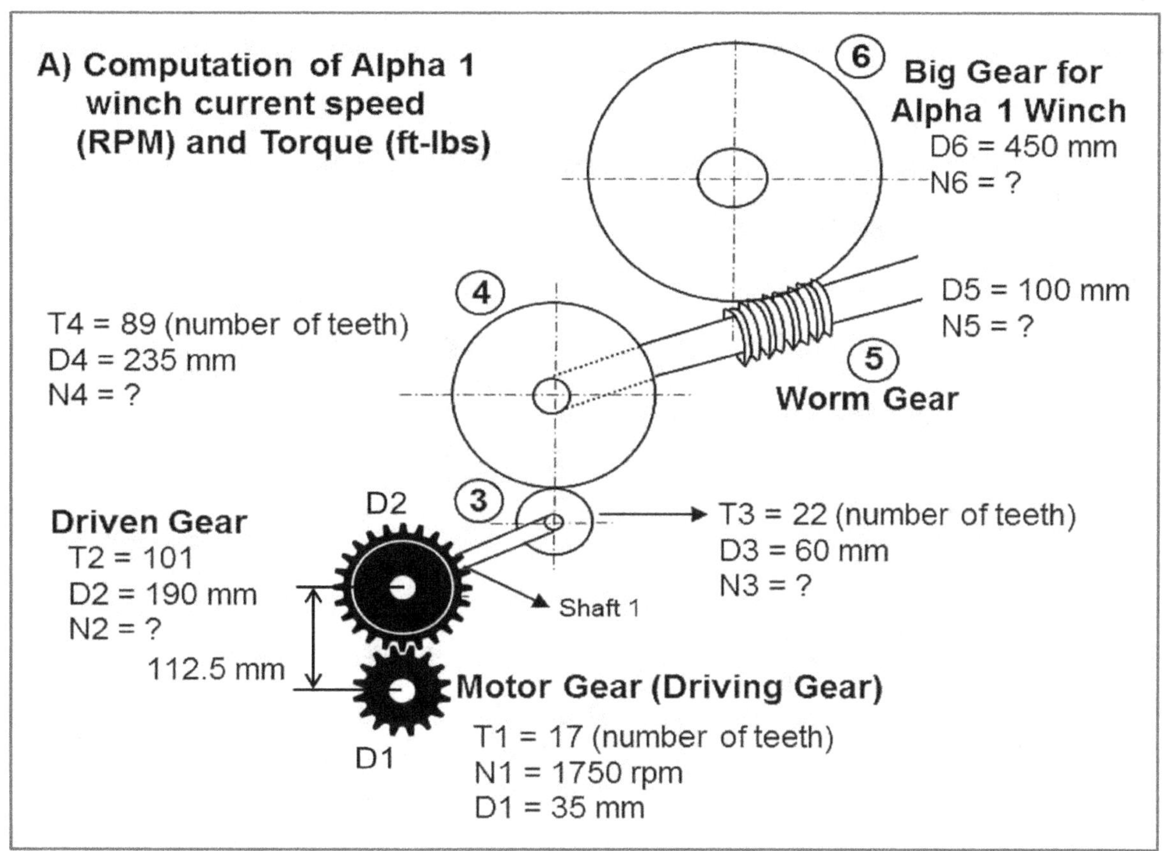

Figure 7.3: Alpha 1 Anchor Winch Gear Teeth Dimension and Diameter (not scaled)

Computing for N2
D1N1 = D2N2
N2 = D1N1/D2 = 35(1750)/190
N2 = 322.37 rpm

Computing for N3
N2 is equal to N3 since they are coupled
N2 has the same shaft, hence N3 =322.37 rpm

Computing for N4
D3N3 = D4N4
N4 = D3N3/D4 = 60(322.37)/235
N4 = 82.31 rpm

Computing for N5
N4 is equal to N5 since they are coupled
to the same shaft, hence
N5 = 82.31 rpm

Computing the speed of N6 will give the RPM speed of the spool
D5N5 = D6N6
N6 = D5N5/D6 = 100(82.31)/450
N6 = 18.29 (0.90)

Current Torque of Alpha 1 winch
Torque = HP (5252)/RPM
Torque = 10 (5252)/16.46

N_6 = 16.46 rpm   Torque = 3,190.76 ft-lbs
Gear Ration of Alpha Winch ratio = 1:5.43

Option 1: Changing the Gear Ratio to 1: 1 @ increase 30 hp

Computing for N2   **Computing for N3**
$D_1N_1 = D_2N_2$ N2 is equal to N3 since they
$N_2 = D_1N_1/D_2 = 112.5(1750)/112.5$ are couple to the same shaft
N_2 = 1750 rpm   therefore N_3 = 1750 rpm

Computing for N4   **Computing for N5**
$D_3N_3 = D_4N_4$ N4 is equal to N5 since they
$N_4 = D_3N_3/D_4 = 60(1750)/235$ are couple to the same shaft
N_4 = 446.81 rpm   therefore, N_5 = 446.81 rpm

Computing the speed of N6 (Large Gear) will give the speed in RPM of the spool
$D_5N_5 = D_6N_6$ Current Torque of Alpha 1 winch
Torque = HP (5252)/RPM   Torque = 30 (5252)/89.36
$N_6 = D_5N_5/D_6 = 100(446.81)/450$ Torque = 1,763.50 ft-lbs
N_6 = 99.29 (0.90)
N_6 = 89.36 rpm
Gear Ratio of Alpha 1 winch 1:1

Option 2: Changing the Gear Ratio to 1: 2 @ increase 30 hp

Computing for N2   **Computing for N3**
$D_1N_1 = D_2N_2$ N2 is equal to N3 since they
$N_2 = D_1N_1/D_2 = 75(1750)/150$ are couple to the same shaft
N_2 = 875 rpm   therefore N_3 = 875 rpm

Computing for N4   **Computing for N5**
$D_3N_3 = D_4N_4$ N4 is equal to N5 since they
$N_4 = D_3N_3/D_4 = 60(875)/235$ are couple to the same shaft
N_4 = 223.40 rpm   therefore N_5 = 223.40 rpm

Computing the speed of N6 (Large Gear) will give the speed in RPM of the spool
$D_5N_5 = D_6N_6$ Current Torque of Alpha 1 winch
$N_6 = D_5N_5/D_6 = 100(223.40)/450$ Torque = HP (5252)/RPM
N_6 = 49.64 (0.90)   Torque = 30 (5252)/44.70
N_6 = 44.70 rpm   Torque = 3,524.83 ft-lbs
Gear Ratio of Alpha 1 winch ratio = 1:2

Option 3: Changing the Gear Ratio to 1: 1.5 @ increase 30 hp

Computing for N2   **Computing for N3**

D1N1 = D2N2
N2 = D1N1/D2 = 90(1750)/135
N2 = 1166.67 rpm

N2 is equal to N3 since they
are couple to the same shaft
therefore N3 = 1166.67 rpm

Computing for N4
D3N3 = D4N4
N4 = D3N3/D4 = 60(1166.67)/235
N4 = 297.87 rpm

Computing for N5
N4 is equal to N5 since they
are couple to the same shaft
therefore N5 = 297.87 rpm

Computing the speed of N6 (Large Gear) will give the speed in RPM of the spool

D5N5 = D6N6
N6 = D5N5/D6 = 100(297.87)/450
N6 = 66.19 (0.90)
N6 = 59.57 rpm
Gear Ratio of Alpha 1 winch ratio = 1:1.5

Current Torque of Alpha 1 winch
Torque = 30 (5252)/59.57
Torque = 2,644.95 ft-lbs

Option 4: Changing the Gear Ratio to 1: 25 @ increase 30 hp

Computing for N2
D1N1 = D2N2
N2 = D1N1/D2 = 100(1750)/125
N2 = 1400 rpm

Computing for N3
N2 is equal to N3 since they
are couple to the same shaft
therefore N3 = 1400 rpm

Computing for N4
D3N3 = D4N4
N4 = D3N3/D4 = 60(1400)/235
N4 = 357.45 rpm

Computing for N5
N4 is equal to N5 since they
are couple to the same shaft
therefore N5 = 357.45 rpm

Computing the speed of N6 (Large Gear) will give the speed in RPM of the spool

D5N5 = D6N6
N6 = D5N5/D6 = 100(357.45)/450
N6 = 79.43 (0.90)
N6 = 71.50 rpm
Gear Ratio of Alpha 1 winch ratio = 1:1.25

Current Torque of Alpha 1 winch
Torque = HP (5252)/RPM
Torque = 30 (5252)/71.50
Torque = 2,203.54 ft-lbs

Option 5: Changing the Gear Ratio to 1: 3 @ increase 30 hp

Computing for N2
D1N1 = D2N2
N2 = D1N1/D2 = 100(1750)/130
N2 = 1346.15 rpm

Computing for N3
N2 is equal to N3 since they
are couple to the same shaft
therefore N3 = 1346.15 rpm

Computing for N4
D3N3 = D4N4
N4 = D3N3/D4 = 60(1346.15)/235

Computing for N5
N4 is equal to N5 since they
are couple to the same shaft

N4 = 343.7 rpm   therefore N5 = 343.7 rpm

Computing the speed of N6 (Large Gear) will give the speed in RPM of the spool

D5N5 = D6N6	Current Torque of Alpha 1 winch
N6 = D5N5/D6 = 100(343.7)/450	Torque = HP (5252)/RPM
N6 = 76.37 (0.90)	Torque = 30 (5252)/68.73
N6 = 68.73 rpm	Torque = 2,292.45 ft-lbs
Gear Ratio of Alpha 1 winch ratio = 1:1.25	

Computing the time for the net to reach the fishing boat at a distance of 1 mile
T1N1 = T2N2
T2 = T1N1/N2
T2 = (5.34 hours) (16.46 rpm)/89.36 rpm
T2 = 0.98 hours

The diameter of both the motor gear (D1), which is the driving gear and the driven gear (D2), will be modified. I have provided five options regarding the size of the driving and driven gear, which were D1 and D2, respectively. We also calculated the time for the net to reach the ship when released up to one mile in length. The original design of the winch will take 5.34 hours to reach the ship when it was released one mile from the ship. Even the dumbest fish caught in the net will have plenty of time to escape.

OPTION	D1 (mm)	D2 (mm)	Gear Ratio	Torque foot-lbs	RPM	Time Hours (1 mile)
Original	35	190	1:5.43	3,190.76	26.46	5.34 hours
Option 1	112.5	112.5	1:1	1,760.50	89.36	0.98 hours
Option 2	75	150	1:2	3,524.83	44.70	1.97 hours
Option 3	90	135	1:5	3,644.95	59.57	1.47 hours
Option 4	100	125	1:1.25	2,203.54	71.70	1.23 hours
Option 5	100	135	1:1.30	2,292.45	68.73	1.27 hours

Figure 7.4: Summary of Options for Gear Modification for D1 and D2

After one week, I submitted my calculation to my friend John. I provided him several options as to the size of the gear ratio (figure 7.4). He told me that he would be going for the ratio of 1:25 or option 4 with an rpm of 71.70. Based on the calculation, if this gear ratio will be used, then the time for the fishing net with a distance of one mile to return to the ship will be reduced from 5.34 hours to 1.23 hours. Finally, he went to a machine shop to have the gear fabricated, which were the driven and driving gear, fabricated to the exact diameter he wanted for gear D1 which is the driving gear, and gear D2 for the driven gear. The machine shop fabricated the gear, and after a couple of weeks, it was done. They installed the revised gears on the winch and have it tested. During their trial run, John called me; I remember I was in Malaysia conducting some maintenance training, and he said that if I could come. I said I was out of the country. He said that the revised gears have been tested, and the crew and other shipping boats were amazed by the new speed of the anchor winch. He told me that they caught so many fish during their first run using the modified gears. The crew went home and feasted with their families that night, and I think they ate as many fish as they can. I was happy to hear about that. I think in this case, the reliability of the anchor winch to deliver and catch fish was increased, and I rest my case on that. The winch was originally designed for the anchor of the ship and modified it to catch fish, and it did catch some fish and I think that's all I have to say about that.

- *While others say that maintenance cannot improve the reliability of the equipment but it can only sustain it. My take on this is two-fold. It depends on how you define maintenance. While I agree that considering maintenance as a task will only sustain reliability, however, considering maintenance as people and human beings can definitely improve the reliability of the equipment without a doubt.* From Rolly Angeles

7.4: Why Reliability is Everybody's Responsibility

Technically, if we use Bazovsky's definition and formula, we can increase the reliability of our equipment and assets. This is a wrong concept since the inherent design reliability of any equipment is given at 100% on the assumption that the equipment is running without any single failure, breakdown, equipment loss, and downtime. With this, breakdown losses are reduced, and reliability percentage increase. If the reliability of the equipment was at 60% last month and this month it becomes 85%, it does not mean that reliability actually improved. The reality is we are not actually increasing the reliability in this case since it is already given at 100% at its design stage, and we are just sustaining it by avoiding unnecessary failures in the equipment.

Achieving 100%, reliability in the real world is not possible since there are many losses to address. One of them is the Planned Shutdown. We cannot operate the equipment without any form of Preventive Maintenance. When we perform Preventive Maintenance, we need to shut down the equipment. Equipment reliability can only be increased by going beyond maintenance and modifying the capacity of the equipment. Once the capacity of the equipment is modified, the reliability increase, but again, the maximum value for reliability we can get is 100%. This means that we cannot go beyond this unless a second modification takes place. Running the equipment at 100% reliability during its entire lifespan is simply an impossible task as there are equipment losses and downtime, both planned and unplanned, that cannot be eliminated. Hence, if we have a piece of equipment that can produce 24,000 units per day or 720,000 units per month, ideally in a perfect plant, still, in the real world, that will not happen, as we need to factor in the losses and downtime in the equipment.

1 day = 24 hours/day x 1000 units/hour = 24,000 units/day
1 month = 24 hours/day x 1000 units/hour x 30 days/month = 720,000 units/month

However, if the equipment is a non-dedicated machine where it is designed to produce other products the machine needs to undergo a changeover where the equipment needs to be stopped. This could not be done in a split of a second, since there will be some changeover or conversion that will be done on the equipment that will include changing the parameters, and changing some jigs and dies to produce the next product. This process, called set-up, changeover, or conversion, will be done on the equipment when the equipment is stopped. The same goes true when we conduct some Preventive Maintenance on this machine. This means that the inherent design reliability cannot be surpassed. We cannot reach a perfect output on these machines.

In the case of the anchor winch, where I modified the size of the gears, the rpm of the anchor winch increased from 16.46 to 71.50, and the time for the net to reach the shipping boat when released one mile was shortened from 5.34 to 1.23 hours. Comparing the old design to the modified design, the reliability improved by 441% by comparing it to the original design of the winch. Once again, this time, the winch can only run at a maximum of 100% reliability, which is at 71.50 rpm. This time we are now basing it on the new speed of the winch with the new design. Again, we cannot surpass the reliability of the new design here unless a second modification will be done to further increase the speed of the winch. This is what it meant that reliability couldn't be increased by any means of doing maintenance. Still, initially, modification can improve the capacity of the equipment temporarily if we compared it with the previous design of the equipment.

Here is the difficult part. Reliability is not only limited to our physical equipment and assets. If we speak about asset management, this will include everything, including our company's name, logo, and trademark. In this case, reliability is no longer controlled nor managed by maintenance. It now becomes a C-Level responsibility, and the plant needs to remain in business. As said previously, reliability is the responsibility of everyone in the entire organization because reliability has something to do with everything we do. This will include having a reliable way for human resources in the process of hiring the right people, locating reliable suppliers by purchasing, having a reliable storeroom that can be dependable when maintenance needs parts, on having reliable operators in the plant, from the computer or laptop we use. I can go on and on with this one. Everything needs to be reliable, that is why reliability is not just a maintenance responsibility. Still, it is everybody's responsibility in the plant, since these functions need to deliver whether the people involved are directly or indirectly related to the outcome of the equipment to generate revenue. If any of these functions failed to deliver then operations will definitely halt. Reliability is just like a chain since the strength of the chain entirely depends on its weakest link. Just like an organization where each function represents a link, the strength of the whole organization entirely depends on its weakest function.

7.5: What Happens When Organizational Silos are Removed

Every organization, company, or industry has its own distinct culture. By definition, culture can be said as the way of doing things around here. It has something to do about their common values, shared, beliefs shared things, shared sayings shared doings, and shared feelings that had been provided to us by those old-timers in industries. According to Schein, culture is the pattern of basic assumptions that a given group has invented, discovered, or developed in learning to cope with its problems of external adaptation and internal integration that has worked well enough to be considered valid and therefore taught to new members as the correct way to perceive, think and feed concerning these problems.

Figure 7.5: Conflicting Goals and Objectives among Organizational Functions

A culture is like a fingerprint, and there are no two human beings with the same fingerprint and a human being have 10 fingers, which means that no finger has exactly the same exact fingerprint. Just like in industries, every industry has its own organizational function from the Human Resources, Warehouse, MRO Storeroom, Finance and Accounting, Production, Maintenance, Safety, Health and Environment, and others, which represent the fingers. This means that even if these people work in the same industry, their culture might be unique to other functions of the entire organization since this is how they think and react to situations. This individual organizational culture creates division or silos despite having a common Vision and Mission of the industry. The problem is that there are cases where the Vision and Mission of these functions contradict if not come into conflict with the vision and mission of the other organizational functions. If the direction of the Finance and Purchasing people is towards cost cutting and the direction of the entire maintenance workforce is to preserve their equipment and assets, then the entire maintenance function will be affected and may not deliver their directive.

Let us be realistic at this point, the industry is built so that it can generate revenue which in turn converts into profit for the industry. Industries need to sell their products or services for them to exist and remain in the competition. These products or services rendered by industries provide will be produced by the equipment. Although both operations and maintenance will be directly involved in the equipment, other functions of the organization will also be indirectly involved. The sad fact is industries create their own problems through their rules, policies, procedures, and standards that they generate and there are many cases where their policies generate problems, especially for those who are directly involved with their equipment and assets. When silos exist, then there is always the doubt that equipment will perform its intended function and that is to produce either a product or service for their clients, which in return provides them the revenue and profit. Silos create a separation and a loss of chance to collaborate with one another to become better. If silos exist in an organization, then expect operators just to operate, equipment to fail, maintenance to repair, and if the part is not around, maintenance to cannibalize parts somewhere since the storeroom does not stock the item, and a lot of other problems to occur. Industries depend on their equipment and assets to function so that they can earn revenue and that can only happen if their equipment is operating. While others talk about managing assets, the truth is that assets, machines, and equipment can only be managed if they are well maintained. It does not take a rocket scientist to figure this out. All we need is just a modicum of common sense.

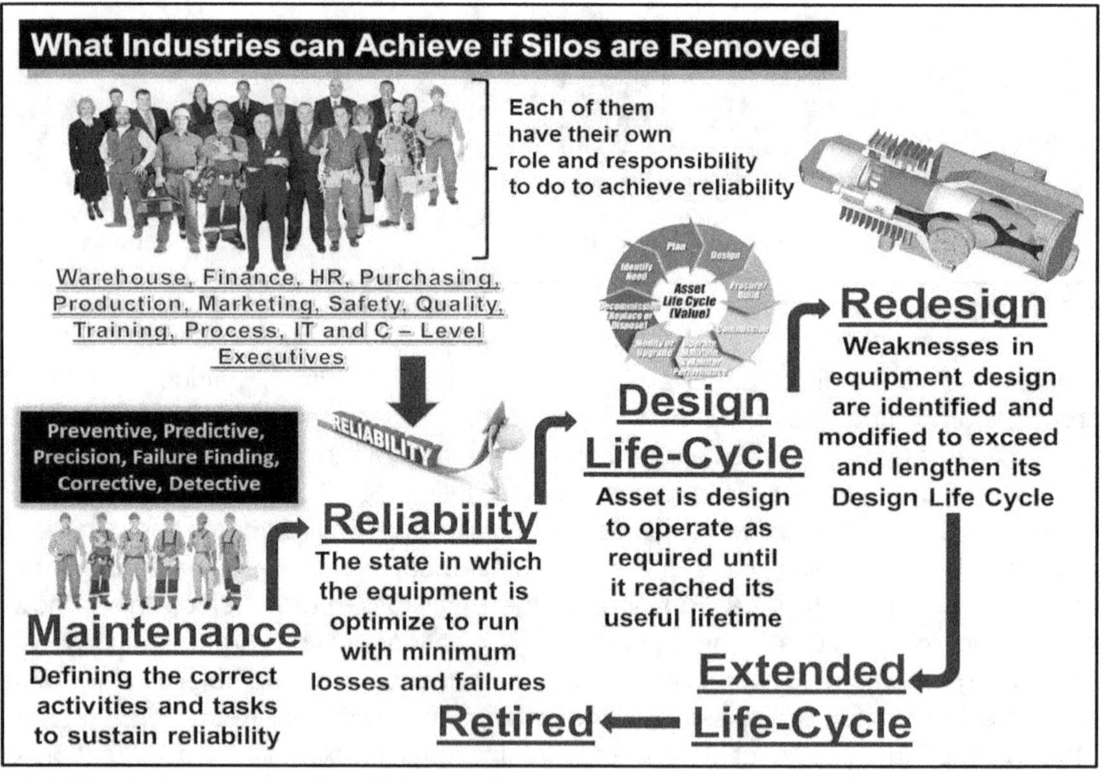

Figure 7.6: What Industries can Achieve if Silos are Removed

7.6: The Three Forces in Industries

The goal of any industry on the planet is to earn revenue and money and that should be first and the priority. If Safety will be the first then let us make earning revenue for the industry a second priority. I agree 100 percent that safety is important and should not be compromised for profit, but like Quality and Reliability each of them is as important as Safety and none of them should be the first. What is important is that besides teaching their employees about safety and quality, maintenance people should start to educate their people on what reliability is all about, and the role of each and everyone in the plant. Quality, safety, and reliability people should work as one and not in isolation so that these people can have a better chance of dealing with human errors, which are the primary cause of industrial accidents and disasters. Providing stricter safety rules and procedures for everyone to comply with will not address human errors. What I believe is making the equipment safe to operate is not synonymous with making that equipment reliable, and having products of high quality, but what I believe is that if the equipment is reliable, then it is safe to operate for the operators. What is lacking between these people is constant communication and utmost cooperation so that we can optimize the reliability of our equipment and assets. No disrespect to Safety and Quality people, but what I believe is that it is time for maintenance to teach these good people what reliability is all about and why each of these functions should know their role and responsibility in having reliable equipment and assets.

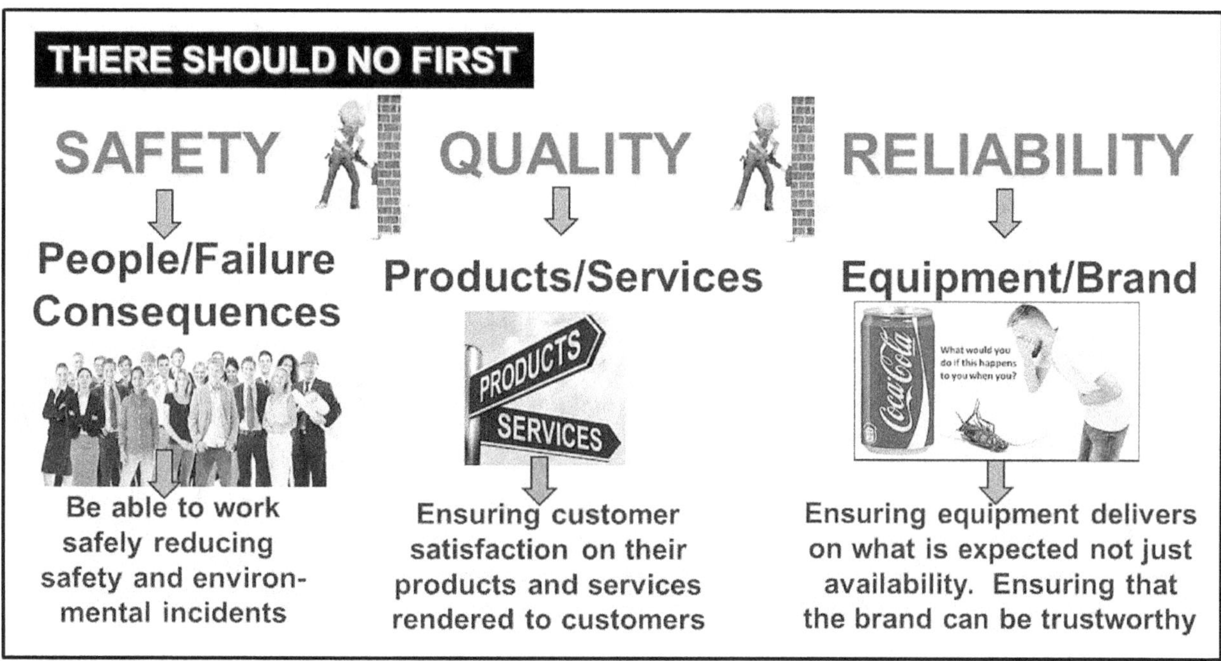

Figure 7.7: Why There Should Be No First

Maintenance includes the activities needed to preserve, conserve and sustain our equipment and assets. Safety, Quality, and Reliability is the effect of doing the correct maintenance tasks on the equipment and assets. This can only happen if these three forces do not work in isolation but in constant communication and treat each of these priorities with equal weight and importance. That is why among the three, no one is first as each of these priorities is connected and equally important.

What we want in industries is for these three champions to apply the concept of the La Cosa Nostra of Mafia that whenever there are disputes among the Bosses, a Sit-Down is arranged for these bosses to settle their disputes. Well of course, after that, they kill each other. We just apply the first concept of the Sit Down and disregard the killing. If these three champions can just settle, collaborate, consolidate, and cooperate with each other. Then what I believe is that you have a much greater force to reckon with.

Chapter 8

Considering Redesign to Improve Reliability

> *Let us not take for granted that if the equipment does not suffer from any equipment failures, then it is considered reliable. I have experienced many equipment, machines and assets that seldom fail, but experienced a lot of equipment losses such as defects, minor stoppages, design speed loss, and conversion. The word reliability means can it be trusted to deliver and perform its functions?*

8.1: Difference Between the East and the West Approach to Improvements

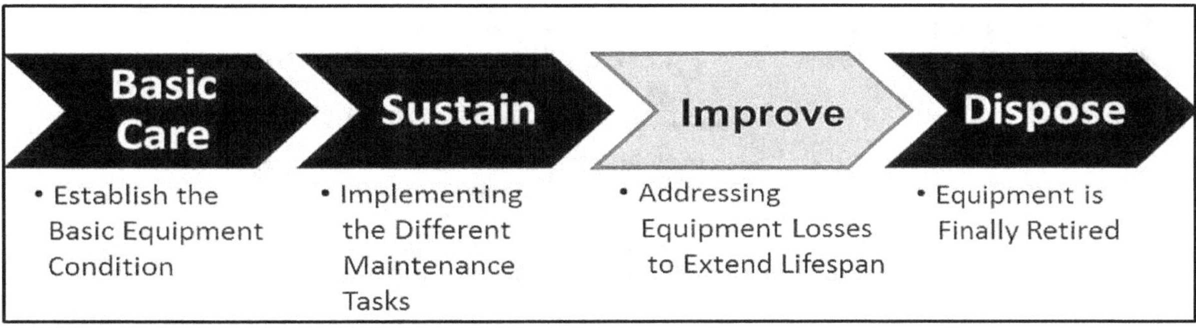

Figure 8.1: Extending Equipment's Lifespan

This chapter is about extending equipment's life cycle, but first, let us discuss the different approaches between the east and the west on improvement. There are two types of improvement, **kaizen, and innovation**. The kaizen improvement originated in Japan, the east, while innovation is mostly preferred by the western countries. The Kaizen methodology is the simple belief that any product, process, service, or anything devised by man can continuously be improved. This is merely the concept of continuous improvement, which means improving incrementally, or small improvements continuously. It contradicts Fredrick Taylor's belief that there is only one best way to do anything. It is the simple truth behind Japan's economic miracle, and the reason the Japanese have become masters of flexible manufacturing. The Kaizen message is that not a single day should go by without small incremental improvements done in the company. The Kaizen Methodology refers to small but consistent and incremental improvements.

While the US industries continue to focus more on big volumes of mass production in which the way to victory was to have smart people manage organizations that produce goods in large batches at lower costs leading to high profits. On the other hand, the Japanese competitors developed a new and much more powerful paradigm, where winning organizations are those that listen to the voices of their people, customer, design products and services that meet or exceed their customer's expectations, and continuously improve all the organizational process that lead to customer satisfaction. In March of 1980, Richard W. Anderson, General Manager of Hewlett-Packard Data System Division, issued a chilling report that had a dramatic impact on their industry. HP had inspected and tested 300,000 units of 16k ram chips from one Japanese and three US manufacturers. He discovered that at incoming inspection Japanese had a failure rate of zero, while the American chips had between 11 to 19 failures per 1000 after 1000 hours of use. The failure rate of Japanese chips was between 1-2 per 1000 against 27 per 1000 for US chips.

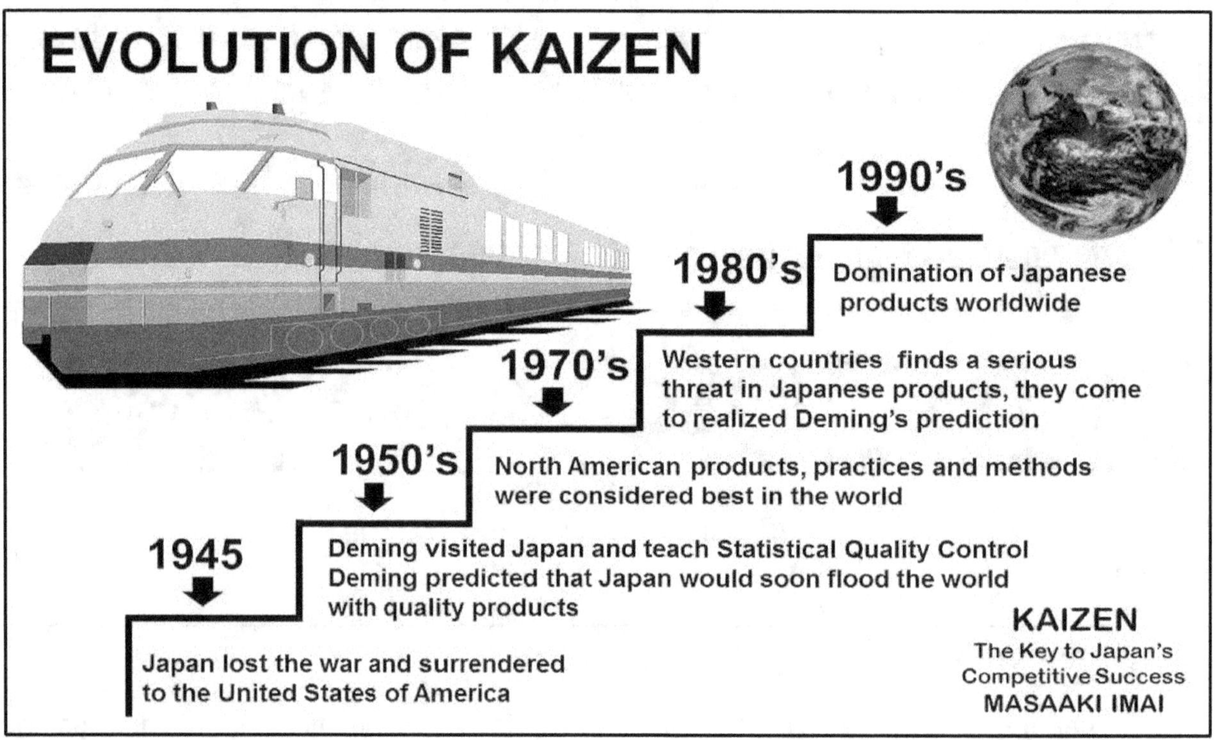

Figure 8.2: Evolution of Kaizen

When Japan surrendered to the US in 1945, Japan was in complete devastation. Out of humanity, the US sends an unknown and modest person by the name of W. Edward Deming. His role was to assist the Japanese Government to recover from its current situation and provided a restructuring plan. Hence, Dr. Deming taught the importance and principles of quality, and Japan listened. One Japanese student asks Deming, how long would it take to shift the perception of the world from the existing paradigm that Japan produced cheap products to one that can produce high-quality products? Dr. Deming told the group that if they would follow his directions, they could achieve the desired outcome in five years. As Dr. Deming told it, they surprised me because they did it in four years. Dr. Deming was invited back to Japan repeatedly where he became a revered counselor.

Because of his efforts, he was awarded the Second Order of the Sacred Treasure by the former Emperor Hirohito. Japanese scientists and engineers named the famed Deming Prize after him. It is bestowed upon organizations that apply and achieve stringent quality-performance criteria in Japan. Japan may have actually lost the battle but it was the United States that lost the war in the economy. The irony is that Deming become so popular in Japan but in his own country, he was a completely unknown and unheard person. The truth is that the staggering success of Japan was built upon the principles of an American citizen ignored in his own country.

On the other side, we have innovation, which is more of a breakthrough improvement in which its impact could be felt immediately. Kaizen requires a mindset; that is why all employees must contribute and suggest something which is a way of life for most Japanese manufacturing firms. It does not matter whether its impact is big or small. It's a part of their day-to-day culture and philosophy. I was employed many years ago in a Japanese firm, and as far as my poor memory can recollect, they would prefer to use the word "better" than the word "best." Their reason is simple, Japanese people challenge themselves that the best can still be made better.

Perhaps kaizen, also known as continuous improvement, has something to do with the Rolling Stones' famous song, "I Can't Get No Satisfaction," as the band itself, the older they get, the better they become. Kaizen's true meaning is somewhat difficult to absorb and comprehend since this is about mindset. Most Japanese operators and maintenance will perform small incremental improvements in their equipment or asset since their belief is that the older the equipment gets, the more it should operate smoothly. This mindset is very different from the Philippines and other countries in Asia (as I cannot speak for the rest of the countries) since the older the equipment gets, the more problematic it becomes.

Comparison	Kaizen	Innovation
• Effect	Long-term and lasting	Short-term but dramatic
• Pace	Step by step, incremental	Great leap, a big step
• How often it is done	Many times continuously	One time
• Timeframe	Continuous and incremental	Intermittent and non-incremental
• Change	Gradual and constant	Abrupt and volatile
• Involvement	Everyone	Few selected with the capability
• Approach	Team approach	Individual's creativity
• Mode	Maintenance Improvement	Rebuild, redesign and modification
• Spark	Conventional know-how and state of the art	A technological breakthrough, new invention, and design theories
• Ideas	Information is open and shared	Classified, confidential, trade-secret
• Requirement	Requires little or no investment but great effort to maintain	Requires large investment or capital but little effort to maintain
• Effort Oriented	People	Technology
• Responsible	Everyone in the organization	Selected few

Figure 8.3: Comparison between Kaizen and Innovation

Innovation, on the other part, refers to something new. It is considered a change made to an existing product, service, idea, or field. We can say that the first telephone invented by Elisha Gray and Antonio Meucci was an invention; the mobile phone, today with so many apps and features, can be considered an innovation. Not everyone can innovate, unlike kaizen, which is for everyone in the organization. Figure 8.3 is a cross-comparison between kaizen and innovation.

Let me share this short story about a Japanese manufacturing soap factory to explain the difference between kaizen and innovation. The final process in this soap manufacturing will be packaging the soap and placing it in a box distributed to the different retailers and supermarkets. The problem started when one of the customers who purchased this soap only contains the box with no soap inside. A few days followed, and several customers also complained of the same problem. One customer even posts it on his Facebook Social Media about the box without the soap inside and named the company, which went viral with more than one million views. It really damaged the reputation of the company. The factory's Top Management was alarmed and called for a meeting with the Head of Engineering, Quality, Operations, and Process and other big heads in the plant to address the problem. After several days of brainstorming, the factory decided to invest in an expensive x-ray machine, which will be included in the packaging process. One operator was also assigned at the x-ray to inspect all soaps 100% on her shift. The plant operates for three shifts, so there are three operators assigned to do the task. There will be one operator per shift assigned to the x-ray machine, just like in the security area in an airport before you board the plane. The soap factory management team also purchased an automated weighing system that includes all the soap in the box that should be measured with + 5 /- 5 grams. The operations team again assigns three people, one operator per shift, to do the tasks.

The plan was executed with the x-ray machine and automated weighing system, but still, there are a few customers who purchased the soap without the item but only the box. Although the complaints were dramatically reduced, it was not 100% foolproof. The six operators assigned were doing their tasks, which were assigned to them even if it bored them to death. But, hey, they also need to take a break and eat their lunch. This means that both the x-ray and the weighing system were unmanned when these operators eat their lunch. Finally, one operator proposed his suggestion to his supervisor to borrow one of the blowers from their facilities and run it continuously, blowing away those empty boxes that contain no soap inside. When this was implemented, the six operators were relocated to other parts of the production, both the x-ray machine and the automated weighing scale were never used again. It was just a simple suggestion from an operator, which they implemented, and the problem of the missing soap was finally solved. In Kaizen, they make their improvements as easy and as simple as possible.

Japanese are very simple people. I remembered one time when our JIPM consultant visited us as part of his scheduled consultancy with the plant. There were several teams from Autonomous Maintenance who was scheduled to be interviewed by the JIPM Consultant. Since the JIPM Consultant does not speak the English language, he was accompanied by a Japanese interpreter and both of them speak Kanji. One of the teams was presenting and one of their recent activities was replacing the door of the machine, as it is full of scratches. The Japanese consultant was asking for the old door that was replaced and the team showed it to him. A table was near the door of the machine. The JIPM Consultant asked the interpreter to help him move the table so that when the door swings, it will not be hit by the table anymore. Finally, after moving the table, a couple of feet away, the consultant asked the team how much the cost of the new door and the AM team replied that it cost around 12,000 PHP as it was made of thick acrylic material. The consultant asked the team how much it cost to move the table a couple of feet away so that it does not hit the door anymore and the team members looked at each other and were blown away by the remarks of the JIPM Consultant. That was how simple the JIPM consultant was.

• *When NASA began the launch of astronauts into space, they found out that pens wouldn't work at zero gravity since the ink would not flow down to the writing surface. It took them one decade and an estimated amount of US $12,000.00 million to solve this problem. NASA Engineering and scientists developed a pen that worked at zero gravity, upside-down, underwater, on practically any surface, including crystal, and in a temperature range from below freezing to over 300 ° C. What did the Russians do? They used a pencil.*

Kaizen means improvement. Moreover, it means continuing improvement in personal life, social life, and working life. When applied to the workplace Kaizen involves everyone in the organization, including managers and workers. According to Willam Manly, Senior Vice President of Cabot Corporation, he quotes that I thought the Japanese have two religions in Japan, Buddhism, and Shintoism, but now I find a third, which is Kaizen. Kaizen resorts to positive thinking, where the belief is that we can turn each problem into an opportunity for improvement, and the starting point in an improvement is to identify the problem. Here are some important things to reflect on about Kaizen;

• It reflects on gradual, unending, continuous improvement
• Not a day should go by without some kind of improvement made somewhere in the company.
• Quality in its broadest sense is anything that can be improved, and the means to justify improvement is Kaizen.
• Make a continuous effort to establish a system to support the P-Criteria.
• There must be a close connection, continuous interaction, and communication between the design stage, development stage, production stage, and marketing stage.
• Quality first, not profit first. If we take care of Quality, then the profit will follow.
• Remember that the next station or process is your customer.
• Follow the PDCA (Plan, Do Check, Act) for a continuous improvement cycle.
• Kaizen must involve everyone in the organization from top management down to the shop floor.
• Although Kaizen is culture-based, it is not culture-bound which means that it can work in any organization once you learn to accept its principles

If asked which is better Kaizen or Innovation. This can be compared to the race between the rabbit and the turtle in which the lesson is slow and steady wins the race. Although innovation is a leap-frog approach, in most cases this is just a one-time event, while Kaizen is an incremental and small improvement but it is continuous since Kaizen is considered a way of life for the Japanese people. This means the improvement will be done repeatedly. As Buzz Light year from Toy Story said, to infinity and beyond.

8.2: Equipment Redesign Explained

Equipment redesign or modification means changing something from the original specification of the equipment or asset to address a particular or existing problem. This can involve changing the shape, size, measurements, or strength of the material of a particular part, spare, or item. Redesign or modification may also include an engineering change, a change in the process, adding something to the equipment, or replacing or changing something to address a particular problem. This may also involve changing a process or any standard procedure performed on the equipment or changing the equipment or component itself. All equipment has its original inherent reliability, which means that the equipment or asset is only designed to function on which it is intended to do. A pump with a rated capacity of 500 gallons per minute cannot discharge beyond that capacity since this is what it was designed for. Increasing the capacity means either modifying the speed or completely changing the component with a higher rated capacity. Increasing the reliability of the equipment cannot be done by any means of performing maintenance tasks. This means that no amount of maintenance tasks can increase nor improve the inherent reliability of equipment or asset since this is how it was designed to operate. Increasing the reliability of a piece of equipment can only be done by going beyond maintenance, which is either redesigning or modifying the asset. Remember that the goal of any maintenance task is simply to sustain or preserve the equipment. Going beyond that means walking the thin line and performing a redesign or modification.

During the entire life span of the equipment, there will be problems experienced on the equipment that should be addressed by a particular group to counteract the problem.

8.3: What Modification and Redesign Includes

There are many reasons why equipment is redesigned or modified during its entire lifespan. According to FMEA and FMECA, reducing the RPN or Risk Priority Number for failures with high severity can only be done through redesign, or modification, while RCM states that a failure with safety or environmental consequences cannot be allowed to fail. This means that run to fail is not an option. If there were no particular maintenance tasks that can address a particular failure mode that has safety or environmental consequences, then this would be subject to redesign or modification. In FMEA a high severity ranking of 10 means that the failure can affect the operator, plant, and maintenance safety or may have non-compliance with any known government regulation and provisions. This means that if the risks are high and the consequences of the failure are unacceptable, and no maintenance tasks can address the failure, then the default is to redesign or modify. Here are some considerations and questions to consider if redesign or modification will be a feasible option to undertake in the equipment. If you answer yes to all of these questions, then redesign, or modification is feasible and valid.

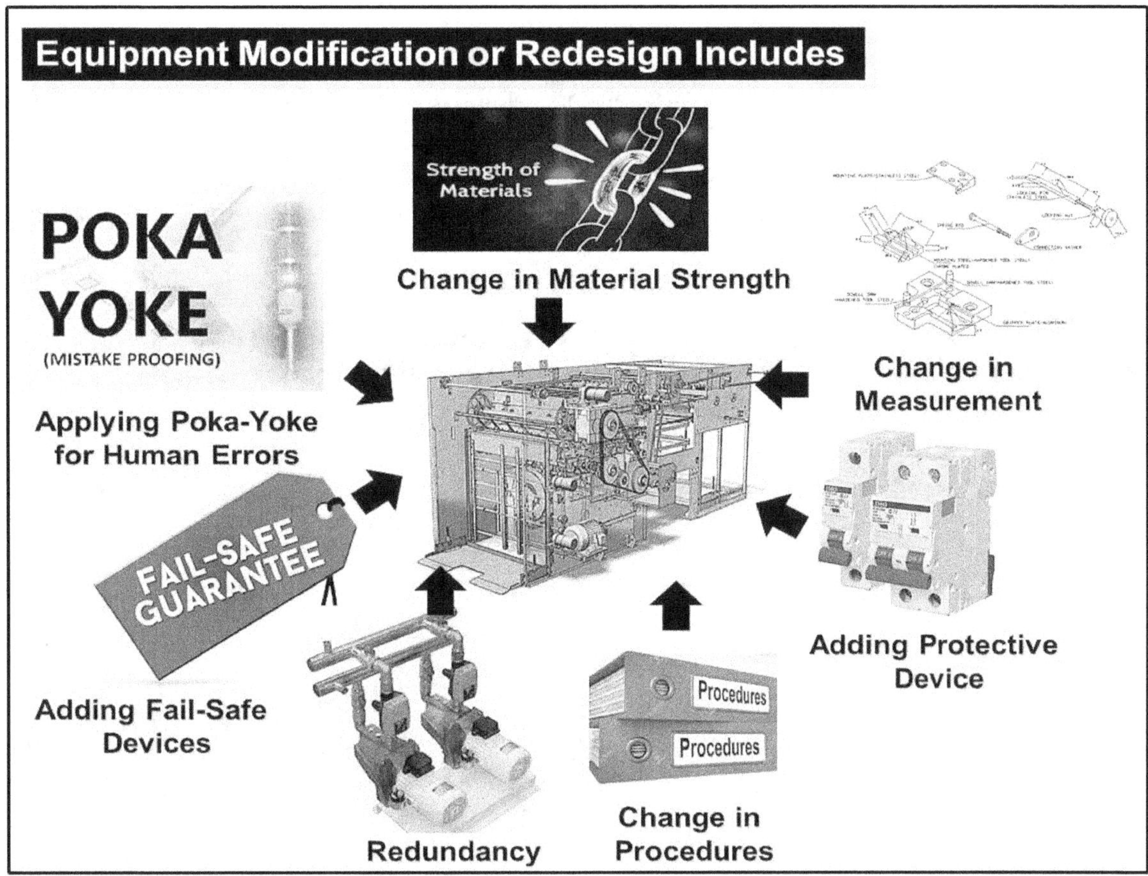

Figure 8.4: Modification and Redesign Includes

• Does the failure involve major operational consequences?
• Is the cost of breakdown maintenance high?
• Is there no backup or redundancy in place?
• Are there specific costs that can be eliminated by the design change?
• Does the design have no harmful effects, which can be generated afterward?
• Is there a feasibility study on the payback or return on investment on the cost of modification?
• Can redesign or modification eliminate or reduce the consequences of the failure?
• Is the asset to stay for a long time and will not be decommissioned in the next few months?

Redesign and modification are done on the equipment to address a particular problem or address a particular design weakness on the equipment, machine, or asset. These modifications or redesign may be in the form of the following;

1. Adding Protective Devices: These controls can either prevent or predict that a failure mode is on the verge of occurring or detect that an asset is about to fail or break in operations. These controls must detect the failure before it happens and not in the situation where it is already happening. A low-level alarm of the main fuel tank will sound if the level in the tank is at 2000 liters. When the fuel ran out of the tank, an alarm will sound for everyone to realize that the tank is already emptied. These design machinery control are our defenses against failures, which can come in different forms. These controls should be in place especially when the impact and consequences of the failure are of very high criticality. The majority of these controls and defenses include protective devices. They are placed on the equipment to protect something, which is called the protected function. A functionality inspection or Failure Finding task is performed on these protective devices to determine if they are still functional or not and to avoid the chances of a multiple-failure. Multiple failures occur when both the protective device and protected function are both in a failed state.

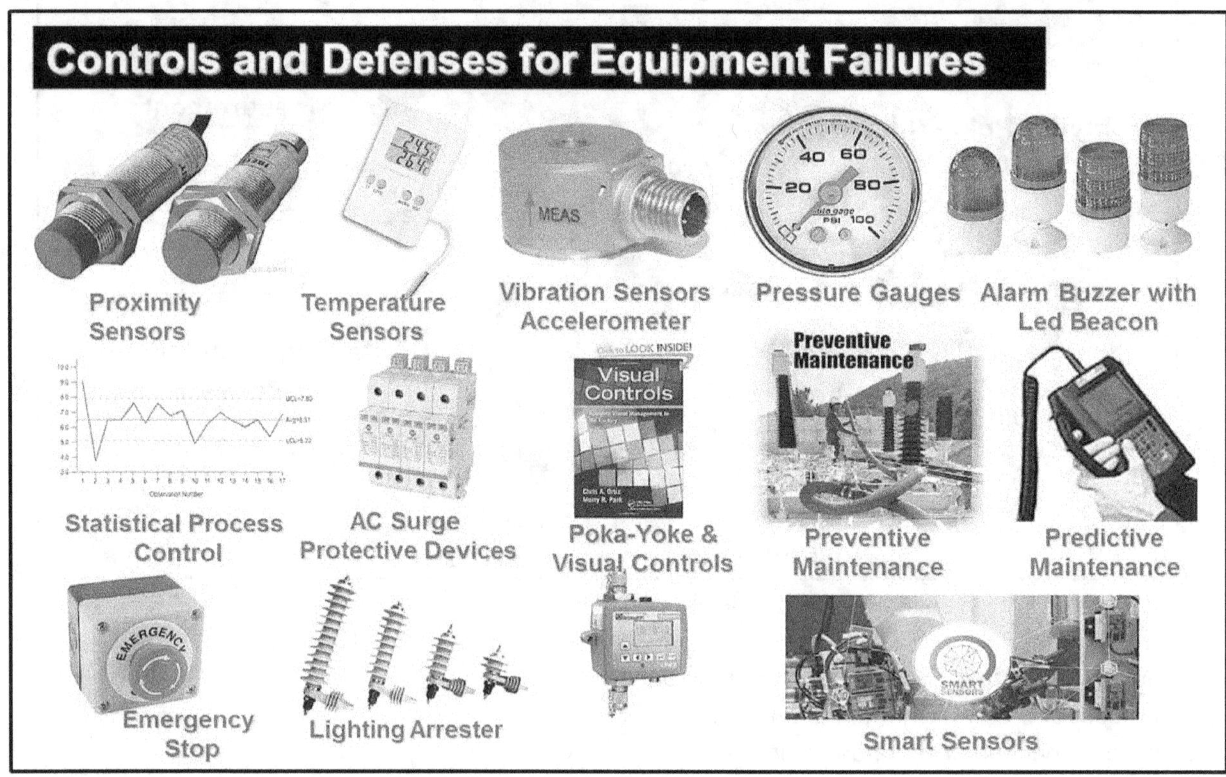

Figure 8.5: Controls and Defenses for Equipment Failures

2. Change in Material Strength: Changing the strength of the material usually means changing the material to a harder one. This is done to increase the lifespan of the part or item.   The strength of a material is its ability to withstand an applied load without failure or plastic deformation. The field of strength of materials deals with forces and deformations that result from their acting on a material. A load applied to a mechanical member will induce internal forces within the member called stresses when those forces are expressed on a unit basis. The stresses acting on the material cause deformation of the material in various manners including breaking them completely. Deformation of the material is called strain when those deformations are placed on a unit basis. Careful study is required when changing the material to a harder one since increasing the hardness of the material will make the material part more resistant to dulling of the sharp edges and can increase the chance of brittle fracture.

3. Implementing Poka-Yoke: Poka-Yoke is a Japanese term that means mistake proofing. Examples include elevator alarms, having a different socket and electrical outlet for 110/220 volt or auto volt appliances, low battery indicators, fuel indicators in the car, child lock doors, and so on. Poka-yoke can be any mechanism used by industries, especially manufacturing processes, that helps an equipment operator avoid errors and mistakes. Its purpose is to eliminate product defects by preventing, correcting, or drawing attention to human errors as they occur. It was developed by Shigeo Shingo as part of the Toyota Production System. The term Poka-yoke was applied by Shigeo Shingo in the 1960s to industrial processes in which its primary purpose is to prevent human errors. Poka-yoke aims to design the process to detect and correct mistakes immediately, thereby eliminating defects at their point of origin.

4. Adding a Fail-Safe Device: A fail-safe is a design feature that in the event of a failure or breakdown, will respond in a way that will cause minimal or no harm to the equipment, environment, or to the operator. Fail-safe means that a device will not endanger lives or property when it fails. This will cause a piece of machinery or asset to revert to a safe condition in the event of a failure or malfunction. Unlike inherent safety to a particular hazard, a system being fail-safe does not mean that the failure is eliminated, but rather the system's design can prevent or mitigate unsafe consequences in the event of a failure. That is, when a fail-safe system fails, it remains at least as safe as it was before the failure.

5. Adding a Redundancy: Redundancy means duplicating the system or component. In these cases, failures may be allowed or tolerated through redundant or duplicated functions at some point. The presence of redundancy or alternative means of production is a feature of the operating context, which must be considered in detail when defining the functions of the asset in its present operating context. Oftentimes termed as standby-unit, even with the same equipment type, standby units have different degrees of maintenance requirements as the duty unit, and most failures for standby units remain hidden. Standby equipment is installed to ensure the availability of process systems or sub-systems at a high level. The common operating practice is to run the standby equipment intermittently with the duty equipment but it is not recommended to run both the duty and stand unit at the same interval. One should run longer, say five months for the duty and one month for the standby.

6. Changing Procedures: Changes in procedures and standards will also be included in the redesign or modification process. There are several reasons for changing the procedure or standards such as updating obsolete procedures or upgrading to a newer version, addressing a particular human error, or adding an additional item that is not part of the existing procedure.

7. Change in Dimension or Measurements: Modification or redesign may also include changing a particular item or part's existing dimension and measurements. An example of this is a modification of a pick and place arm due to frequent turnover of strips during a change of magazine for a plating machine causing mispick. The Y-Aligner arm was modified by extending its arm to 3 millimeters to cater to the upcoming strips so it can hold it better and avoid the strips to fall, which cause machine errors and minor stoppages

8. Increasing the Design Speed or Capacity of the Equipment: Improving the design speed loss means increasing the speed or capacity of the equipment. This is explained in detail in the next section on equipment losses.

8.4: Reasons for Moving Beyond Maintenance and Redesign

The purpose of performing redesign or modification is first, to address existing losses on the equipment, and second, to extend the Life Cycle of the equipment. However, for teams conducting the RCM analysis, it is not recommended to discuss the subject of redesign during the RCM meetings; for the reasons that first, it will only increase the time spent in doing the RCM analysis, and second, the team doing the RCM analysis may not be the exact people who will be doing the redesign. Granted that the plant is implementing TPM Planned Maintenance, the team executing Phase 2, Lengthening Equipment Lifespan by Addressing Design Weaknesses would be responsible for redesigning and modifying the equipment. However, suppose there are certain maintenance activities or tasks that must be done before the redesign will be completed; temporarily, it should reflect in the maintenance tasks. This will only be removed once the redesign had been completed and implemented in the equipment and asset. In conducting RCM, if a redesign is agreed to address a specific failure mode, just write the word redesign or modification in the proposed tasks and move on to the next failure mode. Just inform the team doing Phase 2 on Planned Maintenance. Once the RCM analysis is completed, summarize all the failure modes that defaulted to redesign and submit it to the engineering department. Here are some possible reasons why redesign or modification is warranted on the equipment.

1. To Reduce the Consequences of Failure: When dealing with failures with safety or environmental consequence, if the level of risks associated with the failure will lead to these unacceptable consequences, then any means necessary to eliminate, reduce or mitigate the failure should be done. If performing any form of maintenance tasks will not address these failure consequences, then our last resort is to modify or redesign.

2. To Make Hidden Failures Evident: When the impact of multiple failures is simply unacceptable, we can modify or add another protective device to make the failure evident. Multiple failures are where both the protected function fails because the protective device is in a failed state. However, this may not be a fool-proof solution since adding another protective device to make the existing protective device evident will only make the additional protective device hidden. In figure 8.7, a triple redundancy Over Speed Device is placed to indicate if the speed of the turbine increase. In this case, we are not certain if one or two Over speed device is still working or not, hence a beacon is placed to detect if the Over speed device is still working or not. For as long as the beacon is lit, the Over speed device is working. However, in this case, the failure of the beacon can be considered hidden, hence an alarm is installed to detect if the beacon is working or not. In this case, the failure of the beacon now becomes evident, but the failure of the alarm becomes hidden. What I am saying is that there will always be hidden failures in the equipment. The only way to capture these hidden failures is to conduct a functionality inspection for these protective devices to determine if they are still functioning or not.

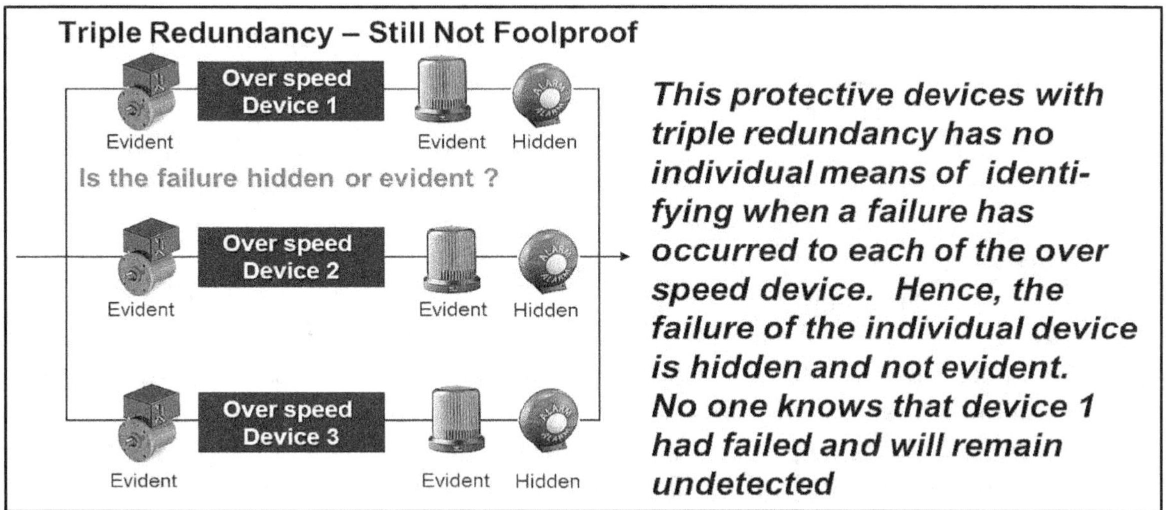

Figure 8.6: Failures of Protective Devices are Hidden

3. To Avoid Recurring Failures: Recurring failures are considered chronic by nature. This means that chronic failures occur because of a complex combination of causes. Although the cost of chronic failures is less compared to sporadic failures, however, in the long run, it may cost more than sporadic failures because of their recurrence. To address chronic failures is to address each cause independently. A tool designed for chronic failures and defects will be P-M Analysis.

4. Reduce Equipment Downtime: Except for Design Speed Loss, most of these losses on the equipment will result in downtime. This means that when the equipment is down, it is not producing any revenue and profit for the industry. For example, redesigning the process of changeover or set-up on the equipment can reduce the set-up time from 2 hours to 0.5 hours on the equipment. This means an additional 1.5 hours of uptime and revenue can be generated on the equipment by reducing the set-up time.

5. To Lengthen Parts Lifetime: Modifications and redesign can be targeted to parts and items with an inherently short lifespan. The goal is to extend the life span of these parts with inherent design weaknesses. This is usually the goal of Phase 2 on Planned Maintenance, which will be discussed in detail in the next chapter.

6. If No Maintenance Tasks Can Address a Particular Failure Mode: If there are no specific maintenance tasks that can address a particular failure mode, then this can be addressed through redesign or modification. In fact, in RCM, redesign, and modification is part of the default tasks, and are usually the last option on the RCM decision diagram, which means that if all maintenance tasks are exhausted and will not address a particular failure mode, then this is where redesign or modification is feasible and recommended to adopt.

8.5: Contradicting Beliefs on TPM and RCM on Improvements

RCM starts by determining the asset functions and failure modes, while TPM believes that the first step is to address the equipment's basic condition. Many failures start with small things and it is these small things that are often left neglected that caused big problems in the end. Big problems are just an accumulation of small problems. TPM believes that big failures can be prevented if we address the basics first. In fact, some of the failure modes can be reduced if basic equipment condition is well established on the equipment.

TPM believes that to advance to any improvement initiative, basic equipment conditions must be carried out first on the equipment. This basic equipment condition includes keeping the equipment clean, proper lubrication, equipment with complete bolts with the right torque, and no leaks of whatever form. Meaning, why don't we address first the basics and complete all the screws and bolts that have been missing in the equipment before performing any vibration analysis on it? On the other hand, RCM states that the first step is to change the way people think and apply this change thought to their equipment. I like what RCM is saying because even before the basic equipment conditions should be well established both operators and maintenance should change the way they think about their equipment and build this new way of thinking as part of their culture. Both operators and maintenance should first understand what harm it could do to their equipment if these basics will be neglected on the equipment.

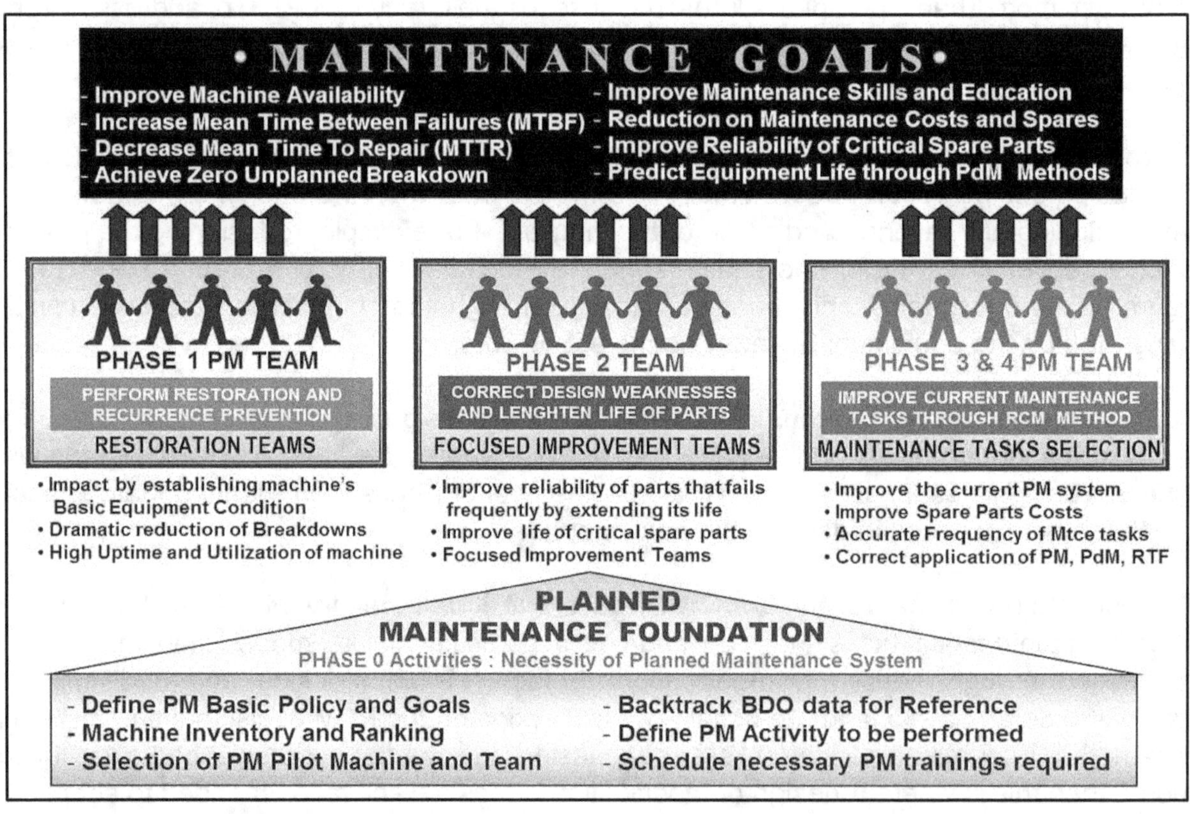

Figure 8.7: The Goal of Planned Maintenance

Although I like to limit this discussion to the equipment side, TPM focuses on addressing the six big equipment losses. By reducing these equipment losses, OEE will improve. On the other hand, RCM focuses both on the primary and secondary functions of the equipment. It highlights the importance of secondary functions of the equipment and that it allows us to understand that there are cases where the failure of secondary functions poses a bigger threat rather than the failure of the primary function. Imagine multiple failures occurring because the protective device, which is a secondary function, is in a failed state.

I believe that this is where both TPM and RCM have some forms of contradiction. TPM is heavy on improvements. It believes that the equipment should be continuously improved, while on the other hand RCM focuses its strength on adopting a structured approach to maintenance in addressing all possible failure modes and adopting the most feasible maintenance tasks to use with the aid of a decision diagram or an algorithm. The word itself, Reliability-Centered Maintenance, means that if we centered our focus on maintenance, perhaps we can improve the reliability of the equipment. In this case, the team selects the most appropriate tasks for each failure mode based on its consequences.

According to John Moubrey, there is a thin line between improving and maintaining the equipment. There are times that we modify the equipment only to learn that the equipment will be decommissioned because of obsolescence. Most improvements take time to accomplish. Some may even take several months, As Moubray said, a person who is on duty today has to maintain the equipment as it exists today and not what it should be there in the future. Therefore, before considering improving and redesigning the equipment, have we asked ourselves if the asset is here to stay for a long time or will be decommissioned. I have seen many improvements being wasted on their equipment because the people who perform the improvement do not know that their equipment will soon be decommissioned. Hence, before trying to improve or redesign the equipment we must have a concrete answer to these questions.

8.6: Equipment Downtime Explained

Every industry has one thing in common, they have assets, equipment, and machines to maintain. Industries acquire equipment and assets to produce revenue, whether they may be manufacturing goods or delivering services. The available time in one year is given as 365 days or 8760 hours. Some industries operate for 365 days a year, while other industries operate less than 365 days a year. If an industry operates let's say 320 days in a year, this can be said as the industry's available time. However, the available time is simply the sum of the equipment's Uptime and Downtime as reflected in figure 8.8.

Uptime is when the equipment is running and utilized, while downtime is when the equipment is not running or generating any revenue. However, downtime can be classified into two, which can be either Planned or Unplanned Downtime. Planned Downtime is also considered as Non-Machine Related Downtime while Unplanned Downtime can also be considered machine-related downtime. Machine-related downtime is caused by machine losses. This can be caused by machine breakdown, set-up, and conversion, quality defects, start-up losses, infant mortality failures, minor stoppages, and changing cutting blades. Non-machine downtime is caused by other factors such as performing Scheduled

Preventive Maintenance, operator's break time, meetings, no raw materials, no operator, and others.

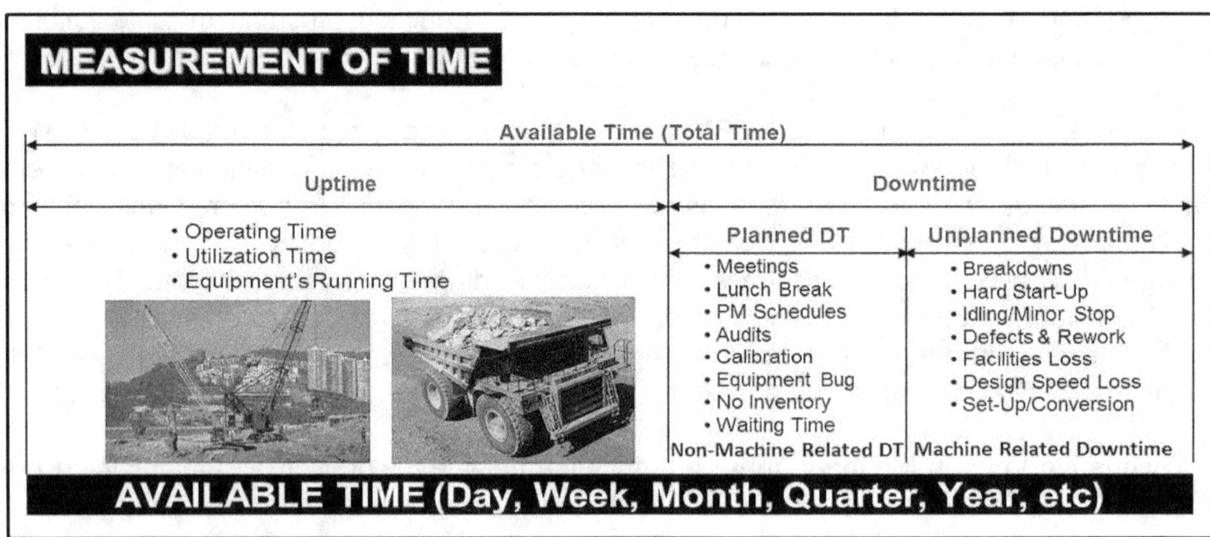

Figure 8.8: Measurement of Time

Planned Downtime	Uplanned Downtime
Non-Machine Related Downtime	**Machine Related Downtime**
- Shutdown Loss	- Breakdown Losses
- Preventive Maintenance Activities	- Set-up and Changeover
- Scheduled Outage	- Start-up Losses
- Operator's Breaktime	- Changing Blades, cutters
- No Operator	- Quality Defects where operator
- No Raw Material to Process	stopped the machine
- Christmas and Holidays	- Idling and Minor Stoppages
- Meetings and General Assembly	- All Function Loss Breakdown
- Machine Audits	
- Predictive Maintenance Activities	
- TPM Activities such as Initial Cleaning	
- Operator Cleaning Activities	

Figure 8.9: Agree on What to Include as Planned and Unplanned Downtime

Although these losses cannot be zeroed out, reducing these losses is also one key factor in lengthening equipment's life cycle. What is important for the maintenance function and those involved is to once again apply the concept of the La Cosa Nostra, sit down and discuss what will be included as Planned and Unplanned downtime. This is very important since neglecting to do this can compromise the indices and KPIs maintenance is tracking.

8.7: Different Equipment Losses Explained

As discussed in the previous section, reliability is the probability that an item will operate without failure throughout a specified interval. According to Bazovsky, when a piece of equipment works well and works whenever called upon to do the job for which it was designed, such equipment is said to be reliable. Satisfactory performance without breakdowns while in use and readiness to perform at the design time is the criteria of equipment's reliability. The measure of a piece of equipment's reliability is the frequency at which failures occur in time. If there are no failures, the equipment is 100% reliable. Again, Bazovsky's claim may only be true if and only if the equipment problem will only be limited to breakdown. You see, failures and breakdowns are just a subset of a bigger problem called equipment losses. If I may explain this in the simplest way I can, just imagine yourself ordering a pizza in a pizza parlor. The standard cut of a pizza will be in eight slices. This means that failures and breakdowns are just one slice of the whole pizza pie. My point is that to consider the equipment reliable, we should not only be speaking in terms of failures and breakdowns but the entire equipment losses. A piece of equipment can experience no failure but the speed may be reduced or minor stoppages are rampant on the equipment. Breakdowns and failures are not the only cause of machine downtime.

We simply cannot declare equipment is reliable if it is suffering too many defects or if the change over time in the equipment is excessively long. Equipment losses are much broader in scope since equipment failures are just a part of the losses. Addressing all the possible losses on the equipment or asset will require different strategies and methodologies way beyond just improving the way we maintain the equipment. The key is to identify the losses the equipment is suffering and adopt the most feasible method to address these equipment losses. However, even if we addressed all these losses in the equipment, there are still other problems that can be generated by human error, facilities requirements, or even the environment itself. These problems are way beyond the equipment itself. When all these equipment losses and other problems on the equipment have been addressed and taken care of, then only I can declare the equipment is indeed reliable which means that the asset can be trusted to perform as it is intended to do. Let us not take for granted that if the equipment does not suffer from any equipment failures, then it is considered reliable. I have experienced many equipment and machines that seldom fail, but experienced many equipment losses such as defects, minor stoppages, and conversion. The word reliability means that the equipment can be trusted to deliver and perform its intended functions.

Let us discuss the different equipment losses that can affect its performance and reliability. Each of these equipment losses experienced on the equipment will require different methodologies indicated in figure 8.10. Addressing each of these losses will require different strategies as indicated in figure 8.11. A cross-functional Focused Improvement or Kobetsu-Kaizen team will be responsible for these losses except for breakdown losses.

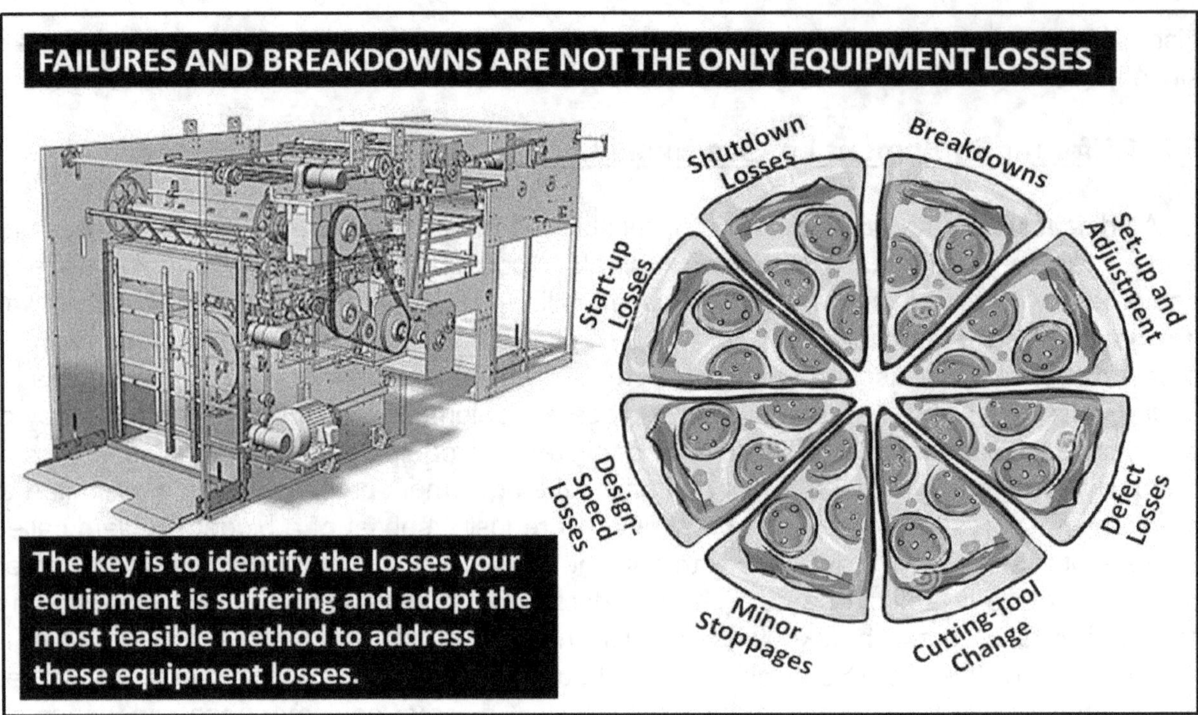

Figure 8.10: Failures and Breakdowns are Not the Only Equipment Losses

8 Equipment Losses	Strategies to Address this Losses	What to Measure
Breakdown Loss	• TPM Planned Maintenance and RCM	Availability, BDO, MTBF, Downtime, Reliability
Set-up, Changeover	• SMED (Single Minute Exchange of Dies)	MTTS. Downtime
Minor Stoppages	• Autonomous Maintenance	MTBA, MTTA
Start-up Losses	• Both RCM and Precision Maintenance	PM Effectiveness
Design Speed Loss	• PM Phase 2, LCM, IFCA or EEM	Efficiency, MTCE Cost
Quality Defects	• P-M Analysis, Common Analytical Problem Solving Tools	Yield, % Defects
Cutting-Tool Change	• Modification, Redesign	Downtime, MTCE Cost
Shutdown Loss	• Both RCM and Precision Maintenance	PM Effectiveness, BDO

Figure 8.11: How to Address the Different Equipment Losses

1. Breakdown Loss: Also called equipment failure. This is a type of equipment loss that can stop the equipment since its primary function failed. There are two types of breakdowns; first, we have the function reduction breakdown where a specific function failed, but the equipment can still operate. Let us say that you are driving your car, and the AC is not working; the car can still travel since it still satisfies its primary function, which is to travel from one point to another. Although there will be lesser comfort as the AC is in a failed state. These are breakdowns that account for the largest proportion of plant failures.

Although they are not recorded as a breakdown, there will be cases where the failure of a function reduction breakdown or secondary failure is far more dangerous than the failure of a primary function, or function loss breakdown. The second type of failure is a function-loss breakdown where something failed, and the equipment completely stopped and can no longer be used completely. These are losses in which production is stopped. Other industries termed this as unplanned downtime or breakdown maintenance. Breakdowns cause delivery delays, quality problems, longer cycle time, and an increase in the cost of doing maintenance. A single breakdown can create havoc throughout the plant. Analysis of the breakdown and failure can be attributed mostly to human problems, basic neglect of design problems, lack of operator skills as well as lack of repair skills that is why a basic understanding of the failure must be addressed thoroughly. Breakdowns are caused by many factors, and most of the time, slight deteriorations are overlooked, which contributes highly to equipment breakdowns.

Reducing Breakdown Losses: Breakdowns are caused by all sorts of factors but we usually notice only the big problems or sporadic failures and overlook the slight defects that also contribute to them. Obviously, the big problems deserve attention but slight defects deserve equal time and attention since big failures start from small things such as dirt, dust, contamination, and debris. Many breakdowns happen simply because minor things such as loose or missing bolts, abrasion, and contaminants are being ignored. When we start considering these small things and give them our utmost attention and priority, only then we can start to reduce these unplanned breakdowns in the equipment. One of the holistic methods to reduce equipment breakdown losses is the implementation of the four phases of Planned Maintenance, which is explained, in the next chapter. The goal of Planned Maintenance is two-fold. The first two Phases will attempt to improve the equipment by reducing unplanned breakdowns through restoration and addressing parts with inherent design weaknesses, while the next two Phases will aim to improve the system on how maintenance is carried out on the equipment to create a sustainable process. This is where Reliability-Centered Maintenance can be integrated into Phases 3 and 4 of Planned Maintenance. Equipment is designed to function under given operating conditions within certain tolerances. Vendors and OEMs often times develop a generic task on the condition that their equipment will be operated with the same parameters. What is important is to understand how your asset will be operated and what conditions will it be operating. Traditionally, when a breakdown occurs, maintenance will restore broken parts unknowingly that a possibility of secondary failure takes place as a result of the first failure since there is a possibility that other parts can be affected. Maintenance must also understand when it is time to conduct a Root Cause Failure Analysis investigation so that failures would not repeat themselves. The main objective of conducting root cause is for the industries to learn from the things that go wrong regardless of whether the consequences of the failure are big or small. Maintenance people must not only be capable of repairing but analyzing failures.

2. Set-Up Loss, Conversion, or Changeover: In several manufacturing industries, some machines are not dedicated and produce different types of products. Hence, when it is time to change from one product to another, it will require the time to remove dies and jigs for one product, clean up, prepare dies and jigs for the next product, reassemble the equipment, adjust the machine, perform trial runs and make further adjustments until the product of acceptable quality is now achieved from the equipment. It begins when the production of one product is completed and ends when standard quality is attained on the next product being processed.

Reducing The Time for Set-up and Conversion: Shigeo Shingo's (SMED) Single Minute Exchange of Dies deals with several techniques to reduce the set-up time and adjustments without reducing its accuracy. According to him, the standard set-up time in manufacturing must be around 10 minutes and below. The challenge is reducing the set-up and conversion time in manufacturing. According to Shigeo Shingo, he observed that set-up time would compose of the following activities:

• Preparation of materials, jigs, tools, and fittings will be at 20% of the set-up time.
• Removal and attachment of jigs, tools and dies will be at 20% of the setup time.
• Centering and dimensioning will be around 10% of the setup time.
• Trial Processing and adjustments will consume around 50% of the setup time.

Shigeo Shingo recommends that the first step in improving the set-up time is to distinguish activities that can be performed while the equipment is running from those activities that can be performed only when the equipment will be shut down. It is important to list what will be the activities that will be included in the external from the internal setup. External set-up is those activities that can be performed while the machine is still running while an internal set-up is those activities that can be performed only when the machine had been totally shut down and is now ready for conversion. The goal of set-up time is to minimize the time to perform it, and one of the techniques is to list down all the steps performed in doing the set-up and conversion. Once the lists on the external and internal set-ups are completed, try to find out some internal setup that can be done when the machine is still running on its last lot before conversion. One book I recommend to reduce the setup and conversion time will be SMED or Single Minute Exchange of Dies by Shigeo Shingo. Here are some basic tips to reduce set-up time on the equipment.

• Simplify clamping mechanism by using quick-fitting jigs.
• Adapt parallel operations; two people working together can perform a set-up faster.
• Optimize the number of workers and division of labor, especially on complicated setups.
• What type of preparations needs to be made in advance?
• What tools must be on hand?
• Are the jigs and tools to be installed in good condition?
• What type of workbench is needed?
• Where are the dies and jigs placed after removal?
• How will they be transported?
• What types of parts are necessary during conversion?
• How many people are needed to perform the set-up?

3. Idling and Minor Stoppages: As equipment becomes more complex and automated, more losses are attributed to idling and minor stoppages. Some companies term this as errors or assists. The Japanese call this Chokotei. First, let us define this type of loss. A minor stoppage occurs when a failure or an error in automatic handling, processing, or assembly of parts and workpieces, or a piece of equipment stopped due to the occurrence of quality-related abnormality. This type of loss happens mostly in automated processes in manufacturing industries. A minor stoppage is different from a breakdown. This is when the workpiece flow stops, the operator resets the machine, the operator reactivates the process, and the machine runs once again. There is no actual breakdown that took place. The problem with this type of loss is the number of occurrences or frequency is much greater than the actual breakdown, and most of the time, this type of error is left unrecorded since the time to fix it can be done by just resetting the start buttons. Most of these types of short stoppages are left unrecorded since the time spend to record and write down the error will be longer than the time to actually fix it. Idling and minor stoppages sometimes are mistaken for breakdowns, which must not be the case, and it must be completely separated from breakdowns even if the waiting time for the technician is long. This is especially important when calculating indices and metrics on maintenance that involved time such as MTBF, Availability, Utilization, and others.

Reducing Minor Stoppages: JIPM (Japan Institute of Plant Maintenance) experts believe that having equipment free from dirt and dust can reduce minor stoppages by 20 to 60%. More complicated minor stoppages that cannot be reduced through cleaning must be addressed by a cross-functional Focused Improvement team composed of a cross selection of people from different departments knowledgeable on the problem. Some examples of idling and minor stoppages include workpiece jamming, resetting of sensors, an error reading on the monitor of computerized equipment where the machine is stopped temporarily, the operator resets, and the machine starts running again. These problems occur more frequently in automated equipment. In addressing minor stoppages, it is important to observe what is happening. These can be done by carefully observing the equipment until a minor stoppage occurs and by conducting an MTBA Analysis. Since errors and assists are more frequent than failures and breakdowns in automated and highly complex equipment, and with the absence of software and CMMS system, an MTBA Snapshot can be taken from the equipment. Here are the steps to perform an MTBA Snapshot.

• Step 1: Prepare an MTBA Snapshot Form
• Step 2: Select a piece of equipment that have a high frequency of assists and errors
• Step 3: Sectionalize the equipment per station or sub-assembly and record all possible assists that
  can occur on each sub-assembly. Make sure to code each error as in figure 8.12.
• Step 4: Perform an MTBF Snapshot on the equipment. The minimum snapshot time should be 2
  hours. The longer the MTBA Snapshot, the better. Write the duration, frequency, and type of
  assists that occurred on the machine
• Step 5: Generate corrective actions and modifications. Correct actions should be SMART. Specific,
  Measurable, Attainable, Reasonable and Time-Bounded
• Step 6: Horizontal Replication of the MTBA improvement for similar equipment that possesses the
  same problem. Note that if there is similar equipment that does not possess the same problem, it is
  recommended not to fan out or replicate the modification for these stable machines.

Code	Minor Stoppages Errors	Code	Minor Stoppages Errors
PPM	Pick and Place Malfunction	CE	Conveyor Error
MP	Mispick	PF	Product Falling
ML	Misloading	SE	Software Errors
MNC	Magazine Need to Continue	SM	Sensor Malfunction
PJ	Product Jamming	CS	Clogged Sensor due to Dirt
EL	Elevator Jamming	CC	Clogged Chute
BE	Buffer Error	ICA	Insufficient Compressed Air

Figure 8.12: Sample Codes for Minor Stoppages

4. Design Speed Loss: This is a type of equipment loss on production, which is caused by the difference between the design or theoretical speed and the actual operating speed of the equipment. Lack of care at the design stage of the equipment may result in speed reduction. There are cases where the equipment is operated below its operating speed limit since quality defects and breakdowns are frequently encountered if the machine will run on its design speed. Equipment may be run at less than the design or ideal speed for a variety of reasons, which may contribute to mechanical problems, quality and defect problems, history of past problems, and sometimes not knowing what is the optimal speed of the equipment. On the other hand, deliberately increasing the operating speed can actually contribute to quality-related problems by revealing latent defects in the equipment. This type of loss will not experience any form of downtime but the performance efficiency of the equipment will be affected.

Improving the Design Speed Loss: When the equipment was newly commissioned in the plant, it is operating accordingly to its design speed. As time passed by and as equipment operates continuously, the speed of the equipment is reduced. This reduction in the equipment's operating speed refers to the difference between the actual operating speed and the design or ideal speed of the equipment. There are many reasons why this happens. Design speed may be reduced due to problems with quality or due to some mechanical problems in the past, resulting in short parts lifespan. Another reason is that increasing the speed of the equipment will contribute to quality problems. If this happens, maintenance usually will try to reduce the running speed of the machine; however, in most cases, production will increase the speed once more to deal with production losses only to provide more serious downtime on the equipment. Our goal here is to eliminate the gap between the actual speed and the design speed of the equipment.

• Level 1: Achieve Standard Operating Speed for each product
• Level 2: Increase Standard Operating Speed for each product
• Level 3: Achieve Design Speed
• Level 4: Surpass Design Speed

Maintenance people need to know what parts or spares will be affected when the speed of the equipment is increased. Study the part concerning its design, shape, the strength of materials, dimensions, and other factors. TPM Planned Maintenance Phase 2 on Lengthening Equipment Lifespan of parts with inherent design weaknesses and Focused

Improvement activities will initially help achieve Level 2 on achieving the standard operating speed until the design speed is achieved on the equipment. Further modification and redesign from both the Planned Maintenance and Focused Improvement pillar of TPM can further challenge the team to surpass the actual design speed of the equipment whenever necessary. This one will be a bit more difficult but still possible in certain cases.

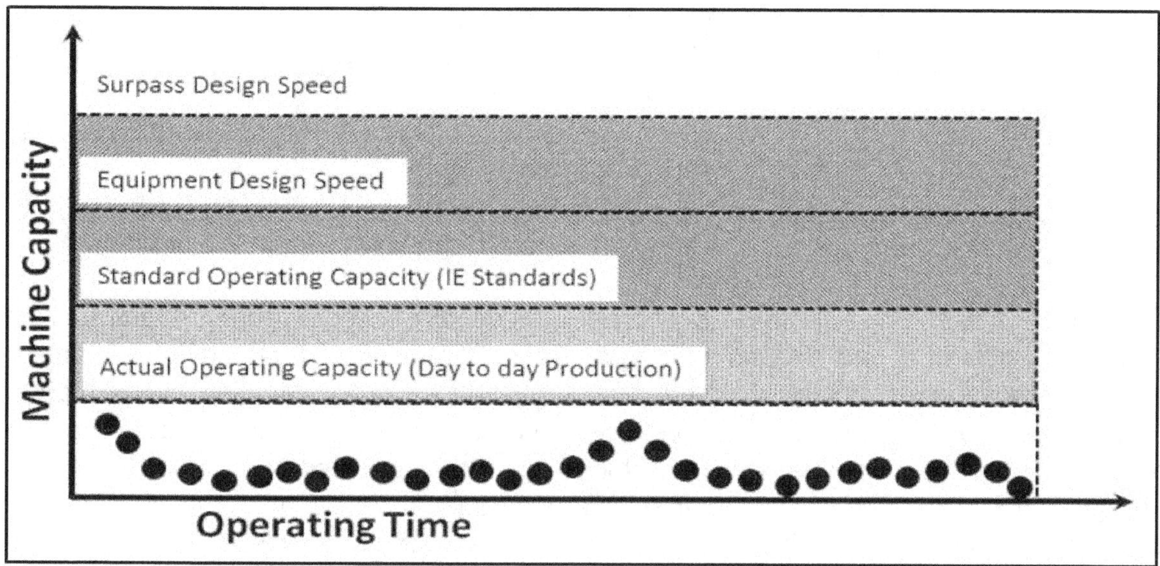

Figure 8.13: Surpassing the Design Speed

5. Tool and Cutting Blade Change: The time loss incurred in swapping any consumable tooling item when it has become worn out, ineffective, or severely damaged. This type of equipment loss is caused by changing the cutting blade due to premature breakage or caused by changing the cutting blade when the service life of the grinding stone, cutter, or bites has been reached. Normally a fixed number of times is set, but if the tool breaks before its lifespan, then we can investigate the cause of the breakage and modify it.

Improving Tool and Cutting Blade Change: Providing a Tool Algo can provide a precise interval for changing these cutting tools. Tool Monitoring Algo we used in the semiconductor end of Line process is used to dictate precisely when to change the dies and punches for a Trim and Form machine. A Tool Algo is an instrument that monitors and records the number of strokes stored on the main computer. This means that if the number of cycles for a die and punch is at 350,000 strokes when the punch reaches 350,000 strokes, it will automatically stop. The tool algo can also be programmed so that when the punch and die reach 320,000 strokes, it will provide a signal to the operator that the time to replace is near. This will allow the operator to prepare a work request to the maintenance that the end of life for the die and punch is closing in. Another way is to increase the hardness of the material can improve the lifespan of the material of the cutting tool, however, this may also cause a new problem since the harder the material, the more prone it is to a brittle fracture.

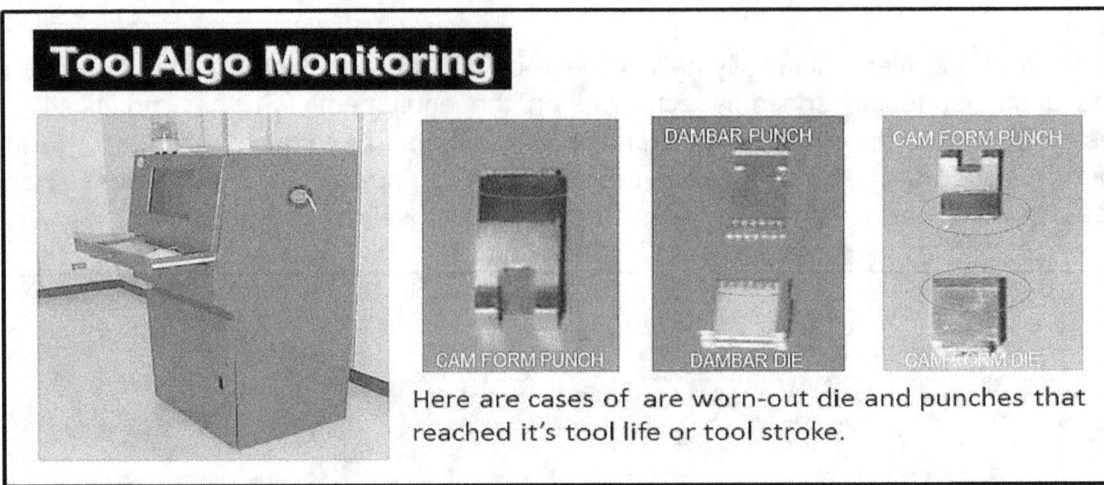

Figure 8.14: Tool Algo Monitoring

6. Start-Up Loss: Start-up loss is a type of loss that occurs during starting up the equipment. Start-up loss means that the material loss caused at the initial stages of production was caused during the period from the start-up of production to the stabilized production stage. Its frequency depends on several factors, such as unstable machining conditions, poor maintenance, and operator skills. Start-up losses can also occur after a poor set-up or conversion or after an intensive overhaul performed on the equipment due to Preventive Maintenance activities. It takes some time for the machine to stabilize. Start-up loss is difficult to identify. Their scope includes the stability of processing conditions, workers' skills, the loss incurred by test operations, the warming up period, and other factors. A thorough understanding of why this type of loss occurs can be addressed by implementing Reliability-Centered Maintenance. Other causes of start-up failures may also be due to infant mortality failure.

Reducing Start-Up Losses: Start-up losses can include warming up the process and infant mortality failures that are experienced right after the equipment has undergone major Preventive Maintenance. Removing intrusive maintenance and integrating Precision Maintenance into Preventive Maintenance can reduce human errors that contribute to infant mortality failures.

7. Shutdown Loss: This is the time when the equipment is shut down for maintenance. However, shut down related work generally affects the operating time of the equipment. Shutdown-related work must be regarded as a loss and reduction of shutdown losses, which can be done by reviewing the maintenance tasks performed on the equipment. Intrusive and unnecessary maintenance tasks and those tasks that do not address any failure mode should be removed from the Preventive Maintenance tasks. Implementing Reliability-Centered Maintenance can reduce the time spent on shutdown losses on the equipment due to the removal of unnecessary or intrusive maintenance. This type of loss is incurred by deliberately shutting down the equipment for a scheduled Preventive Maintenance shutdown due to routine maintenance, scheduled overhauls, and replacement, inspections, scheduled outages, or turnaround activities. Among the other equipment losses, shutdown loss is considered a Planned Downtime.

Shortening Shutdown Period: Reducing Shutdown during Scheduled Preventive Maintenance and scheduled outage means reviewing all the activities to be done on the equipment and reducing intrusive maintenance tasks. These intrusive tasks include all those activities that are deemed unnecessary or any duplicated tasks. The goal of any maintenance task is to address a particular failure mode. Another important factor to consider in shortening the shutdown period is by reviewing the intervals of performing such tasks ensuring that the tasks are done not too soon or too late.

8. Defect and Rework Loss: This is a type of loss that is caused when defects are found, and the product has to be reworked. In general, these defects are likely to be considered a waste that should be disposed of. Product defects are more difficult to address since the majority of these defects are chronic by nature. The problem can come from anywhere including human error, raw materials, the machine itself, the process, or even the environment. An Ishikawa or Fishbone Diagram may be adopted to determine the different possible causes and address them one by one. Included in this type of loss are products being shipped back due to customer complaints. Products are returned and reworked in the production area. Quality defects and rework losses are caused by malfunctioning production equipment. In general, sporadic defects are easily corrected, while chronic defects require a thorough investigation and innovative remedial action. The conditions surrounding and causing the defect must be determined and then effectively controlled. Tools such as P-M Analysis are suitable for dealing with chronic problems. It aims at reducing defects to zero; however, P-M Analysis must only be used after extensive use of conventional problem-solving tools had been used, and around 1 to 5% of the quality problems still exist. This is where P-M Analysis must be used ideally. Sporadic defects are easy to solve, but it is more difficult to solve chronic defects, and most of the causes of defects and rework losses are chronic in nature. Chronic defects occur due to complex causes and a combination of causes. This means that there is more than one cause. Addressing Chronic defects means identifying all the possible causes and correcting each of them respectively.

Reducing Defects and Rework Losses: One tool recommended to address chronic defects is P-M Analysis. P stands for both physical and phenomenon, while M represents the 4M and mechanism. Physical analysis physically analyzes chronic losses according to the inherent principles and natural laws that govern them. Its principle will determine in precise physical terms what happens when a machine breaks down or produces defects and how it happens. This methodology clarifies the mechanics of their occurrence and the conditions that must be controlled to prevent them. This tool analyzes chronic problems such as defects and failures according to the machine's operating principles. In adopting P-M Analysis, it is recommended to first implement a conventional improvement approach such as the use of the conventional analytical problem-solving tools, and when the defect rate is reduced to a minimum, then this is when to apply P-M Analysis to reduce the chronic defects to zero.

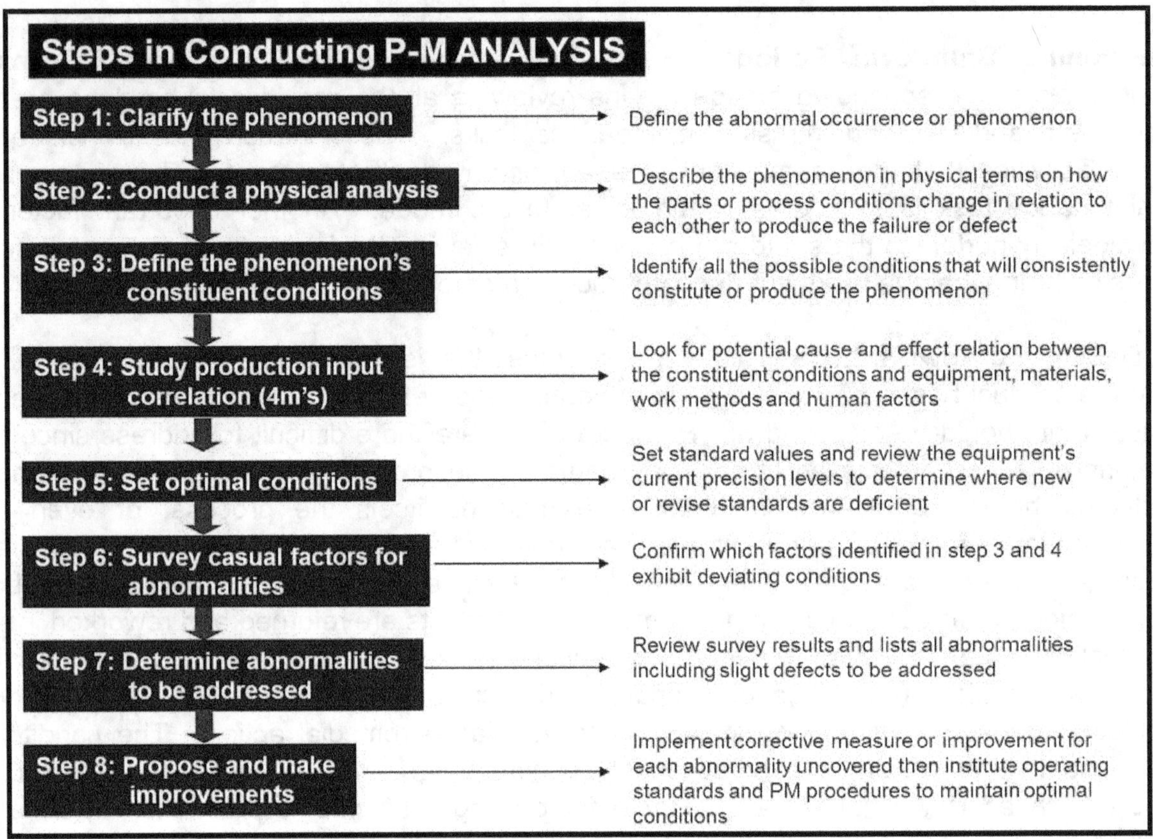

Figure 8.15: Steps in Conducting a P-M Analysis

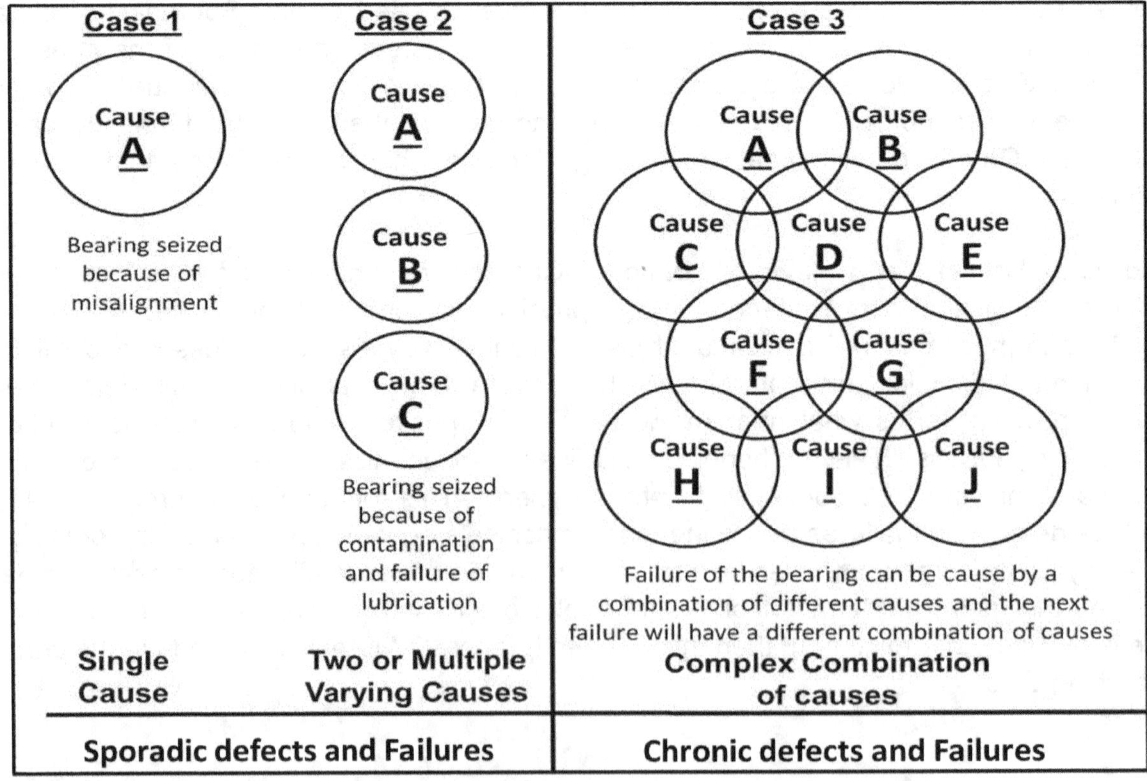

Figure 8.16: Sporadic and Chronic Failures and Defects

8.8: Overall Equipment Effectiveness Explained

These equipment losses discussed in the previous section can be measured through an indicator called OEE or Overall Equipment Effectiveness. OEE has been first introduced in the book of Seiichi Nakajima, An Introduction to TPM, Total Productive Maintenance. Every industry has a product or service to render, but more importantly, it is the customer that spells the difference between why a company exists and still remains in business. Today, industries adopt the law of the jungle and that is to survive or be left behind. Therefore, to survive in this tough business competition, we need our customers to stay and remain loyal, and to do this, the industry needs to satisfy the following requirements.

1) Those that can produce the lowest possible costs.
2) Those that can produce the highest quality product or service with the best customer satisfaction.
3) Those that can produce the fastest delivery of them all.
4) Those that can make decisions adapt and innovate to rapid changes in the market trend.

If any of these cannot be provided by the industry, then there will be repercussions, risks, and the possibility of the industry shutting down permanently. OEE is the primary measure of TPM. Although this is a consolidated KPI, which means that maintenance is not only the sole person responsible for improving OEE as it contains several equipment losses, and Planned Maintenance will only be focusing on breakdown losses, which are just one of the major equipment losses. It will identify the bottleneck equipment critical for the improvement of the overall equipment's productivity.

[10]OEE indicates the relative productivity of a piece of equipment compared to its theoretical performance. It is used to prioritize equipment to improve productivity. OEE contains three components, which include Availability, Performance Rate, and Quality Rate. For those implementing TPM (Total Productive Maintenance), OEE will be the primary measure of performance. It is a function of the complete manufacturing system, process line, or individual piece of equipment. It will measure the effective utilization of capital assets by expressing the impact of equipment-related losses, which includes the following:

• Machine downtime caused by breakdowns
• Time required for Set-Up, Conversion, and Adjustments
• Time lost due to inefficient start-up
• Time lost due to tooling
• Time lost due to Minor Stoppages
• Time lost due to Shutdown losses
• Production lost due to Design Speed Loss
• Production lost due to defective products
• Production lost due to rework and reprocessing

---

[10] Charles Robinson and Andrew Ginder, *Implementing TPM – The North American Experience,* (Productivity Press Inc, 1995), Page 125

OEE is composed of three components, Availability, Performance Rate, and Quality Rate. This means that there are three functions involved in this situation, Availability for maintenance, Performance Efficiency for Operations, and Quality Rate for the Quality function. This means that improving the machine's OEE cannot be done solely by just a single function like maintenance, which some books claim. Improving OEE is a consolidated effort of these major functions with the help of other functions as well. This means that if the MRO Storeroom cannot provide the parts for maintenance, then downtime caused by breakdown will be prolonged. If the raw materials supplied are of poor quality, then the number of defective products will increase which can contribute to a low yield. Except for breakdown losses, which are addressed by the TPM Planned Maintenance pillar, typically, a cross-functional team called Focused Improvement or Kobetsu Kaizen will be deployed to improve the OEE by addressing the losses experienced by the equipment.

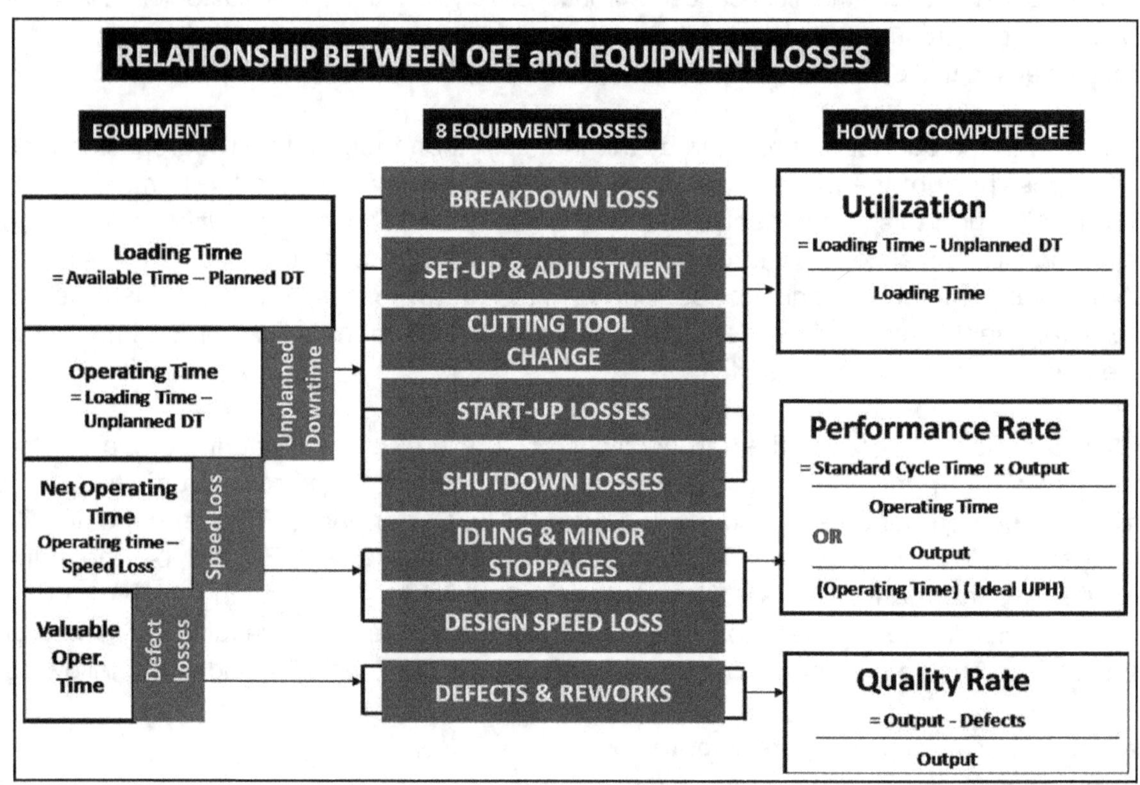

Figure 8.17: Relationship Between OEE and Equipment Losses

Since OEE is focused on the equipment's primary function, which is the output, it is also important to check the secondary functions of the equipment if they are working or not. Although JIPM states that to pass the TPM requirements, an 85% OEE should be achieved. It does not mean that we have achieved success. Remember that OEE is composed of three components and that these three components must be satisfied. Every piece of equipment consists of both primary and secondary functions that are likewise important and needs to be satisfied as well, and there are cases where the failure of secondary functions is as important or even more important than the failure of the primary function itself. This means that equipment can have a high OEE even if it has a severe oil leak or its protective devices malfunction. Hence, both primary and secondary functions of the equipment must be satisfied in this case. Except for breakdown losses, which will be handled by the Planned Maintenance, the best people to impact OEE will be the Focused Improvement pillar of TPM since they will be addressing the different losses on the equipment. When calculating a meaningful OEE, consider the following;

• Make a thorough classification of what constitutes a Planned and Unplanned Downtime
• Decide the formula to be used for availability and be consistent.
• Decide the ideal or rated UPH or cycle time to be used on each individual product especially if the machine is not dedicated
• Since OEE is focused on the primary function, which is the output, it is also important to check the secondary functions of the equipment
• When calculating the OEE of a line or process, just focus on the bottleneck equipment

8.9: The Concept of Focused Improvement

One of the pillars of TPM that address these equipment losses except for breakdown losses will be the Focused Improvement pillar. By definition, Focused Improvement can be considered as the various improvement activities used to accomplish maximum efficiency of individual facilities, equipment, and manufacturing processes as well as the entire plant by thoroughly eliminating losses and improving its performance. *From JIPM Materials.* This includes all activities that maximize the Overall Equipment Effectiveness of equipment processes and plants through uncompromising elimination of improvement losses and the improvement of performance. *From Tokutaro Suzuki*

The Focused Improvement team comprises a cross-selection of individuals who are both knowledgeable on the losses to be improved which can consist of anyone in the entire organization. The team can include maintenance, engineering, process, operations, quality, or anyone who is well familiar with the problem. The goal of the Focused Improvement team is to maximize the Overall Equipment Effectiveness (OEE) by identifying the losses and challenging the team to reduce these eight major equipment losses encountered in their equipment. Focused Improvement teams are like an elite force or special ops in the military that have a specific mission, which is to eliminate equipment losses in their equipment and assets. They can be said as the cream of the crop. These teams must be heavily equipped and trained with several problem-solving and analytical tools to analyze the problem and how to address these equipment losses. Initially, the team will observe their equipment regarding the losses to be addressed. Understand fully the significance of these losses and the rationale for improving equipment's OEE. Gather data on failures, defects, downtime, and other losses and plot them over time. The Focused Improvement team needs to be familiar with the different analytical problem-solving tools for analyzing and reducing these losses on their machines and equipment.

8.9.1: Purpose of this Focused Improvement Guidelines

The purpose of conducting Focused Improvement on industries is to identify the different losses the equipment is suffering and to organize a cross-functional selection of people that will address these losses. Focused Improvement (Japanese termed this as Kobetsu-Kaizen) will include all activities that maximize the Overall Equipment Effectiveness (OEE) of the equipment through the reduction or elimination of these losses and performance. Not all losses will be experienced by every type of industry. The purpose of organizing a Focused Improvement team is to address any of the 16 big losses experienced by the industry on their equipment and assets.

8.9.2: Scope of this Focused Improvement Guidelines

The scope of conducting Focused Improvement activities will apply to critical equipment, machines, and assets in the plant. New equipment and assets will not be covered initially in the Focused Improvement activities as the warranty of these equipment provided by the OEM may be affected due to some redesign and modification changes that will be done by the Focused Improvement teams.

8.9.3: Objective of this Focused Improvement Guidelines

The objective of conducting a Focused Improvement or Kobetsu-kaizen is to assemble a cross-selection of people from the different functions of the organization such as maintenance, quality, reliability, safety, process, operations, and production as well as other departments knowledgeable on the problem or losses identified which will be addressed by the team. The goal of the Focused Improvement team is to eliminate or reduce these equipment losses. Note that breakdown loss will be excluded from the Focused Improvement activities since this will be taken care of by the Planned Maintenance teams. For equipment improvements, another objective is to create a loop to EEM/IFCA (Early Equipment Management or Initial Flow Control Activities) so that improvements can be prioritized according to their importance. The reason for this is to create a loop so that when the industry decides to purchase additional repeat order equipment in the future, these improvements and modifications will be discussed with the vendors and OEM to be included in the new equipment purchase to avoid similar problems that were previously experienced in their existing equipment.

8.9.4: How to Conduct Focused Improvement

Here is a detailed step-by-step guide for conducting Focused Improvement activities to address equipment, manpower, and other losses.

1. Form the Focused Improvement Team: Membership of the Focused Improvement team will compose of a cross-selection of people from different functions of the organization knowledgeable about the problem to be addressed. Minimum of 4 but not to exceed 8 members will compose the Focused Improvement team. Determine how frequently the team will meet until its completion. The Focused Improvement team should be knowledgeable on at least a minimum of three analytical problem-solving techniques. Recommended analytical problems solving tools for Focused Improvement will include Root Cause Failure Analysis, FMEA/FMECA, Ishikawa or Fishbone Diagram, P-M Analysis for chronic problems, and Pareto Analysis, Poka-Yoke, process mapping, and the seven QC tools.   The type of problem-solving or analytical tool to be used by the Focused Improvement team will depend on the type of problem being addressed.

2. Decide on the Improvement Topic: The improvement topic to be addressed by the Focused Improvement team should be focused on the reduction or elimination of any of the 16 big losses currently existing in their equipment, machine, and assets.   Register the improvement topic with the TPM Office.

3. Understand the Situation: The Focused Improvement team should analyze and understand the current situation and identify the major losses and bottlenecks in the overall process.   Losses may either be for equipment, manpower, or other losses.   Provide initial measures and KPIs as a baseline to set up targets.   The Focused Improvement team should backtrack equipment's initial KPI and measurements to track to determine the success or failure of their effort.   One of the measurements they will be tracking will be the OEE or Overall Equipment Effectiveness for addressing equipment losses.

4. Expose and Eliminate Abnormalities: For equipment losses, construct a picture of the optimal conditions of the equipment or process.   Before applying any basic analytical techniques, the team should address all minor flaws, deviations, abnormalities, irregularities, and defects to ensure that basic equipment condition is well in place in the equipment.

5. Analyze the Causes of the Problem: Focused Improvement team can use some basic analytical problem-solving tools to analyze the possible or likely causes of the problem depending on the type of losses to be addressed.   There will also be cases where each loss will require two or more analytical problem-solving tools.

6. Plan for the Improvement: During the development of proposals for improvement, formulate several alternatives.   Never dismiss any ideas at this stage.   For best results, do not limit participation to one or two members of the Focused Improvement team rather all members should participate in this initiative.   If a budget is required, the Focused Improvement team needs to prepare a feasibility or economic study and calculate the return on investment (ROI) to have the budget approved by their top management.   The Focused Improvement team needs to guard carefully those improvements that may result in fresh problems with the equipment.   For equipment-related losses, the team will mostly result to redesign, or modification.   Also note that if a spare or part will be involved in the redesign or modification, the Focused Improvement team needs to check the actual inventory of the original parts in the storeroom and make an economic decision on whether or not to consume these parts first or proceed with the modification or redesign.   If the team decides to proceed with the redesign or modification, then these original parts will become obsolete and the storeroom people and purchasing people should be informed so that these parts will be removed from the storeroom.

7. Implement the Improvement and Evaluate the Results: Once the plan had been completed, create a prototype and implement the improvement.   Evaluate the results of the improvement on the equipment.   If the improvement is successful, it will be replicated to other equipment with the same operating context or problem. It is not recommended to replicate or fan out the improvement to similar equipment, which does not possess the same problem. Check whether targets have been achieved, if not, the team will again proceed to Step 5 and analyze the causes once again.

8. Monitor the Results and KPI: KPI should be based on PQCDSM. (Production, Quality, Costs, Delivery, Safety, and Morale) If equipment losses are being addressed, the Focused Improvement team should monitor the OEE of the equipment (Overall Equipment Effectiveness). If the indices and KPI improves, the Focused Improvement team needs to develop new procedures or engineering changes whenever necessary and disseminate this information to concerned areas and people in the plant.

9. Sustain the Improvement: Implement standardization and measures needed for preventing the recurrence of the problem. Draw up control standards to sustain results. Formulate work standards or routine maintenance tasks to sustain these improvements.

10. Horizontal Replication: Fan-out or horizontally replicate successful improvements generated by the Focused Improvement team to another process, equipment, or system with the same operating context or problems. Similar equipment which does not possess any problems whatsoever should be excluded from the horizontal replication plan.

11. Create a Loop Back to Initial Flow Control Activities (IFCA): Once the improvement had been concluded, create a loop back to IFCA (Initial Flow Control Activities) or EEM (Early Equipment Management) for the possibility of purchasing additional equipment in the future so that the improvement will already be included before purchasing the equipment with the vendor. The issue of the patent should be discussed with the OEM or Vendor.

12. Team Celebration: If the team is successful in the reduction, mitigation, or elimination of the losses identified in their Focused Improvement activities, a simple celebration will commence and the team disbands and goes back to their original duties in the plant.

8.10: Industries Problems Can Go Beyond Equipment Losses

Although the industry's problems may not necessarily be isolated alone to equipment losses. Total Productive Maintenance or TPM identifies the 16 big losses, which also include manpower and facility losses. There are losses still beyond what has been covered which may include the environment and others that may be beyond the scope of this book. This means if the production environment is too humid or corrosive, then it can have an impact on the equipment especially if electronic boards and parts are present. We also have to consider the force majeure or Acts of God beyond what we can imagine.

Usually, these 16 big losses TPM is mentioning are those losses that happen in the industry and can be controlled by the people themselves. However, some problems can be generated outside the industry that the customer or clients experienced themselves. This is just to oversee how complicated things can be in industries. For example, if a product being manufactured provide some health issues to its clients, then that client has the right to sue that industry. What if you are drinking your favorite canned soft drink and suddenly it tastes different, and out of your curiosity you open up the canned soft drink and found two dead cockroaches inside, what are you going to do? What if a certain pharmaceutical industry's products become contaminated which caused the death of some of its clients?

The thing is there are also other problems industries may face which cannot be addressed by any means just like during the recent pandemic where the government thereby ordered industries to close down due to the lockdown to prevent the spread of the virus. As of this writing, a new virus called **Monkeypox** is now emerging and is now present in 12 countries and may provide the same outcome. I hope I am wrong on this. All of these problems can compromise the existence of any business and must also be addressed independently. However, the scope of this book once again will just be limited to the subject of equipment losses and how to deal with them respectively. Improvement in equipment performance and reducing breakdowns can be done by addressing the basic equipment condition.

Figure 8.18: The Mission of Focused Improvement; Destruct and Destroy the Losses

Chapter 9

Addressing Equipment's Design Weaknesses

> *Equipment is not designed perfectly by the OEM. If God created these machines, then they will be perfect. There will always be flaws that need to be understood, and addressed. The OEM may have designed the equipment, but the people in the plant who operate and maintain their equipment will be the ones who will experience these flaws and not the OEM, which means that they will be responsible in addressing them.*

9.1: Understanding the Design Weaknesses of the Equipment

One of the important facets of doing maintenance is about improving and the possibility of extending the life cycle of their equipment and assets. Every equipment has a defined design weakness. There is no such thing as perfect equipment by design. Perhaps if God made the equipment, then I believe that it will be perfect, but that is not the case. What is important is that maintenance must identify these parts in the equipment with a short inherent lifespan. These parts are those that frequently fail in the equipment. Once these parts are identified, maintenance must challenge themselves to increase the lifespan of these parts through improvement, redesign, or modification.

A cross selection of technical people can study the strength of materials, dimensions, and measurements of the original part, then provide a modified prototype model and evaluate it in the equipment. If the lifespan of the modified part has exceeded the original part, then this new design will replace the original part in the equipment. Another important reminder for people involved in modification or redesign is for them to communicate the modification with their storekeeper so that replenishment of the old part under modification should discontinue, especially if the part or item is held in stock inside the storeroom. Remember that once a part had been modified, the original part becomes obsolete. Feasibility and economic study need to be conducted on whether the original part with a short lifespan will still be consumed or will be replaced immediately by the new design. If the decision is to discontinue, then these original parts inside the storeroom will become obsolete. Most modifications and redesign performed on maintenance are geared toward increasing the lifespan of these parts. Another important factor to consider is to change the interval of replacing them in the Preventive Maintenance lists.

9.2: Standard Four Phases of Planned Maintenance

By definition, Planned Maintenance is the deliberate activity of building and continuously improving such a maintenance system. This definition was given by Tokutaro Suzuki in his book TPM in Process Industries. It can also be defined as the maintenance activities performed on a pre-determined schedule of activities. This definition was given by Charles Robinson and Andrew Ginder in their book Implementing TPM, The North American Experience. It is also one of the major pillars of TPM that aims to optimize equipment reliability by achieving a Predictive and Proactive stage in maintenance. In Japanese, Planned Maintenance means Keikaku-Hozen. The objective of Planned Maintenance is to take care of the equipment's total life cycle at the most reasonable cost while improving the skills of the maintenance people as well. Although there are many versions of how industries may adopt the Planned Maintenance pillar, whatever phases or steps are adopted to comply with JIPM consultants or TPM standards, will end up in doing the generic four basic phases of Planned Maintenance. Reliability-Centered Maintenance (RCM) will fit perfectly in the higher Phases of Planned Maintenance.

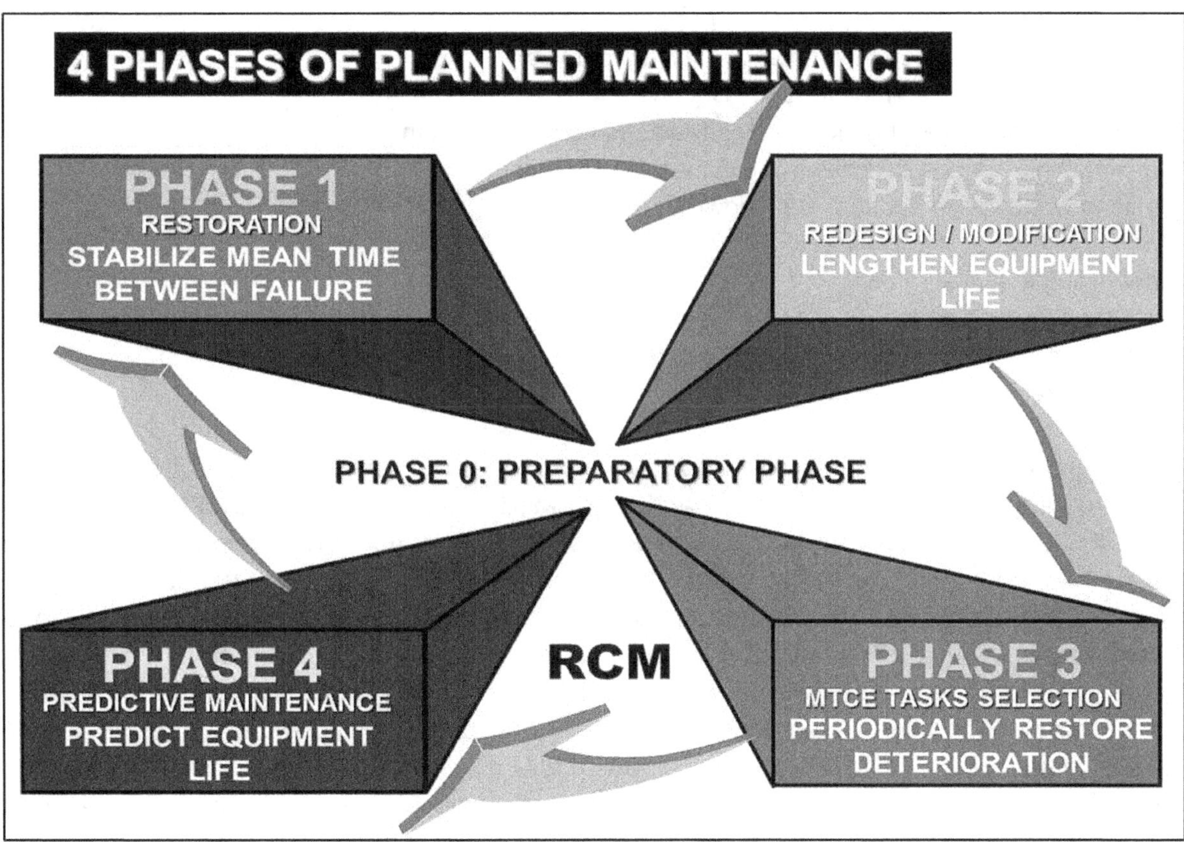

Figure 9.1: 4 Phases of Planned Maintenance

Phase 0: Preparatory Phase: Planned Maintenance will start through a phase called the preparatory phase before moving to the first phase of Planned Maintenance, which is all about equipment restoration. To begin the journey on Planned Maintenance, the first step will be for the maintenance people to undergo some basic training on Planned Maintenance.

If the plant is active in TPM, a Planned Maintenance facilitator will be responsible for conducting this training, or other third-party consultants can be hired to deliver this kind of training. Planned Maintenance Phase 0 details the requirements needed to set up an effective Planned Maintenance structure in the plant. These are the basic requirements needed for a strong foundation. The preparatory phase usually takes from 3 to 6 months or more depending on the size, the number of equipment, assets, systems, sub-systems, components, and manpower in the plant. It is important not to ignore this phase; otherwise, building a Planned Maintenance system without establishing the basic foundation is like building a house on sand. It is best for an industry starting on Planned Maintenance either to hire people internally or externally to act as a full-time Planned Maintenance facilitator. This person will be responsible for developing the nitty-gritty details of implementing this pillar. This person will act as the ambassador to the Planned Maintenance teams. I prefer hiring since there are actually a few people who can give consultancy in this regard. The important thing is that maintenance people must own Planned Maintenance, and it should not be treated as a separate program for the maintenance function. The main activities of Planned Maintenance Phase 0 include;

• Establishing the Planned Maintenance Vision and Mission
• Organizing the Planned Maintenance Council
• Developing and Projecting the Planned Maintenance Master Plan
• Defining the roles and objectives of Planned Maintenance
• Defining what will constitute a breakdown or not
• Preparing the machine inventory
• Conducting machine ranking
• Pilot machine selection
• Getting the initial MTBF/breakdown occurrence of the team's pilot machine
• Establish training needs for the PM team

Misnomer: Some consultants and books will insist that all breakdowns can be zeroed out, which is quite unlikely. Breakdowns can be grouped into both Planned and Unplanned and the reduction of breakdowns is actually focused on Unplanned Breakdowns. I was asked many times in my class, are zero breakdowns really possible? My response is if I sell you a drug and I tell you that if you take this medication once a day, then you will be immortal. Would you believe me? In addition, the student will reply, of course not. Then that's my answer. Then a follow-up question will be, but TPM (Total Productive Maintenance) says it can be done. When we experience no failures, and zero breakdowns, it does not mean nothing really fails. First, if we replace a part before it fails because of wear-out mode, then this will not be declared a breakdown, but technically, it failed. Second, we are not eliminating the breakdown, but we are just prolonging the equipment's MTBF. Third, if zero breakdowns are really possible, then we cannot proceed with the higher Phases of Planned Maintenance, as we need a failure to happen to identify the parts with inherent design weaknesses as those parts are candidates for improvement and modification. I can vividly recall our previous JIPM Consultant saying you need to have a breakdown before proceeding to Phase 2. Fourth, what TPM is simply claiming is having a Zero Unplanned

Breakdown for a temporary period. If zero failures and breakdowns are really possible, then you and I will be jobless.

Phase 1: Establishing Basic Equipment Condition through Restoration: If both Autonomous Maintenance and Planned Maintenance begin to accept the concept that equipment is always a shared responsibility for both of them. These two TPM pillars will address abnormalities and restore equipment back to its original basic condition. As Autonomous Maintenance activities focused more on the exterior part of the equipment, the Planned Maintenance pillar addresses the interior parts of the equipment. The goal of establishing Basic Equipment Conditions hand in hand with their equipment is to transition from a stage where accelerated deterioration is rampant to a stage of natural deterioration for parts that are age-related in nature where they tend to reach their natural lifespan. The objective of this phase of Planned Maintenance is to reduce the number of unplanned breakdowns experienced on the equipment; hence, when Phase 1 activities are done on Planned Maintenance, we are dramatically improving the equipment in terms of reduction in unplanned breakdowns.

Phase 2: Addressing Design Weaknesses through Improvement: Once accelerated deterioration has been reduced, there are parts inside the equipment that will wear out and suffer from natural deterioration, which means that the part has reached its natural lifespan. There are spare parts that will deteriorate naturally. The team will expose and analyze several parts of the equipment with an inherently short lifespan and correct design weaknesses by improving its parts dimension, the strength of materials, construction, dimensions, measurements, and so on. A Maintenance Prevention design is usually used for this activity and later on looped back to the Initial Flow Control Activity Pillar or Early Equipment Management so that when the company decides to purchase additional equipment in the future, these improvements are discussed with the designers to be included in the new equipment purchase. A cycle must be established where MP Design improvements must be looped back to IFCA or Initial Flow Control Activities of TPM. Correcting design weaknesses can prevent major breakdowns from recurring unexpectedly. Teams are trained on several analytical problem tools such as P-M Analysis for a more detailed approach to dealing with chronic failures and breakdowns. As most of these breakdowns are caused by human errors. Both operators and maintenance must upgrade their skills to reduce human errors. The application of Poka-Yoke solution may solve human errors but not necessarily improve the level of understanding of the mistake, or error caused by the person involved. Again, for Phase 2, maintenance should be trained in problem-solving techniques and analytical tools since we are improving the equipment by addressing these parts with design weaknesses.

Phase 3: Periodically Restore Deterioration with the Aid of a Decision Diagram: Once the parts with design weaknesses have been addressed in Phase 2, their lifespan will increase; hence, we need to have a thorough review of how we are going to maintain it. This is accomplished with the aid of a maintenance algorithm. RCM refers to this as the decision diagram. A thorough study of the different maintenance tasks must be done to establish the correct maintenance tasks and intervals. Since the basics had already been established, we can perform this phase through the use of Reliability-Centered Maintenance

(RCM). Not all tasks need to undergo Preventive Maintenance. This is accomplished by having a thorough understanding of the failure modes that affect the asset. Each part will have its own unique failure characteristic pattern. The key in this activity is to understand the six failure patterns so that we can derive the correct maintenance tasks for each failure mode through the aid of an algorithm or decision diagram. The decision diagram will aid the team in making a decision on the best maintenance tasks to carry out to address every single possible failure mode that can affect the equipment. In this activity, we are no longer interested in reducing the number of breakdowns, but rather we are now focusing on how to improve the way we maintain our assets and equipment. Which tasks should undergo Preventive, Predictive, Run to Fail, Failure Finding Tasks, and Modification? The decision as to which tasks to undertake will depend on the consequences of failure itself.

Phase 4: Predict Equipment Life: In Phase 4, we now introduce the concept of Predictive Maintenance or Condition-Based monitoring techniques, although this is similar to using the human senses in a much higher perspective with the aid of sophisticated precision instruments. Several parts of the equipment can be predicted through the use of these specialized diagnostic instruments. This is done by checking the condition of the equipment rather than basing maintenance on schedule. The key to using this technique is to determine the P-F interval. P stands for potential failure and F for functional failure. If failure shows signs or symptoms that they are on the verge of failing, then these parts will be a good candidate for using Predictive Maintenance. For example, a bearing may produce noise, an increase in temperature, or an increase in vibration. These are symptoms or potential failures indicating that a failure is likely to occur or is on the verge of occurring. Therefore, maintenance can schedule an inspection of the equipment before the actual failure can take place. One of the advantages of using Predictive Maintenance is that we only replace them if they are on the verge of failing. If nothing is wrong, then the decision is to allow the equipment to continue running.

9.3: Details of Implementing Phase 2 of Planned Maintenance

One of the important facets of doing maintenance is not only about sustaining the equipment but it is also about improving the life cycle of their equipment and assets. Every equipment has a defined design weakness. There is no such thing as perfect equipment by design. What is important is that maintenance must identify these parts in the equipment with short inherent lifespan. Once these parts are identified, maintenance will challenge themselves to increase the lifespan of these parts through improvement, redesign, or modification. A cross selection of technical people can study the strength of materials, dimensions, and measurements, of the original part, provide a modified prototype model and evaluate it in the equipment. If the lifespan of the modified part exceeds, then this new design will replace the original part in the equipment.

Lengthening the lifespan of these spares, and items will be the key to extending the life cycle of the equipment. History records, spare withdrawals, and breakdown reports can be used to identify these parts and items with a short inherent lifespan. Most modifications and redesign performed on Planned Maintenance are aimed at increasing the lifespan of these parts. If the decision is to proceed with the modification, purchasing people should be notified to stop the reordering of the existing part. Since the interval had been increased, the replenishment period should also be considered which entirely depends on the lead time for the new modified part to arrive. If the existing design has a lifespan of one month and the new design has a lifespan of six months, then the replenishment time for purchasing should be adjusted. The Preventive Maintenance interval for replacement should also be changed as the lifespan increase.

In phase 1, a parallel activity must be conducted with Autonomous Maintenance on addressing the basic equipment condition in which the main goal is to bring the equipment back to its original basic condition. In Phase 2, addressing equipment design weaknesses are performed parallel to improving the analytical skills of maintenance in addressing recurring failures. The objective of this Phase is to challenge the team to further improve the lifespan of parts that frequently fails, and address failures attributed to human errors both from operation and maintenance. This time from accelerated deterioration to natural deterioration and finally the challenge is to extend the part's lifespan.

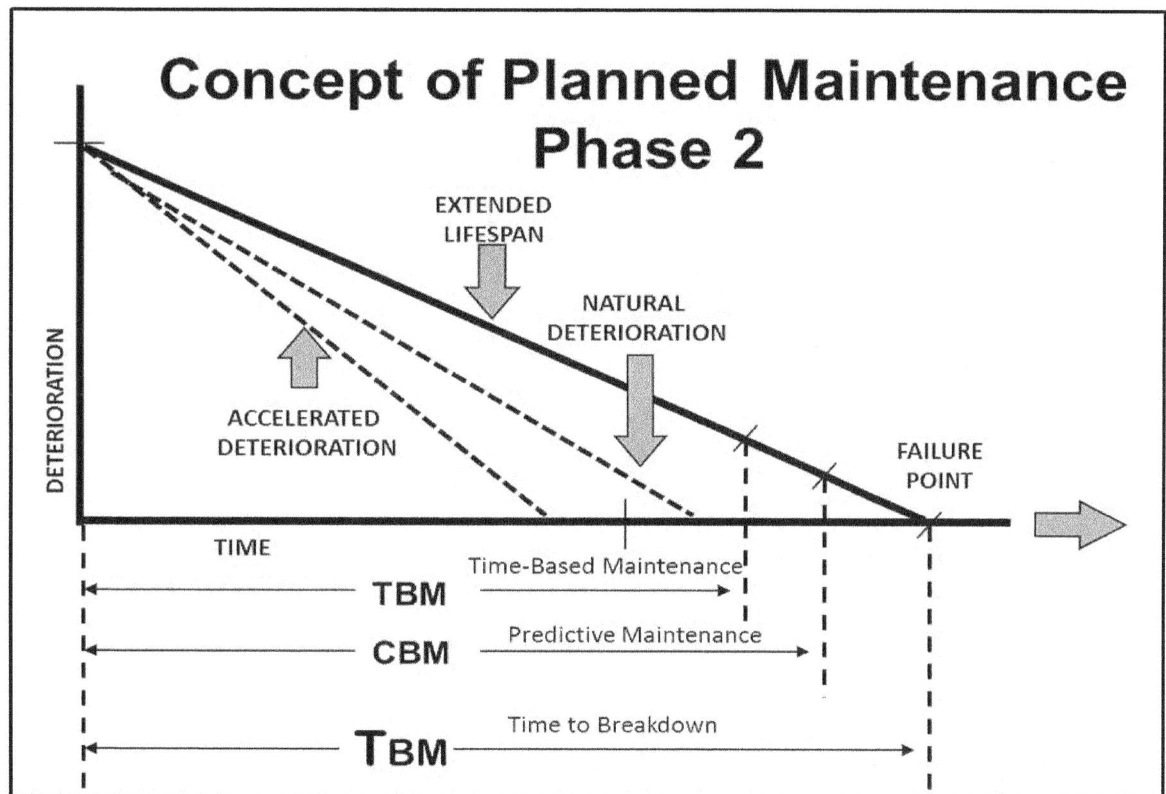

Figure 9.2: Concept of Planned Maintenance Phase 2

Candidates for Addressing Design Weaknesses will Include;
• Parts with an inherently short lifespan
• Spare parts that frequently fail
• Fast-moving spare parts
• Parts that do not reach their natural lifespan

- Parts that adhere to wear and tear
- Parts with material-related problems
- Parts with Safety or Environmental Consequences
- Parts that frequently caused minor stoppages
- Parts or items where the lifespan can be increased

Below are the details on how Planned Maintenance Phase 2 is done, which aims to identify equipment design weaknesses by lengthening its lifespan. Figure 9.3 will be the Phase 2 Roadmap of Activities that will be used by the Planned Maintenance teams.

Planned Maintenance Phase 2 Roadmap of Activities				Place Company Logo	QUALITY — NO EXCUSE
For Pilot Equipment					
AREA: _______________		EQUIPMENT : _______________		Teamname : _______________	
LEADER:					
OBJECTIVE : Lengthen equipment life by correcting design weaknesses and address recurring breakdowns permanently					
NO.	ACTIVITY DESCRIPTION	Code	FORMS USED	WORKWEEK	
1	Update MTBF / BDO data on Phase 1 of Planned Maintenance and register the team for Phase 2	S1-01 / S1-01a	MTBF / BDO Analysis Chart / PM Registration Form	Plan / Actual	
2	Cover the following training's : - Planned Maintenance Phase 2 - Root Cause Failure Analysis and FMEA	S2-02a	Training (8 hrs) / Training Monitoring & Skills Assessm	Plan / Actual	
3	Check recurrence of Breakdown and identify the sub-assembly and corresponding part	S1-01	MTBF / BDO Analysis Chart	Plan / Actual	
4	Draw and identify spare parts that frequently fail or have a short lifespan on the equipment	S4-05	Equipment Weaknesses lists	Plan / Actual	
5	Prioritize improvement and rank lists according to its severity and consequences	S4-07	Component / Spares Priority Lists	Plan / Actual	
6	Analyze rootcause of design weakness, detail the analysis per part	S4-08	P-M Analysis Form	Plan / Actual	
7	Recommend for Improvement in Design (Change in material, shape, dimensions etc.,)	S4-09	Design Improvement Recommendation / MP Design Form	Plan / Actual	
8	Proceed with the improved design and validate (Prepare prototype whenever necessary)	S4-09	Design Improvement Recommendation / MP Design Form	Plan / Actual	
9	Proceed with the other Identified Equipment Weaknesses Lists and perform activities 6 - 8	-------		Plan / Actual	
10	Make necessary changes in procedure, PM interval, and prepare OPL or GTF for operators and maintenance	S2-10	Good to Find or OPL	Plan / Actual	
11	Monitor cost impact and usage of part / component with improved design and check problems if any	S4- 13	Modified Spares Parts Cost Comparison	Plan / Actual	
12	Summarize Improvement Activity and update MTBF/BDO and other indices	S4-14	Improvement Activity Summary / MTBF/BDO Analysis Chart	Plan / Actual	
13	Update PM activity board and perform self-audit on on Phase 2 of Planned Maintenance and proceed to Audit	----	Step 4 Audit Guidelines	Plan / Actual	
14	Identify Fan-Out to similar machine type (Note : Identify only the machine, do not replicate yet)	S4-16f	Horizontal Replication Form	Plan / Actual	

LEGEND : P = PLAN   A = ACTUAL   Note : Team must have Minutes of Meeting for the activities and Attendance Check for the members   Rev. 06, 04/20/22

Figure 9.3: Planned Maintenance Phase 2 Roadmap

1. Update Current Indices and KPI in Phase 1: Once the team passed the audit process on Phase 1 of Planned Maintenance, the team is now ready to proceed with Phase 2. The PM team will update its indices such as MTBF and breakdown occurrences, which should dramatically be reduced, as they restore the equipment in Phase 1.

2. Team Registration: The team will register again for Phase 2 and will be provided a detailed roadmap as in figure 9.3 on how to start and end this activity. The same team that goes through the process of Phase 1 restoration must be the same team that will undergo the activities in Phase 2. Usually, a sticker will be placed on the equipment, which indicates that the machine has already been certified on Phase 1.

3. Undergo the Training on Planned Maintenance on Phase 2: The team on Planned Maintenance should schedule themselves to attend the training on Phase 2 on correcting design weaknesses by lengthening its lifespan. Another important training for the team to undergo in this phase is understanding at least three or more analytical and problem-solving tools such as FMEA/FMECA, P-M Analysis, Pareto Analysis, Fish-Bone Diagram, Poka-Yoke, Fault Tree Analysis, and Root Cause Failure Analysis. The team should likewise understand the concept of Life Cycle Costing. Since Phase 2 of Planned Maintenance is about modification, the team needs to understand how to analyze failures before they can try to redesign or modify the part itself. It should be made clear to the team that modifications and redesign to be done on the equipment should not generate new problems.

4. Check recurring failures and identify parts that fail frequently. The team should analyze the equipment by identifying parts that fail frequently or those with a short inherent lifespan. The goal of this phase of Planned Maintenance will be to identify these parts and challenge the team to increase the lifespan of these parts since no equipment is designed perfectly. There are always parts with weaknesses in design that will fail or break randomly or parts that wear out easily on the equipment. Check history records, spare withdrawals, and other dossiers. What is important for the team is to generate a list of these items or parts that will be subject to modification.

5. Draw the part and identify its weakness in the design: Once these parts have been identified, the team should draw or take photos of these parts and identify the weak points in the design and analyze why these parts fail easily on the equipment. These documents will also be used as a teaching aid to Autonomous Maintenance and operators when they reach the higher steps of their Autonomous Maintenance activities. Operators should also be aware of these parts since they will always be the first to witness the failure before the maintenance. They will always be the first line of defense against any failure that can happen on the equipment and, more importantly, if some of these failures can be detected by the operator's human senses. What is important at this stage is to determine why and how these parts fail, and if they provide symptoms that they are on the verge of failing.

6. Prioritize and rank the list on which part to modify first: List all the weaknesses identified accordingly and prioritize based on the severity and consequences of the failure. Once these parts had been identified and analyzed, the team should prioritize which parts they will modify or redesign first. An FMEA or FMECA can be used at this point to determine the criticality of the part so they can prioritize which of these parts will be modified first on the equipment. Likewise, it is important at this point that the team communicates with both purchasing and the storeroom people to control the ordering of these parts to be modified, especially if they are being stocked in the storeroom. Remember that if these parts are stored in large quantities, a decision has to be made on what to do with these parts because once the modification is fabricated and placed on the equipment, these parts become obsolete parts. Hence, their inventory should be known. Figure 9.4 can be used in prioritizing the improvements and modifications identified.

Criteria : Prioritization of Corrective Actions Implementation	Ranking
Safety/environmental problem consequences or non-compliance to Government Regulation. Unavailable necessary resources or unacceptable cost and time consumption. Having a zero chance of success and 100% probability of undesirable impact.	10
Very remote availability of necessary resources or almost unacceptable cost and time consumption. Almost zero chance of success or almost 100% probability of undesirable impact.	9
Remote availability of necessary resources or near unacceptable cost and time consumption. Have a remote chance of success and near 100% probability of undesirable impact.	8
Very low availability of necessary resources or very high cost and time consumption. Have a very low chance of success or very high probability of undesirable impact.	7
Low availability of necessary resources or high cost and time consumption. Have a low chance of success and high probability of undesirable impact.	6
Rather low availability of necessary resources and rather high cost and/or time consumption. Rather low chance of success and rather high probability of undesirable impact.	5
Moderate availability of necessary resources, cost, time consumption, chance of success and probability of undesirable impact.	4
Rather highly available resources, rather low cost and time consumption, rather high chance of success and rather low probability of undesirable impact.	3
Highly available resources, low cost and time consumption, high chance of success and low probability of undesirable impact.	2
Fully available resources, very low cost and time consumption, near 100% chance of success and near zero probability of undesirable impact.	1

Figure 9.4: Table for Prioritizing Improvements and Modification

7. Analyze the Root Cause and perform failure analysis if necessary: When the list has been provided as to which to prioritize for modification based on its criticality, the team needs to perform a failure analysis on these parts to determine the physical cause of the failure or simply answer how and why these parts and items fail prematurely. It is important to analyze the failure first or conduct some basic analytical problem-solving tools before proceeding with any design improvement changes for the team to understand the reason for the failure. What is important is that modification should not possess or generate any new problems on the equipment itself. The team must understand the reason why the part has a short inherent lifespan.

8. Recommend for the improvement in design: Redesign or modification may be in the form of changing the shape, size, structure, measurements, dimension, or strength of the material of the part. Once the team has identified the weak points in the design of the part, a prototype of the modified part should be developed either by changing its shape, dimensions, the strength of materials, or construction. Remember that the goal of the team will be to exceed the lifespan of the previous existing part.

9. Proceed with the improved design and validate its effectiveness: Once the team has completed the analysis on the modified part, the team needs to fabricate a prototype and actually test it on the equipment. To be successful, the team should note the difference between the lifespan of the modified part and the original part itself. Once the lifespan of the original part has been exceeded, a cost-benefit study should be done at this stage since there will be cases where the cost of the modified part might be more expensive than the original part, but in the long run, would be cheaper since the lifespan of the part has far been exceeded. If the team has been successful in its design changes, it should inform both the storeroom and purchasing people to stop ordering the old part as this will now be considered an obsolete part of the storeroom. The decision here should be supported by a cost-benefit or a feasibility study. If the lifespan has increased, then the team needs to advise the purchasing of changing the replenishment if this is being stocked and likewise the interval for PM replacement will also need to be changed.

10. Proceed with the other identified weakness lists and follow activities 6 to 8: The team should now perform the same activities on the other parts that need to be modified on their equipment by simply proceeding with activities 7 to 9 until they have completed all the parts on the equipment that needs to be modified. An MP or Maintenance Prevention design form will be done indicating the difference between the previous design and the modified part and looped back to the Early Equipment Management or Initial Flow Control Activities of TPM. If the plant decides to purchase additional new equipment in the future, these modifications done to the equipment will be discussed with the designers and OEM of the equipment so that they can add these changes before purchasing the equipment. These activities in Phase 2 will take some time, depending on the number of parts to be modified at this stage.

11. Make necessary changes in procedures, and PM specs and conduct training whenever needed for maintenance and operations: Once all the parts with inherent design weaknesses have been modified, the team needs to adapt changes in their Preventive Maintenance since the interval or frequency of replacement has changed. A One Point Lesson or Good to Find form can be generated at this stage to serve as a teaching aid to operators. Phase 2 teams should also discuss this matter with the Purchasing people especially if these items are currently stored inside the storeroom as the replenishment of these parts will be longer depending on the lead time for the part to arrive.

12. Monitor the impact of improvement concerning cost and summarize the improvement activity: Allow the modified part to run for several months on the equipment and monitor the length of time it has run compared with the previous part. Monitor and observe the modified parts and update indices, most especially on cost, MTBF, and the number of breakdowns at this stage. The breakdowns as a result of Phase 1 activities had dramatically been reduced, and we still expect the breakdowns to be reduced at this phase of Planned Maintenance.

13. Horizontal Replication on other machines with the same problem: Replication of improvement changes and modifications will only be done on the same type of asset or equipment that poses similar symptoms of the problem. This means that the modified part will also be adapted to other similar equipment in the plant that inhibits the same symptoms of the problem. Note that if there are similar pieces of equipment in the plant that does not possess this type of problem, then it is not recommended to place the modified part on that equipment. Proceed with Phase 2 Audit and Certification. The team should complete its documents for Phase 2. The team performs a self-audit at this stage and once they are confident enough then schedules themselves for a Phase 2 audit where the Phase 2 team will try to defend the validity of its improvements and modifications

Most breakdowns occur because maintenance is too busy dealing with the failures themselves. Performing restoration on their equipment can dramatically reduce breakdowns by as much as 50% or even more. Negligence on these basic factors happens simply because people perceived breakdowns as normal and as part of their daily job routines. Restoration plays a crucial factor in Planned Maintenance, and Top Management must not only support this activity but also be committed and own it as well. For expectations, as the team completes Phase 2, breakdowns had been dramatically reduced to a minimum and MTBF had increased. In Phase 3 of Planned Maintenance, the team will finally conduct an activity to determine the correct maintenance strategy and create sustenance for Phases 1 and 2. At this phase, maintenance will need to determine the right maintenance tasks to perform on their equipment by having a decision diagram or an algorithm. This is where RCM will be used. The strategy is to take care of the basics first before implementing RCM.

9.4: Preliminary Requirements for Planned Maintenance Phase 2 Teams

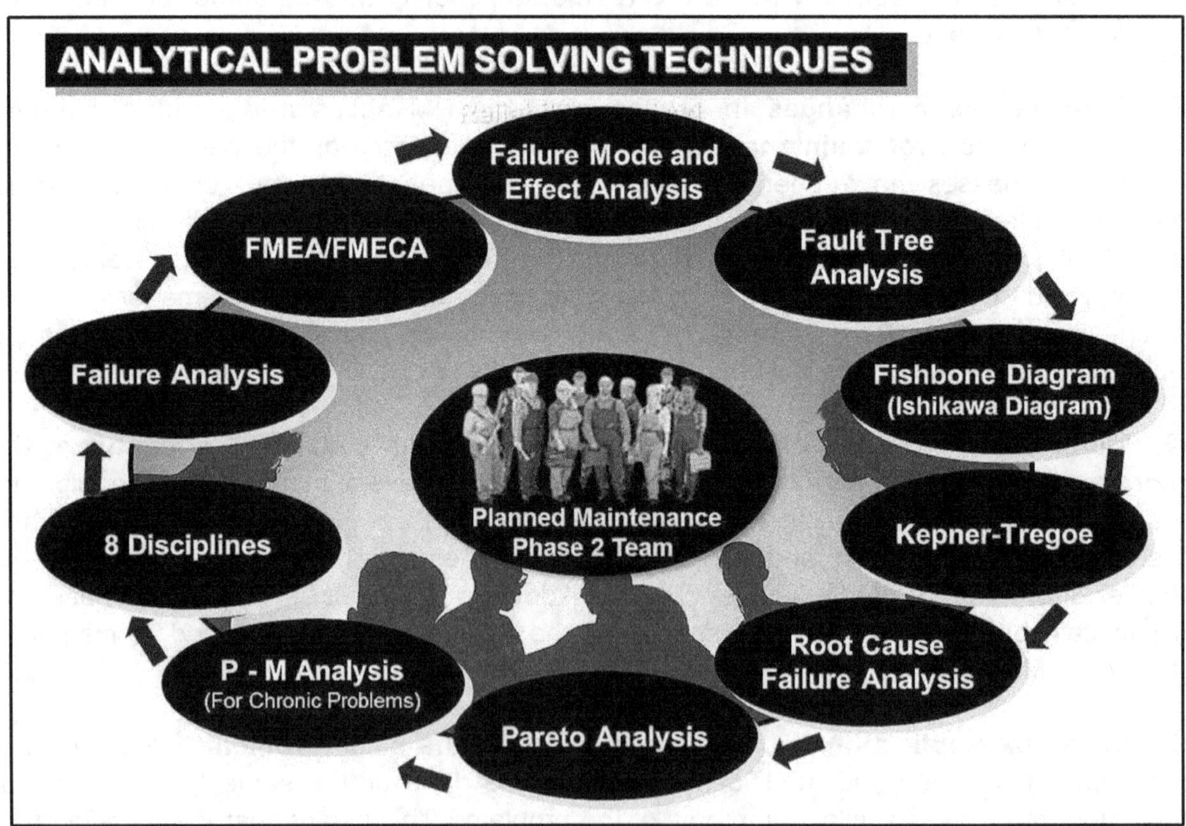

Figure 9.5: Different Problem Solving Tools

A preliminary requirement for Planned Maintenance Phase 2 teams is for the members of the team to be fully trained on at least a minimum of three problem-solving tools since the main activity for Phase 2 is all about improvements, modification, and redesign. The team needs to understand the cause of the problem first before proceeding with the improvement or redesign. The PM Phase 2 teams will be the ones to decide which analytical problem tools they would use for Phase 2 activities. Here are the most common analytical problem-solving tools that the Planned Maintenance team will be using.

Fishbone or Ishikawa Diagram: The cause-and-effect diagram was developed by Kauro Ishikawa of Tokyo University in 1943 and is often called Ishikawa Diagram. They are also known as fishbone diagrams because of their appearance. The cause-and-effect diagrams are used to list all the different probable causes attributed to a problem or an effect. The cause-and-effect diagram can aid in identifying the probable causes why a process goes out of control or a machine fails. An Ishikawa diagram is typically the result of a brainstorming session in which the group members offer ideas on how to improve a product, process, service, failure, or problem. The main goal is represented by the trunk of the diagram, and primary factors are represented as branches. Secondary factors are then added as stems, and so on. Creating the diagram stimulates discussion and often leads to increased understanding of a complex problem. A fishbone diagram is an analytical problem-solving tool that provides a systematic way of looking at the effects and the causes that create or contribute to those effects. Figure 9.6 is an example of an Ishikawa Diagram.

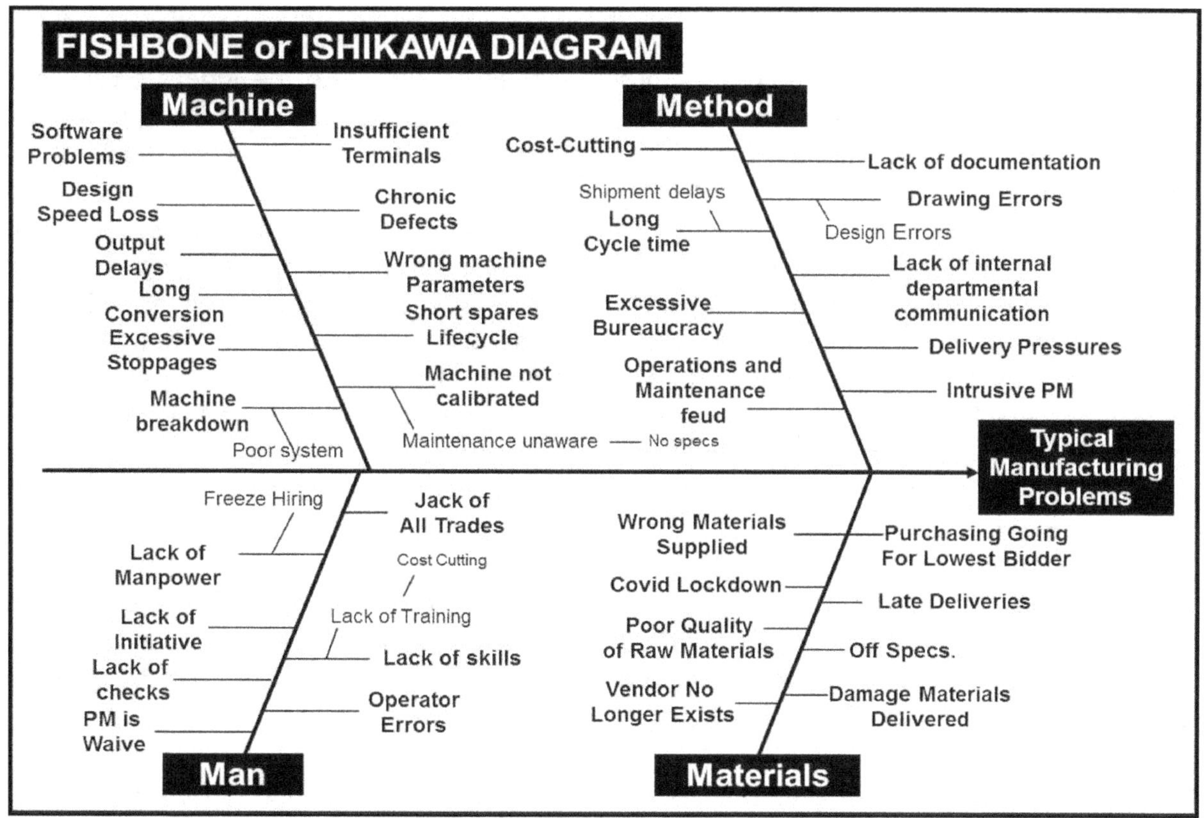

Figure 9.6: How Ishikawa or Fishbone Diagram is Constructed

FMEA/FMECA: Failure modes and effects analysis (FMEA) is a step-by-step approach for identifying all possible failure modes in either the design, equipment, process, system, service, or product. Failure modes mean the ways, or modes, in which the equipment might fail. However, the coverage of this guideline will only be for equipment FMEA. Depending on the type of FMEA, the tables on the severity, occurrence, and detection might differ, respectively, but the process of conducting FMEA will be the same whether this is a machine, service, product FMEA, and so on. Failures are any breakdowns or defects that will affect the equipment and assets of the plant. Failure Mode and Effect Analysis (FMEA) is a proactive tool for evaluating failure modes and their potential probable causes. Note that FMEA/FMECA will be limited only in determining the most likely cause or probable cause and not the Root Cause of the failure itself. It helps prioritize critical failure modes and recommends countermeasures and corrective actions for avoiding catastrophic failures to improve maintainability, safety, quality, and reliability. It is also commonly defined as a systematic process for identifying potential equipment failures before they are likely to occur, with the intent to eliminate or reduce the risk associated with the failure. Equipment or Machinery FMEA/FMECA is a risk analysis tool used to identify all possible failure modes that have happened in the past and the possibility of happening in the future. The FMEA result highlights failure modes with relatively high probability and severity of consequences. This concept will identify the severity, occurrence, and detection of each failure mode and identify possible corrective actions to eliminate, reduce or mitigate each failure mode. Once the corrective action had been completed the RPN or Risks Priority Number should be reduced.

P-M Analysis: This method analyzes chronic losses according to the inherent principles and natural laws that govern them. This method was developed by Mitsugu Kaneda, Shirose Kunio, and Yoshifumi Kimura. P stands for Phenomena and Physical. A phenomenon is a deviation from a normal to an abnormal state. M stands for Mechanism and the 4Ms (man, machine, method, and materials). The mechanism is about understanding the mechanics of the abnormal phenomena or how they are being produced. It also needs to understand the relationship between the abnormal event and the four production inputs or 4Ms. Its principle asks in precise physical terms what happens when a machine breaks down or produces bad parts or defects and how it happens. P-M Analysis clarifies the mechanics of their occurrence and the conditions that must be controlled to prevent them. According to the machine's inherent operating principles, P-M analysis will analyze chronic problems such as defects and failures. It is a logical investigation of phenomena such as defects or breakdowns that explain how the phenomena occur regarding their physical principles. To analyze means to break down a whole into several parts and learn their nature and relationships. Physical Analysis uses a machine's operating principles to clarify how various parts interact to generate abnormal phenomena. It means logically explaining how phenomena occurred. It is a bridge that helps us draw a logical relationship between the phenomenon and its potential causes.

Pareto's 80/20 Rule: Dr. Joseph Juran, one of the pioneers of quality, worked in the US from 1930 to 1940. Juran recognized a universal principle which he called the vital few and trivial many. In an early work, a lack of precision on Juran's part made it appear that he applied Pareto's observations about economics to broader work scope. As a result, Dr. Juran's observation of the vital few and trivial many, where 20 percent of something is always responsible for 80 percent of the results, now became known as the Pareto's Principle, which is the 80/20 Rule. This is also known as the vital few and trivial many. The Pareto Principle can be applied in a wide range of manufacturing, management, and human resources. For instance, the efforts of 20% of an industry's staff could drive 80% of the company's profits. Pareto Analysis will be a good problem-solving tool if data is given

Failure Analysis: If we speak about failure analysis, we stop our probe or investigation on the parts or component level or the failure's physical cause itself. We refer to this as the metallurgical aspect as to why the failure occurs. When we send a failed bearing to the metallurgical laboratory, they will check the raceway, perhaps conclude and write a report that the bearing failed due to fatigue. The physical cause is the reason how the parts failed technically. This is the technical explanation of why things broke or failed. This mostly explains the metallurgical factor of why and how the failure occurred. Failure Analysis is an engineering approach to determining how and why a particular part, equipment, or component fails. The goal of failure analysis is to understand the physical cause of the failure. Failure Analysis will be considered to be the first level of any Root Cause Failure Analysis investigation, which is to determine the physical cause of the failure. This means that failure analysis is just a part of Root Cause Failure Analysis.

Root Cause Failure Analysis: Both RCFA and RCA are performed by determining the problem's physical, human, system, and latent causes. The distinction is on the word failure; when we include the word failure in Root Cause Analysis, we refer to equipment-related failures. However, we need to be specific in this case; RCFA will not only be limited to mechanical failures but will also include electrical and electronic failures for as long as this is equipment-related. Quality defects caused by equipment-related problems such as raw materials will be under RCA. Hence, if we speak about Root Cause Failure Analysis, we are trying to understand why something went wrong with our equipment and assets. It will identify and address the basic source or origin of the problem so that industries can finally learn from the failure itself. RCFA will provide a structured methodology for investigating, categorizing, and eliminating the root cause of equipment-related incidents. Root Cause Failure Analysis is not interested in who caused the problem since there are much deeper causes of why people commit mistakes and errors. The goal of any Root Cause Failure Analysis investigation will be to identify and expose these causes, which are known as the latent cause of the problem. Both RCFA and Failure Analysis are evidence-driven endeavors, which means that the failure investigation will be based solely on the evidence. Hence, one requirement to conduct a root cause investigation is to preserve the physical evidence of the failure. This means that a failure needs to happen first before we can conduct an actual Root Cause Failure Analysis investigation.

Fault Tree Analysis: This method was developed in the 1950s by Boeing's Aerospace Engineer for use in the design process's development stages. It is a mathematical tool and will yield probabilities. The purpose of a Fault Tree Analysis is to predict the probability of a specific failure. Fault Tree Analysis is not actually a Root Cause Analysis tool but rather a design tool. It was first used in 1962 for the US Air Force by Bell Telephone Laboratories on the Minuteman Weapon System. It has been adopted by many industries in the field of reliability engineering. It received extensive coverage at a 1965 System Safety Symposium in Seattle sponsored by Boeing and the University of Washington. Boeing first began using FTA for civil aviation aircraft design in 1966. It provides a basis for analyzing design by understanding the most common failure modes that can affect the design of a system. FTA is analyzed using Boolean logic, which combines a series of lower-level events. This method is mainly used in safety and reliability engineering to understand how a system can fail and identify the best ways to reduce the risk associated with the failure. Boolean symbols and gates are used in FTA to identify the most likely cause of the problem.

8-Disciplines: The 8-Disciplines is an analytical problem-solving tool designed to determine the problem's probable or most likely cause. Again the process of conducting 8-Disciplines is not based on evidence. Hence, we cannot conclude that it will derive the root cause of the problem. This method can be applied to defects and equipment-related failures as well as establishes a permanent corrective action based on data and provides the probable cause of the problem. The corrective and preventive measures generated should address both the defects and system causes. These actions should be proliferated to a family of products or similar processes to prevent a recurrence of the same defect or problem in the future.

Kepner-Tregoe: Kepner-Tregoe is a problem-solving tool derived from its founders, Charles Kepner and Benjamin Tregoe. They authored their first book in 1965, which was titled, The Rational Manager. In 1981, they republished a revision of the book titled, The New Rational Manager. Through the years, the Kepner-Tregoe method researched and identified the troubleshooting skills of people. Through the years, people have learned to deal with the complexity of incidents and problems, which happen, mostly in industries. Often, there is a pressure of time to solve the problems and arrive at a solution to the problem, whether the solution is correct or not. The Kepner Tregoe method or KT method is a problem-solving and analytical tool in which the problem is first disconnected from the decision.

Five Why Analysis: Sakichi Toyoda, the founder of Toyota Industries, developed the 5-why technique in the 1930s which became popular in the 70s and is still being used today by Toyota Motor Corporation to find solutions to their day-to-day problems. The approach uses a systematic questionnaire technique to search for the probable causes of a problem. The team uses a tool by asking why at least five times as you work through the various levels of detail. Once the team finds it difficult to respond to why the probable cause of the problem may have been identified. As you trace the why back to their possible causes, you will find yourself confronting issues that affect the original symptom. To be effective, the team's answer to the 5-whys must steer away from blaming individuals. Blaming people leaves us with no option but to punish them, leaving no room for substantial change. The focus is on the process of the problem and not the person involved, since frequently, answering why will lead you to the person responsible for the problem.

MTBA Snapshot Analysis: This is an improvement technique used to address Minor Stoppages and errors on the equipment. MTBA is the average time the equipment performs its intended function between assists. It is also the productive or the operating time divided by the number of assists. Assists and errors are any unplanned interruption or variance from the specification of equipment that requires human intervention of fewer than 6 minutes. A minor stoppage is an equipment stoppage due to a failure or an error in automatic handling, processing, or assembly of parts and workpieces. It sometimes occurs due to quality-related abnormalities. To carry out an MTBA Snapshot is to initially observe the equipment for a minimum of two hours and record all the errors and assists occurring during that period. Usually, the maintenance will perform the MTBA Snapshot, and likewise, correct the error when it happens and indicate the total time for the error to be corrected as well as the frequency it occurred during the two hours observation time. Minor stoppages and errors usually occur on automated equipment or machines that contains many electronic parts.

MTBF Analysis: An MTBF Analysis can be performed as in figure 9.7 to determine the reason for having a low MTBF and understand the failures involved. In this case, if we want to look for the 2nd line up of equipment which is the Substation, Substation E has the lowest MTBF due to the following failures such as the fuse melts on overload, drained battery, circuit faults, and multiple failures of protective devices. Addressing these failure modes will likewise increase the MTBF of this substation. Note that MTBF Analysis will only be used for failures and breakdowns while MTBA Snapshot will be used for equipment prone to errors and minor stoppages.

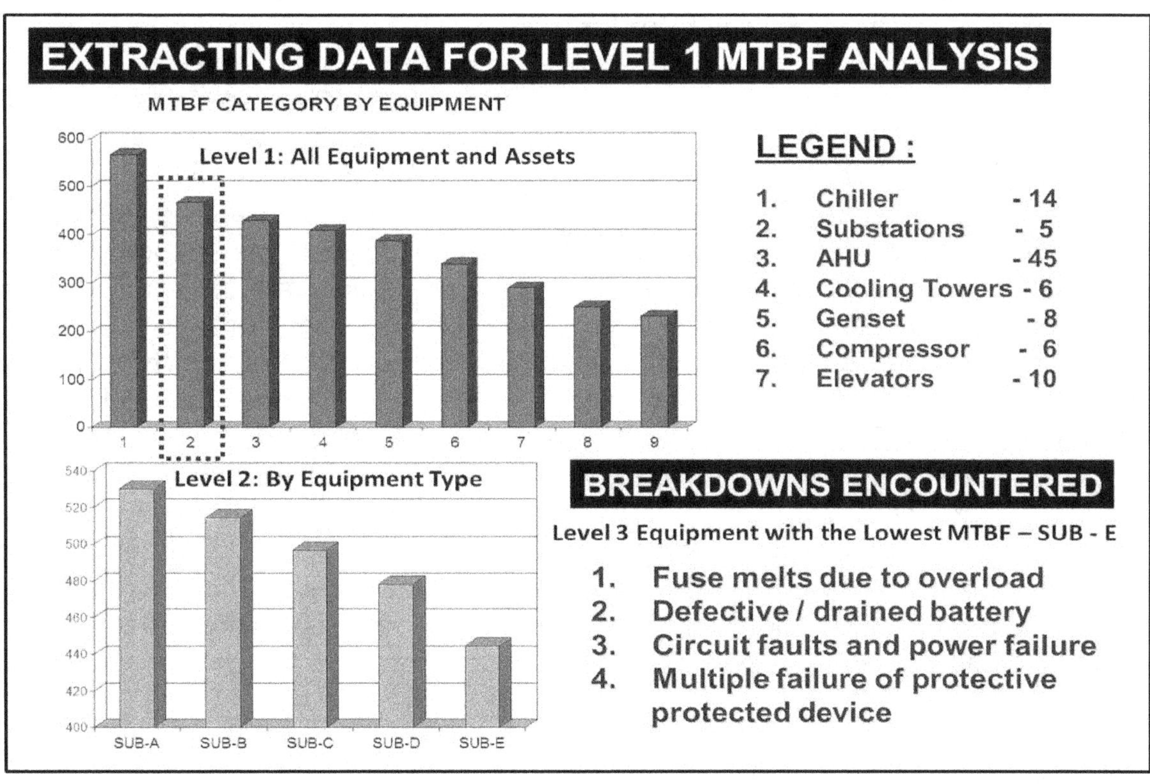

Figure 9.7: MTBF Analysis

9.5: Maintenance Prevention (MP) Design Improvements

[11]According to Japan Institute of Plant Maintenance (JIPM), Maintenance Prevention is defined as the use of the latest maintenance data and technology when planning or building new equipment to promote greater reliability, maintainability, economics, operability, and safety while minimizing maintenance costs and other problems. This means that equipment is designed from the start with easy maintenance and trouble-free operations. When developing new equipment, the design, fabrication, and installation may proceed smoothly with a few problems. Still, the test run and commissioning stages may take some time due to debugging problems during the test period. The key to preventing these problems is to ensure that all processing and operational requirements for the new equipment must be fully incorporated into the equipment requirements used by the OEM. It is better to develop equipment in-house in the plant so that the designers of the equipment understand the operating context and environmental conditions of the plant as well as provide much quicker communication to the users and maintenance of the asset.

MP Design activities reduce the overall maintenance cost and problems by adapting the learning they have experienced on their existing machines. This is done by having maintenance data on their existing equipment and machines, addressing those design weaknesses through improvements from existing assets through modifications and redesign to ensure that new equipment is designed for optimum reliability, easy to maintain, economy, safe, and easy to operate. The MP Design process should improve equipment reliability by investigating weaknesses in the existing equipment and feeding the information back to the designers. Figure 9.8 are samples of MP Design Improvements we generated when we were implementing Planned Maintenance.

Figures 9.9 to 9.10 are a summary of all improvements the Planned Maintenance Pillar generated when I was still employed in the semiconductor industry. We provided a Control Number for each improvement/modification generated. A detailed drawing is provided for every MP Design Improvement. These MP Design improvements are being fan-out or replicated to other similar machines, which possess the same problems. Hence, if we have around 100 wirebond machines and 60 of these machines generate the same problems, then the modification will only be replicated or fan out on these 60 machines. What is important for industries is that all improvements, redesign, or modifications that have been generated should be controlled for every piece of equipment. A copy of the details of the improvement or modification should be provided to the IFCA or EEM Committee. This means that if the plant has some plans to expand in the future and add new machines of the same type, make and model as their existing machines, these improvements will be prioritized according to their importance and discussed with the OEM designers of the equipment to be included when they purchase these machines. This means that the maintenance has experience problems, flaws, breakdowns, and failures of this equipment in the past and has learned from them so equipment and machines that will be purchased in

---

[11] Fumio Gotoh, *Equipment Planning for TPM, Maintenance Prevention Design,*(Portland, Oregon, 1991), Page 72

the future should not possess the same problems they experienced when delivered to the plant.

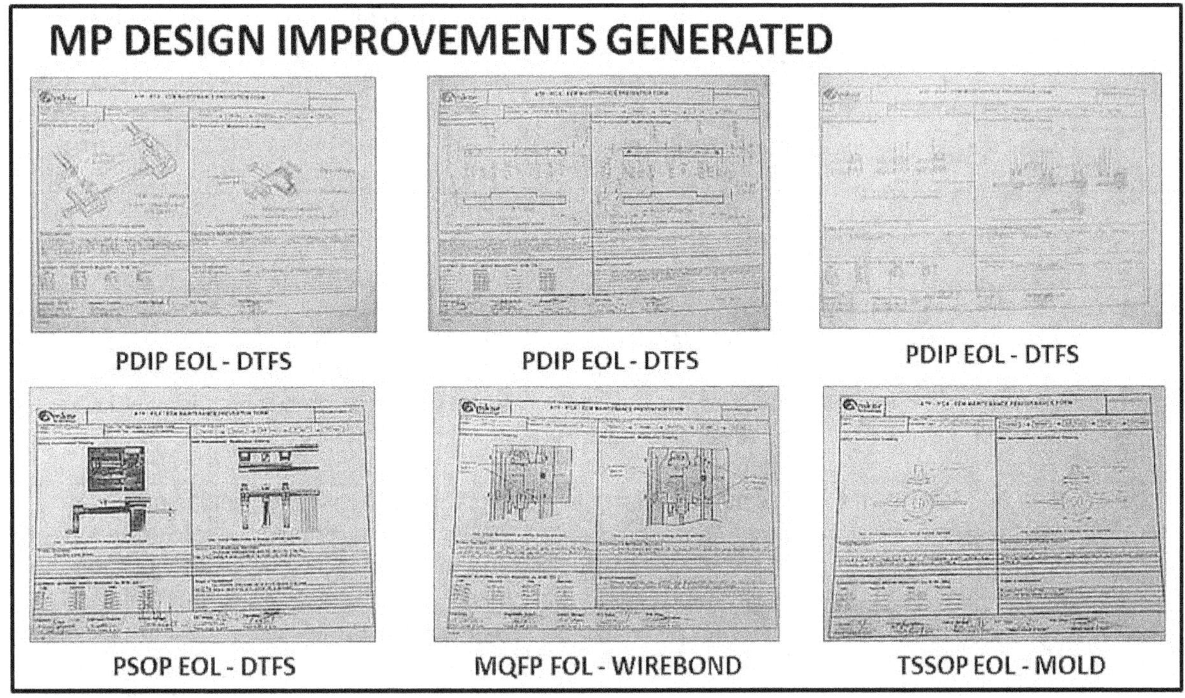

Figure 9.8: Samples of MP Design Improvements Generated

During the development phase, the design, fabrication, and installation stages may proceed with very minimal problems, but during the commissioning phase or in the early phases of operation infant mortality failures and other abnormalities can occur. Failures and breakdowns may be experienced prematurely, which requires repair resulting in a high level of downtime for the new equipment.

[12]The key to preventing such problems is to ensure that all processing and operation requirements of the new equipment are fully incorporated into the equipment requirements used by the equipment design engineers. The best approach is to avoid buying equipment technology developed by another company's engineers and instead develop the equipment in-house. This means that the OEM or vendor of the equipment may be well familiar and versed with the design of the equipment. However, both operations and maintenance are the people that experience the flaws, problems, and failures during their day-to-day operations. They know how the equipment performs unless otherwise, the OEM is the one who also maintains the equipment in the plant. Hence, improvements and modifications were done by maintenance to address these problems on the equipment. These people who maintain and operate the equipment have experienced these problems and if improvements have been carried out, they should be looped back to the plant's Research and Development (RND), or the IFCA or EAM group.

---

[12] [12] Fumio Gotoh, *Equipment Planning for TPM, Maintenance Prevention Design,* (Portland, Oregon, 1991), Page 73

PLANNED MAINTENANCE MP DESIGN IMPROVEMENTS
Our Planned Maintenance – One Mission, One Direction

No.	MP DESIGN CONTROL	EQUIPMENT TYPE	PROBLEMS ENCOUNTERED	MP DESIGN IMPROVEMENTS	DESIGN IMPACT
1	SOT - 001	FICO-TFS	Crumpled strips during the TFS process due to weakness in the rail rack design	Modify SOT-23 SC-70 rear rail track cover	Eliminate downtime and rework due to crumpled strips
2	PSOP - 001	Sunyang Auto TFS	Singulation pusher problem	Modify singulation pusher for Sunyang Auto TFS machine	Minimize frequent replacement of cylinder
3	PSOP - 002	Sunyang Auto TFS	Intermitted breakage of screw causing sensor misalignment and wearing out of main package pusher plate	Place an additional screw holder on the other side of the holder to make it more steady and fixed	Eliminate breakage of the screw and longer lifespan of the package pusher plate
4	PSOP - 003	ASM AD 809	Premature Worn Out Strip Mark Holder	Modify the strip mark holder by integrating it into the main strip mark	Elimination of accelerated deterioration of strip mark holder
5	MQFP - 001	Santec Trim	Frequent damage pin holder due to lead frame indexing stopper assembly mechanical backlash, excessive movement	Fix lead frame indexing stopper	Eliminate problems such as nick and excessive backlash
6	MQFP - 002	Santec Trim	Frequent magazine misalignment during transport	Install set screw on magazine pocket wall	Prevent magazine from jamming and minimize magazine mis-loading error
7	MQFP - 003	Santec Trim	Uncut dambar due to insufficient penetration of dambar punch to debar die caused by a loose camshaft	Create a step on a shaft to prevent turning, and prevent dislocation from its original position	Prevent uncut dambar on the strips
8	MQFP - 004	Santec Trim	Easily worn out lead frame roller and mislocation of spring and jamming of lead frames	Relocate the presence of a check hole where the focus is on the L/F rail's original position	Eliminate jamming of the lead frame as well as minimize chances of crumpled strips
9	MQFP - 005	KNS 1488 Turbo +	Easily worn out elevator screw resulting in magazine drop and unit rejection	Put additional back-up screw-on elevator (L) wall which acts as conversion jig for magazine variation	Eliminate magazine drop and extend the lifespan of the lead screw
10	MQFP - 006	KNS 1488 Turbo +	Exposed Injector blade sensor	Install sensor guard for protection	Zero occurrences of damage to injector blade sensor

Figure 9.9: Samples of MP Design Improvements Generated 1

PLANNED MAINTENANCE MP DESIGN IMPROVEMENTS
Our Planned Maintenance – One Mission, One Direction

No.	MP DESIGN CONTROL	EQUIPMENT TYPE	PROBLEMS ENCOUNTERED	MP DESIGN IMPROVEMENT	DESIGN IMPACT
11	MQFP - 007	KNS 1488 Turbo	Brittle and high-cost material	Change material to Teflon	Low cost and longer lifespan
12	SSOP - 001	Hanmi FS Trim	Accepting misoriented strips	Provide additional pin for lead frame orientation. This will check the orientation even if the vision check fails.	Zero occurrences of inverted strips
13	SSOP - 002	Hanmi FS Trim	Premature worn-out pinhole	Install bearing for pin and shaft and modification of the unit pusher from bushing to bearing type	Prolong lifespan and zero breakdown occurrences on singulation lever and shaft
14	SSOP - 003	HANMI FS Trim	Modification of L/F cut punch remnant guide angle	Modification of angle guide concerning the base from 35 ° to 15°	Eliminate clogging and uncut remnants, which can lead to damage, a broken package of the part.
15	SSOP - 004	ASM AD 809	Falling out of strip mark resulting to illegible mark and inconsistent mark	Modification of new design strip marker assembly	Zero occurrences of illegible and inconsistent marks on the package
16	SSOP - 005	ASM AD 809	Offset camera which resulted in misplaced die and dies crack	Modification of camera bracket	Zero occurrence of misplaced die or crack due to offset camera
17	SSOP - 006	ASM AD 809	Disabled ionizer since it cause machine hangs up and downtime problems	Relocate ionizer from top of the machine to the inside of board compartment and install pressure regulator with gauge	Zero occurrences of machine hang up
18	HERM - 001	SL 9000 Swiss line	Misaligned epoxy dispense boat hole. Does not reach the pitch sensor due to inconsistent indexer roller tension	Replace spring-type indexer tension with a pneumatic cylinder for consistent tension	Maintain 400 to 600 gms indexer roller tension requirement, which prevents indexer error
19	HERM - 002	KNS 1474 Wirebond	Corroded wire path contact pin sensor which tends to cut wire	Modify wire path contact pin sensor	Eliminate non-stop feed motor and cut the wire and good sensing activity to wire path with less cut wire defects
20	HERM - 003	KNS 1474 Wirebond	Broken lead screw and nut	Provide aluminum spacer support on top to support and withstand the heavy weight of the compartmentalized	Eliminate elevator jamming and improve lead screw and nut's lifespan

				magazine	
21	HERM - 004	SL 9000 Swiss Line	Wafer ring jamming due to load and unload due to poor grip of wafer grabber to the wafer ring	Increase the wafer grabber width for better holding grip during the wafer load and unload	Maintain balance, hold of grabber during load, and unload to prevent wafer loss error.
22	TSSOP - 001	Daichi Auto Mold	Insufficient suction of vacuum cap due to the small dimension of the suction area.	Replacement of vacuum cap with a larger suction area,	Extended the lifespan of the suction cap
23	TSSOP - 002	HANMI FS Trim	Unit jamming of offload track due to tube holder area causing breakage of sensor	Provide metal cover for the tube mouth sensor	Reduce frequent breakage of sensor malfunction
24	TSSOP - 003	KNS 1488	The short lifespan of EFO electrode reaches approximately 4000 average hours	Changing the sensor type of maker	Improve the consumption and cost of EFO Electrodes
25	TSSOP - 004	KNS 1488	Exposed sensor assembly that was easily damaged during production and set up	Fabricate a sensor cover that can protect the assembly	Reduce sensor malfunction as it is now well protected
26	CLF - 001	MECO	Intermittent strip fall-out due to low tension of belt clips on the conveyor belt	Modify belt clip from type 304 to type 316	Eliminate product drop and crumpled strips
27	CLF - 002	MECO	Mis-aligned strip in the buffer zone being hit by the Y-aligner arm	Modify y-aligner design extension	Zero Breakdown on strip damage and turnover
28	PDIP - 001	ASM AP-50	Offload error simulation and frequent tube twisting during tube change	Place tube catcher on tube transport assembly	Eliminate twisting of the tube during tube changing
29	PDIP - 002	ASM AP-50	Machine designed only to use one type of tube	Install the set screw to move the track on de-sired height to cater to different types of tubes	Eliminate tube-to-tube transfer of units and flexibility
30	TQFP - 001	APM Auto Trim Form	Frequent error at onload picker vacuum	Spring-loaded lead frame vacuum picker at onload station auto trim	Zero out frequent errors at the onload picker

Figure 9.10: Samples of MP Design Improvements Generated 2

9.6: Case Study 1: PM Phase 2 on Hermetics Analog

This is one of the 21 pilot teams we organized in the year 2000 to 2002 as part of our Overall Planned Maintenance program, which completed Phase 2, which is all about Lengthening Equipment's Lifespan by Addressing Design Weaknesses. All teams passed Phase 2 Certification and as a result, the overall breakdowns of their equipment were reduced dramatically.

	PHASE 2: LENGTHENING EQUIPMENT LIFETIME BY ADDRESSING DESIGN WEAKNESSES PILOT MACHINE TRACKING																							
							Sa Planned Maintenance, Isang Misyon, Isang Direksyon pa rin																	
No.	PM Comm	Reports to	Department	1st PM PILOT	Percent	Plan	WW25	WW25	WW26	WW27		WW28		WW29-30		WW31-32		WW33	WW34		WW35		WW36	REMARKS
					Completion	Actual	Training	1	2	4	5	6	7	8	9	10	11	12	13	14	15	16	17	18	
1	CDELA	RLLAN	CLF	Meco 2 Plating	100.0%	Plan / Actual																			Completed
2	TLIM / KOKO	BMONT	Hermetics (SSEU/Cardo)	Swissline-007 DA-SSEL	100.0%	Plan / Actual																			Completed
3	DBALA	NMARA	MOFP EOL	QPST-001 DTFS	100.0%	Plan / Actual																			Completed
4	RSUAR / CONAR	JESPA	SOIC FOL	SOFD-003 Disco DFD-640	100.0%	Plan / Actual																			Completed
5	JJAVI	CANGC	PSOP FOL	PSAD-006 DA-ASM AD809	100.0%	Plan / Actual																			Completed
6	RONAO	BORLE	TQFP FOL	TQKT-087 WB-KNS1488	100.0%	Plan / Actual																			Completed
7	OPARP	BFAUS	SSOP FOL	SOAD-001 AD809 ASM	100.0%	Plan / Actual																			Completed
8	ZBALO	SCAME	PDIP EOL	PMAP-008 DTFS	100.0%	Plan / Actual																			Completed
9	CGONZ	RNASA	TSSOP EOL	TSAM-003 Mold Dai-ichi	100.0%	Plan / Actual																			Completed
10	CGONZ	RNASA	TSSOP EOL	TSHA-007 Hanmi -F/S	100.0%	Plan / Actual																			Completed
11	JBAYB	SCAME	PDIP FOL	PMAD-023 DTFS	100.0%	Plan / Actual																			Completed
12	GAVYB	VTORI	SSOP EOL	Hanmi #7 DTFS	100.0%	Plan / Actual																			Completed
13	RFABR	BORLE	TSSOP FOL	SOKS-123 WB-KNS1488	100.0%	Plan / Actual																			Completed
14	ROMES	RNASA	TQFP EOL	TQAF-003 DTFS- APM	100.0%	Plan / Actual																			Completed
15	ASOLO	BMONT	Hermetics (Analog)	KNS1474-012 WB-KNS	100.0%	Plan / Actual																			Completed
16	MAPAD	CANGC	PSOP EOL	SOSY-002 DTFS	100.0%	Plan / Actual																			Completed
17	DGALA	NMARA	MQFP FOL	QPKT-087 BATCH 1	100.0%	Plan / Actual																			Completed
18	AJUSAY	DANNR	Facilities	AHU-401	100.0%	Plan / Actual																			Completed
19	RPAND	VGARC	PLCC FOL	PLKS-173 WB-KNS1488	100.0%	Plan / Actual																			Completed
20	BBERN	VGARC	PLCC EOL	HP-82 DTFS	100.0%	Plan / Actual																			Completed
21	CELSO	ETORR	SOT EOL	SMP-005 FICO TFS	100.0%	Plan / Actual																			Completed
				AVERAGE	100.0%																			Note: Dark Shade means delays

Figure 9.11: Planned Maintenance Phase 2 Pilot Teams

One of these teams (Number 15 in figure 7.8) ASOLO or Arnel Solo from the Hermetics Analog division choose Wirebond Machine KNS 1474-012 as their Pilot Machine. Before implementing Planned Maintenance on their Wirebond KNS 1474 machine number 12, the machine experienced many breakdowns, particularly on the bond head, x-y table, work holder, and elevator sub-assembly. The group decided to pursue Phase 2 of Planned Maintenance by identifying parts with inherent design weaknesses and challenging the team to improve the lifespan of these spare parts with recurring breakdowns. They selected P-M Analysis as the tool to address the problems with their equipment.

P-M Analysis Conducted on Wirebond Machine: The failure phenomenon indicates the unmatched design of the load carried and the carrier itself. The machine experienced many premature breakages of the lead screw nut, which leads to quality-related rejects on the product such as rejected packages and deformed wire. The operating principles indicate that a guide nut of the lead screw rod, which is attached to the fork assembly, carried the magazine and brings it up and down and vice-versa. The gradual pitch movement of the lead screw rod on the nut provides sufficient elevator clearance. Their modification includes placing an aluminum spacer to support and withstand the heavy load compartment, which increased the lifespan of the lead screw.

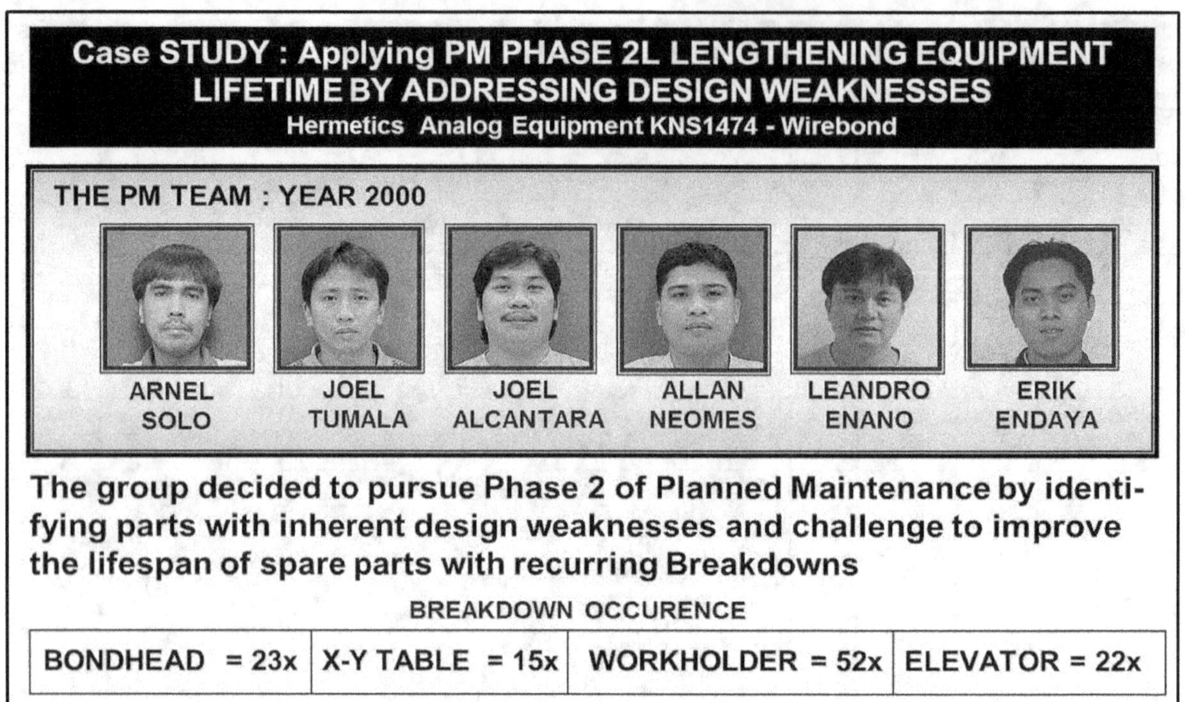

Figure 9.12: PM Hermetics Phase 2 Pilot Teams

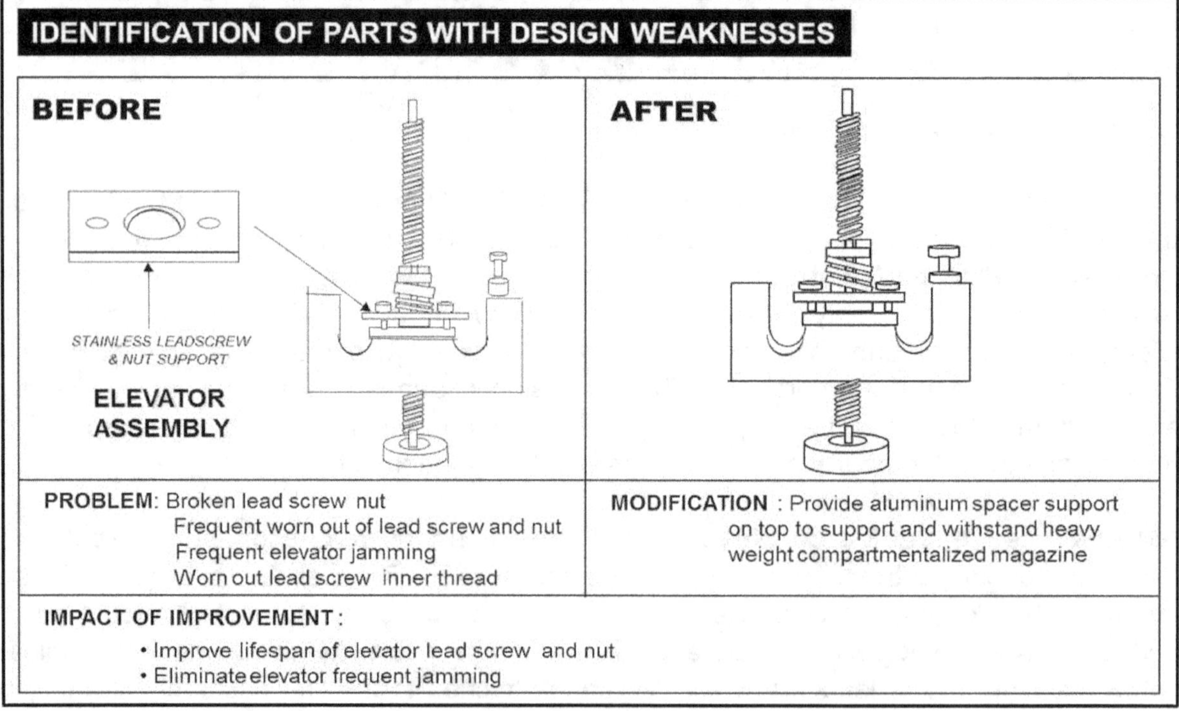

Figure 9.13: PM Phase 2: Modification of Elevator Lead Screw

The physical analysis indicates that the discrepancy in the elevator pitch is critical between the lead screw rod and the nut. The nut cannot resist the heavy weight compartmentalized magazine load that can induce premature breakage of the nut and other machine-related breakdowns. This can also induce depressed wire, product rejection, and other quality-related breakdowns on their KNS Machines. The possible contributing conditions indicate include worn-out lead screw nut thread, dull and worn-out lead screw rod thread, stock-up guide bearing, misaligned lead screw rod, and misaligned elevator guide. The materials design is not for heavy loads. The team findings indicate that the lead screw nut design and durability are insufficient for the load carried. Need to provide the design of the bracket it supported. The failure impact leads to broken lead screw and nut, elevator jamming, pitch slippage due to worn-out thread, worn-out lead screw inner thread, and offset elevator height.

Another problem the team encountered was on the wire patch assembly regarding its contact pin sensor. The problem indicates that the sensor pin was easily corroded since it was exposed to dust which shorten the lifespan of the pin sensor. This will make the sensor more prone to pin loose due to force-fit inserted. The dust accumulated on the pin sensor leads to an unpredictable lifespan due to frequent adjustments due to its sensitivity. Also experienced inconsistent location of the pin because of no locking nut at the base.

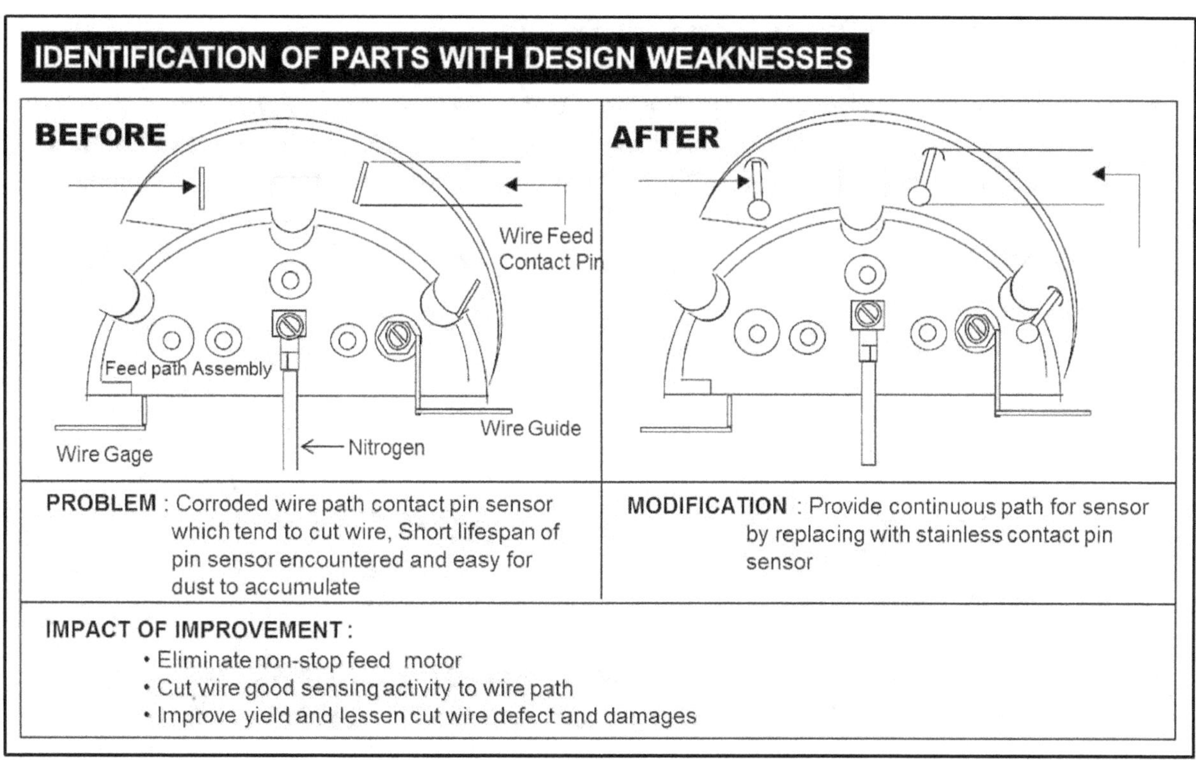

Figure 9.14: PM Phase 2: Modification of Contact Pin Sensor

The team from the KNS Machine Planned Maintenance realized a dramatic reduction in their breakdown occurrences and an increase in their MTBF after modification of design weaknesses in their pilot equipment. The group Passed Phase 2 Certification. Their breakdown occurrences dramatically reduced from 16 as of July 2000, to only 1 as of July 2001 as can be reflected in figure 9.15. Other KNS machines inhibit the same problem, hence, these modifications were replicated to other KNS machines of the same type,

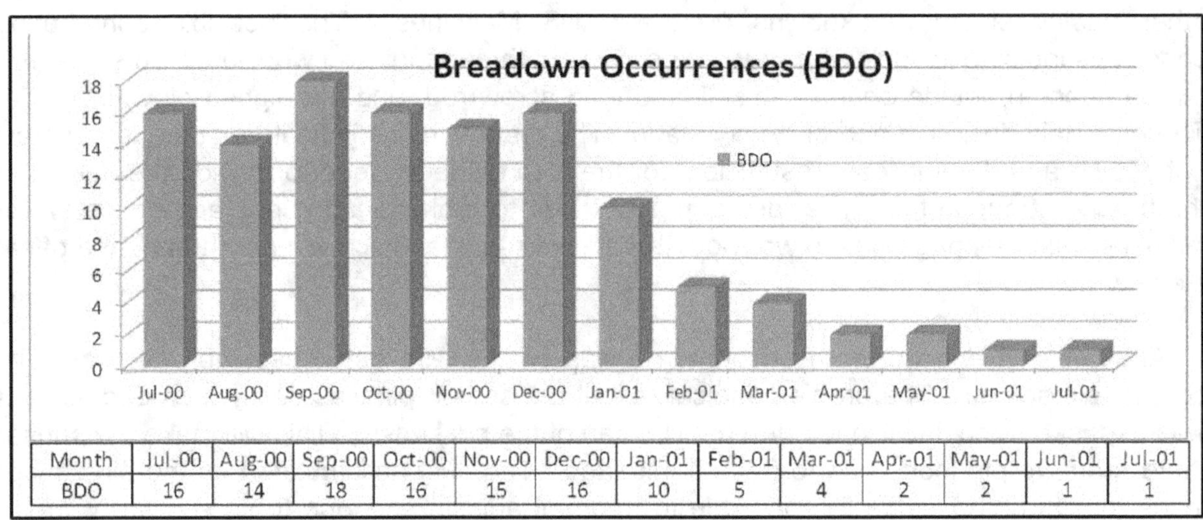

Month	Jul-00	Aug-00	Sep-00	Oct-00	Nov-00	Dec-00	Jan-01	Feb-01	Mar-01	Apr-01	May-01	Jun-01	Jul-01
BDO	16	14	18	16	15	16	10	5	4	2	2	1	1

Figure 9.15: Breakdown Occurrences for the KNS Pilot Team

9.7: Case Study 2: PM Phase 2 on CLF Plating

Another successful Case Study on Planned Maintenance Phase 2 was from the Central Lead Finish (CLF) plating station on their pilot machine Meco 2. Their pilot machine is an automatic load and unloads, strip to strip deflash plus plating machine. Meco machine has several electro-chemical rinsing and drying operation with the capability to process around 2400 strips per hour.

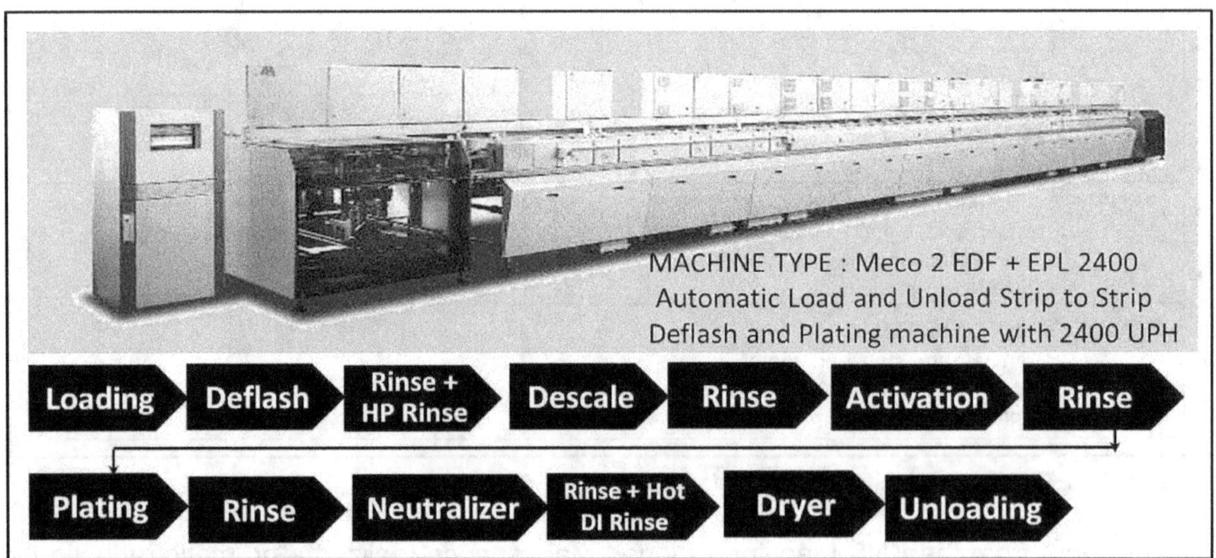

Figure 9.16: Meco 2 CLF Plating Pilot Machine

All their Meco machines were prone to minor stoppages and errors. To address these problems, the Meco team performed MTBA Analysis. Figure 9.18 indicate the improvement in MTBA or Mean Time Between Assists on their Meco 2. Their entire Meco machine was prone to errors, hence, the improvements were replicated to all the rest of the Meco machines achieving the same results as their pilot machine.

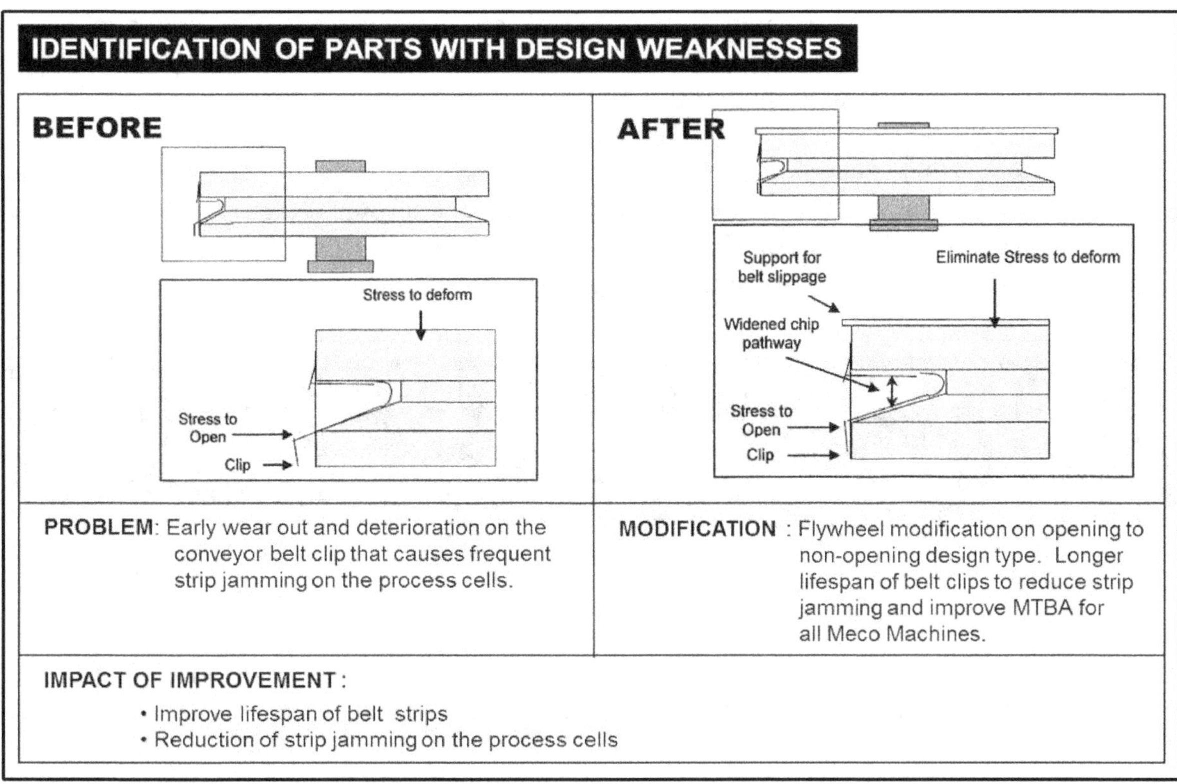

Figure 9.17: PM Phase 2: Modification of Flywheel Belt

CASE STUDY 1 : MTBA SNAPSHOT REPLICATION

MECO	PHASE 1		PHASE 2	
	BEFORE	AFTER	BEFORE	AFTER
	FEB - 98	APRIL - 98	MAY - 98	FEB - 99
MECO 2	16.61 min	22.54 min	45.84 min	115.74 min
MECO 1	11.86 min	67.4 min	118.9 min	117.04 min
MECO 3	15.68 min	51.86 min	51.86 min	87.33 min
MECO 4	10.07 min	39.11 min	48.84 min	105.68 min

MTBA Before and After Data derived from CLF Planned Maintenance Activities 98-99

Following were actual data taken from CLF Plating Meco Machine as part of their improvement activities during their TPM's Planned Maintenance activities, as based from previous records, all MECO and TECHNIQUE machines passed Phase 2 Certification

Figure 9.18: MTBA Snapshot Replicated to Other Meco Machines

Figure 9.18 indicates the summary of all Meco machines' MTBA snapshot, which increased before and after implementing their corrective actions and replicating these improvements with other Meco machines with similar problems

From 1998 to 2001, our Planned Maintenance garnered a total of 9272 improvements in which MQFP FOL was ranked first with a total of 3142 improvements. As advised by our JIPM Consultant that it is mandatory to have 10,000 improvements to qualify for the TPM Excellence Awards which our PM teams are confident to surpass.

OVERALL PM MP DESIGN IMPROVEMENTS

Sa Planned Maintenance, Isang Misyon, Isang Direksyon pa rin . . . . .

| STATION | PM Comm | AREA | MACHINES | | HEAD | TOTAL PM IMPROVEMENTS | | | | | RANK |
			Types	Total	COUNT	1998	1999	2000	2001	Total	
PLCC	RPAND	FOL	29	269	62	0	757	0	0	757	3
	BBERN	EOL	26	117	43	95	54	0	0	149	14
SOT	EBUST	FOL	3	18	9	0	0	127	0	127	17
	CCRUZ	EOL	4	4	6	0	0	6	2	8	22
PSOP	JJAVI	FOL	14	100	28	180	21	45	269	515	5
	MAPAD	EOL	14	29	17	20	6	62	11	99	19
MQFP	DGALA	FOL	6	210	39	2570	479	47	46	3142	1
	DBALA	EOL	6	59	31	40	72	42	15	169	13
SSOP	OPARP	FOL	9	150	29	20	738	10	0	768	2
	GAVYB	EOL	14	33	23	30	47	117	140	334	9
TQFP	RONAO	FOL	9	141	25	166	20	156	2	344	8
	ROMES	EOL	15	32	20	0	56	35	54	145	15
TSSOP	RFABR	FOL	9	218	36	230	0	184	46	460	7
	CGONZ	EOL	17	52	32	0	56	35	54	145	16
SOIC	RSUAR	FOL	20	409	71	0	0	316	164	480	6
	JIMBOY	EOL	22	82	31	0	13	38	30	81	20
PDIP	JBAYB	FOL	14	149	45	0	40	62	0	102	18
	ZBALO	EOL	19	77	32	63	106	72	37	278	10
Hermetics	TUM	SSEL	17	92	34	254	289	157	0	700	4
	ASOLO	Analog	14	36	19	0	0	23	56	79	21
CLF	JERRC	Plating	8	21	30	65	41	38	47	191	11
Facilities	AJUSAY	--	30	199	17	33	146	20	0	199	12
TOTAL			319	2497	679	3766	2941	1592	973	9272	
CUMULATIVE						3766	6707	8299	9272		

Figure 9.19: ATP-Planned Maintenance Total MP Design Improvements Generated

Likewise, we started working on our second Pilot Machine composed of 22 machines across all the divisions and the same results were obtained in which the unplanned breakdowns were dramatically reduced. As we achieved close to zero unplanned breakdowns on the first 22 pilot machines, we also improved the machines' MTBF and downtime trends. A total of 436 machines underwent the Planned Maintenance Activities. The Breakdown Occurrences graph in figure 9.20 speaks for itself. All improvements under Planned Maintenance Phase 2 performed by each department were looped back to the

IFCA or EEM Committee group for control purposes. Hence, if similar equipment will be purchased in the future, improvements and modifications will be discussed with the OEM designers to be built in before being purchased. All improvements generated by the whole Planned Maintenance team were documented, recorded, and controlled.

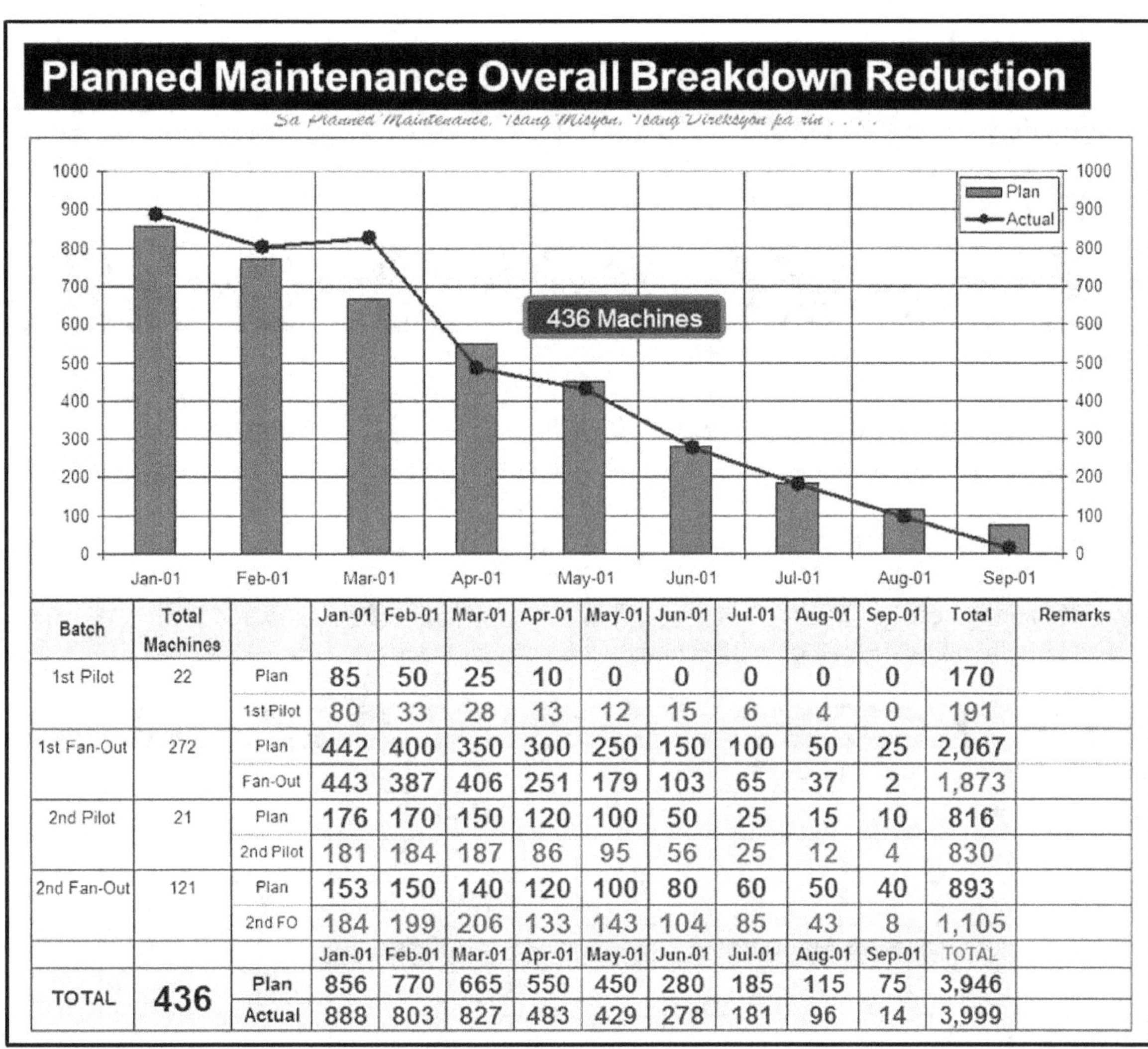

Batch	Total Machines		Jan-01	Feb-01	Mar-01	Apr-01	May-01	Jun-01	Jul-01	Aug-01	Sep-01	Total	Remarks
1st Pilot	22	Plan	85	50	25	10	0	0	0	0	0	170	
		1st Pilot	80	33	28	13	12	15	6	4	0	191	
1st Fan-Out	272	Plan	442	400	350	300	250	150	100	50	25	2,067	
		Fan-Out	443	387	406	251	179	103	65	37	2	1,873	
2nd Pilot	21	Plan	176	170	150	120	100	50	25	15	10	816	
		2nd Pilot	181	184	187	86	95	56	25	12	4	830	
2nd Fan-Out	121	Plan	153	150	140	120	100	80	60	50	40	893	
		2nd FO	184	199	206	133	143	104	85	43	8	1,105	
			Jan-01	Feb-01	Mar-01	Apr-01	May-01	Jun-01	Jul-01	Aug-01	Sep-01	TOTAL	
TOTAL	436	Plan	856	770	665	550	450	280	185	115	75	3,946	
		Actual	888	803	827	483	429	278	181	96	14	3,999	

Figure 9.20: Breakdown Occurrences for the KNS Pilot Team

Likewise, all improvements generated by the Planned Maintenance were also analyzed in terms of their cost as we can see in these cost comparisons between before and after improvement that the cost of modification is more expensive in the short term but the savings generated is felt in the long term basis.

Case Study 1: Resizing the Width of AHU Filter from (24 x 24 x 6) to (24 x 24 x 12)

- Division: Facilities
- PM Phase 2
- Team Leader: Arman Jusay
- Model Machine: AHU-401
- Impact: Lengthen the lifespan of the filter from 1800 to 3600 hours (Note: 1 USD = 51 PHP)

Details	Before	After	Unit
Filter Size (L x H x W)	24 x 24 x 6	24 x 24 x 12	inches
Cost per Unit	$ 47.00	$ 60.00	USD
Total Filters per Air Handling Unit (AHU)	16	16	pieces
Filter Costs per AHU	$ 752.00	$ 960.00	USD
Total Air Handling Units	7	7	units
Average Life in hours	1800	3600	hours
Average Life in years	0.2055	0.41096	years
Cost for 7 AHU per year	$ 25,615.57	$ 16,351.96	USD
Yearly Savings in USD	**$ 9,263.62 per year**		**USD**
Year Savings in PHP	**PHP 472,444.62 per year**		**PHP**

Figure 9.21: Case Study 1: Resizing the Width of the AHU Filter

Case Study 2: Modification of Flywheel Belt for MECO Plating Machine

- Division: Central Lead Finish
- PM Phase 2
- Team Leader: Jojo Santos
- PM Committee: Cesar dela Torre
- Model Machine: Meco 2
- Impact: Lengthen the lifespan of the flywheel belt from 5 to 24 weeks

Details	Before	After	Unit
Replacement Frequency of Belts	4 to 5 weeks	24 weeks	weeks
Cost per 1 set	$ 600.00	$ 2,200.00	USD
Usage per Machine	2	2	pieces
Costs per Machine	$ 1,200.00	$ 4,400.00	USD
Total Fan Out Costs for 4 Machines	$ 4,800.00	$ 17,800.00	USD
Change in Strength of Materials from	Tool Steel	Stainless Steel	material
Average Life in Weeks	4 to 5 Weeks	24 weeks	weeks
Average Life in Years	0.0961538	0.4161538	years
Yearly Savings in USD	**$ 49,475.02 per year**		**USD**
Year Savings in PHP	**PHP 2,523,266.02 per year**		**PHP**

Figure 9.22: Case Study 2: Modification of Flywheel Belt for Meco Machine

Case Study 3: Modification and Localization of Gripper Plate

- Division: Central Lead Finish
- PM Phase 2
- Team Leader: Jojo Santos
- PM Committee: Cesar dela Torre
- Model Machine: Meco 2
- Impact: Lengthen the lifespan of the gripper plate from 30 to 110 days

Details	Before	After	Unit
Costs per clip	$ 0.33	$ 0.33	USD
Usage per Equipment	6,000	6,000	pieces
Costs per Equipment	$ 1,980.00	$ 1,980.00	USD
Total Costs per Year	$ 24,090.00	$ 6,750.00	USD
Total Fan Out Costs for 4 Meco Machines	$ 96,360.00	$ 26,280.00	USD
Average Life in Days	30 days	110 days	days
Average Life in Years	0.082191781	0.301369863	years
Yearly Savings in USD	**$ 70,080.00 per year**		**USD**
Year Savings in PHP	**PHP 3,574,080.00 per year**		**PHP**

Figure 9.23: Case Study 3: Modification and Localization of Gripper Plate

Case Study 4: Modification of blurred illuminations from 1200 milli candela to 4100 Milli-Candela

• Division: MQFP Wirebond

PM Phase 2

• Team Leader: Erwin Almosera

• Model Machine: QPKT – 82 (Wirebond KNS Machine)

• Impact: Modification of the Candela and reducing PRS errors

Details	Before	After	Unit
Costs per assembly	$ 476.00	$ 476.15	USD
Total KNS Machines	149	149	pieces
Total KNS Costs for all 149 Machines	$ 70,924.00	$ 70,946.35	USD
Average Life in Hours	35,040 hours	39,420 hours	USD
Average Life in Years	4 years	4.5 years	days
Yearly Costs in USD ($)	$ 17,731.00	$ 15,655.08	years
Yearly Savings in USD	**$ 2,075.92 per year**		**USD**
Year Savings in PHP	**PHP 105, 871.92 per year**		**PHP**

Figure 9.24: Case Study 4: Modification of Illumination from 12000 to 4100 milli candela

9.8: Performing the Planned Maintenance Audit

Each Phase of Planned Maintenance will require an audit except for Phase 0, which is the Preparatory Phase. The audits are done as the Planned Maintenance completes each phase to check that the team complies with the activities performed on Planned Maintenance and if the teams are ready to challenge the next phase on doing Planned Maintenance. For Phase 1, an initial requirement is that the team should have completed at least 80 % or more of their restoration activities before they can request to be audited or certified. If the team passes the audit process, the equipment will be certified in Phase 1, and the team to proceed to Phase 2 of Planned Maintenance. For Phase 2, at least 90 % of the parts with inherent design weaknesses have been corrected and evaluated on the machine or equipment.

The Planned Maintenance audit is initially done by a TPM Facilitator together with the maintenance managers from other areas of the plant. Maintenance managers should own the Planned Maintenance Audit since they will also benefit from these activities, and they are also in a way responsible for their equipment since they are the owners. During the initial audit, the Planned Maintenance facilitator briefs the maintenance manager on what to look for during the audit process. An audit guideline will be provided detailing what to check. The Planned Maintenance audit will compose of two parts. They will audit both the team and the equipment.

Team Audit: The first part of the audit will be a team presentation, which can be done on the activity board together with the auditors. The team briefs the auditors on their respective activities on Planned Maintenance. The presentation usually takes 30 minutes to 1 hour or more, where the team will highlight their activities. To make sure the team is ready to take on the audit, the Planned Maintenance team performs a self-audit on themselves to gauge their confidence before proceeding with the actual audit. Once they are confident enough, the team will send a request to the TPM Office for an actual schedule for the audit to take place. The objective of doing the audit is to see to it that the team takes pride in their success in reducing the breakdown rate of their equipment before advancing to the next phase of Planned Maintenance.

Equipment Audit: After the team presentation is completed, both the PM auditor together with the team proceed to the equipment, and the auditor checks the validity of the documents concerning the actual condition of the equipment. They will also look for other unwanted items on the machine. If something is found, it will be included in their pending requirements. The team cannot move on to the next phase until the requirements of the audit have been complied with. The maintenance managers performing the audit will ask the teams about the benefits of doing Planned Maintenance and how they were going to sustain the activities, they had performed, so as not to let the equipment deteriorate and be left unattended once again. Standards being set must be performed since this will serve as a means to sustain the equipment, otherwise, if these standards will not be carried out, then everything will just revert back to the way it was before, and nothing will ever change. Sustaining the equipment should be done at all measures.

After some question and answer portion between the team and the Planned Maintenance auditors, the auditor will summarize their findings and inform the team. The audit is now concluded. The team will be graded for their performance. If the team attains the standards and requirements as expected from them, they will deserve a passing mark, which is usually a minimum rating of 85% and above, and they will move on to the next phase of Planned Maintenance. If many items should be addressed by the team that was found during the audit, the PM team may receive a remark of an unconditional passing score unless otherwise all pending requirements are settled on the date both agreed by the auditor and the team. It is highly recommended not to provide a failing mark for the team unless they have not actually done the Planned Maintenance activities, as this can de-motivate the team. Remember that the audit should be done both on the team and on the equipment itself.

9.9: Horizontal Replication of Improvements

Horizontal Replication or in TPM words commonly known as fan-out is the repetition or replication of improvements from the pilot to other similar equipment in the plants. But there is a general rule to follow in this case since even if the equipment is identical in model, vendor, and make, the problem one machine possesses might not be at all present in other equipment, especially if this is a case of human error. I remember during my employment days when I was still working in the semiconductor industry, there was an improvement group called B.K.M (Best Known Method); and the objective of this group is that all improvements, modifications, redesign, or changes made to any equipment should be replicated or fan-out with all equipment of similar types and maker whether they possess the same problem or not. I was totally against this group (since I attended RCM previously before that and learned about the equipment operating context). I used to argue with the implementers of this group a lot. My argument is that if I have 100 similar equipment in the plant of the same type and make and if only 3 of them emit the same problem, my point is that only the three equipment should be modified and not all 100 equipment. In this case, the B.K.M. group insists on carrying the improvement to all 100 pieces of equipment, which is not the right thing to do. I asked this question to a few friends of mine who are also experts in maintenance and reliability and here are their responses.

PERFORMING EQUIPMENT ON IDENTICAL EQUIPMENT

DETERIORATIONS UNCOVERED	1	2	3	4	5	6
• Air leak in the system	●	●	X	X	○	○
• Excessive belt tension	●	●	●	○	●	○
• Excessive vibration on the fan wheel	●	●	●	○	X	○
• Soft foot on the base foundation	●	●	●	X	X	○
• Worn out bearing	●	X	●	X	●	○
• Worn out drive coupling	●	X	○	○	●	○
• Excessive grease in ball bearings	●	○	●	●	○	○
• Excessive misalignment of shaft	●	○	●	○	●	○
• Abnormal end thrust load	●	X	X	X	X	X
• Incorrect direction of rotation	●	X	X	X	X	X

○ ⟶ Restoration still ongoing   ● ⟶ Restoration completed   X ⟶ Deterioration not present

Figure 9.25: Horizontal Replication of Improvements

• Hello Rolly, I agree with you that it is not always a good idea to spend money and make changes just because a similar piece of equipment needs it. Equipment is like our children; they need individual attention but to be held to a standard that meets the needs of the whole. Thank you for your email. *By Nathan Wright (Maintenance Book Author)*

• Rolly, I have never thought about this deep question. I think of the airplane makers, and when they find an improvement, they spread it out to the entire fleet. Besides, when an RCM analysis turns up a failure mode with a dire consequence, all the units of the same type should have the modification. I am happy being wrong on this but would side with quality. ☹ *By Joel Levitt, Maintenance and Reliability Book Author*

• Rolly, your point is well taken, but consider the following, if the identified problem is an inherent design deficiency that would be common to all 100 assets, in this case, the modification should be made to all 100. However, if it is not an inherent design deficiency and the failures are limited to one or two of the assets in specific areas or applications, there is no value and a high potential for creating more problems in modifying all 100. As a general rule, when problem-solving, you determine whether the problem is unique to one asset or to all common or similar assets. In the former, the problem is unique to that asset, and if all similar machines exhibit problems, then the problem is systemic, e.g., caused by something common to all similar assets. I would suggest that you not apply one rule for all cases. Instead, determine to the best of your ability, whether the problem and modification are unique to one or a few assets or systemic and exhibited by all. Quality rollout practice is common, and many consider it best practice. Still, only in cases where the modification is warranted typically, the root cause is an inherent design weakness common to all assets in the family or a systemic (infrastructure) problem. *By R. Keith Mobley, a Book Author,*

Although Joel Levitt's point here disagrees with my thoughts, I fully agree with him since he was talking about the airline industry, and we all know the consequences of failure if a plane crash. I have no argument with this but only for manufacturing equipment and assets where there will be no media, crooked old politicians, insurance people, and lawyers mingling around after your equipment failed. It is just you and your boss. Many years ago, I had a friend and mentor. We used to call him Mang Tibo. In the Philippines, the word Mang means elderly. This person is from the hard knocks. I asked him this question a long time ago if I made an improvement on this equipment to address a specific problem, should I replicate it with the other equipment too. He looked at me for a while and told me a story. He said that he was the bread earner of his family, and he had eight sons. One day, his youngest son got the flu and at the same time have a high fever. Mang Tibo went to a drugstore to get some medication for his son, who was sick. When he went home, he gave his son the medicine. He called all his sons and gave them the same medicine even if they were not sick. The eldest son said that he was not sick, and Mang Tibo said to just follow his orders. After one hour, his second eldest son got some rashes on his skin and has difficulty breathing. He was allergic to the medication given to him. Mang Tibo rushed him to the doctor immediately and was confined for a couple of days. He said, from that moment forward, he learned his lesson the hard way. Mang Tibo never answered my question directly, but his story provided a clear answer to that.

Usually, Quality people do these kinds of things in industries. Their reason is quite obvious as they adhere to the concept of JIC or Just in Case. My question is what if it is a case of JICIDNH? You might be wondering, what on this planet is that ugly acronym? It means, what if "**Just in Case it did not happen**? Here is my concern with the B.K.M. group. Any two machines can be identified if they are of the same type or maker. However, even if the equipment were identical, the operating context may not be exactly and precisely identical in every sense. Different operators can operate the equipment differently and I believed that the operators are not twins and do not have the same DNA. The equipment may not run with the same operating hours and so on. This is what I meant by not exactly having the same operating context statement. The second is the cost of the modification. This means that if the modification will be carried out to the entire 100 pieces of equipment and the cost of modification is $ 1,000 per equipment, then the total costs of modification will be $ 100,000 instead of only $ 3,000.00. I rest my case on this, and I believe that only equipment with the same problem should be included in the modification, for those that are stable, the best thing to do is to leave it alone and I think that's all I have to say about that.

Chapter 10

The Next Challenge - Extending Equipment's Life Cycle

> *In lubrication, many assumed that the word clean and the word new are the same. These are two different words. Just like a child, if the child is fat, it does not necessarily mean that the child is healthy because the word fat and healthy are not the same. In fact, it may be the opposite. For equipment with lubricating systems, one thing is for sure, the cleaner the oil, the longer the life span of the equipment and the lesser the cost of operating and maintaining them.*

10.1: Reducing Oil Contaminants for Longer Equipment Lifespan

One of the biggest problems that shortened the life of our equipment and assets has something to do with oil contamination. First, let us define what oil contamination is. Oil contamination in oil is anything that should not be present in the oil besides its base and additives. These contaminants may be in the form of solids, liquids, or even gases in bubbles that caused foaming. Contamination can enter the system in two ways through ingression, which means that it came from the outside and finds its way inside the equipment or it can be generated inside the system through the wear process. These build-in or generated inside means that it is internally generated and already inside the equipment through the process of wear. Internally generated contaminants are mostly the result of the different wear processes occurring inside our equipment, such as abrasion, adhesion, erosion, fatigue, and corrosion. These build-in contaminants obtained through the wear process are usually in the form of tiny metal debris that can be traced through an oil analysis program called Wear Metal Debris Analysis. Other non-metal elements can also be detected, often caused by the depletion of additives, sludge, gums, and varnishes, which usually start to occur once the oil starts to oxidize.

Contaminants can also be induced by maintenance through top-up or when changing oil itself without proper precautions taken, and ignorance on how oil is being contaminated. As I have covered in my other books, new oil is not necessarily clean as it also contains contaminants. There is a test that already shows that new oil contains contaminants. Even a small amount of contaminants ingested into the system can cause catastrophic failures when the contaminants can be trapped in very tight clearances, which can cause three-body abrasion and generate a new set of contaminants. The process of abrasion causes another batch of contaminants that flows together with the oil. The majority of these oil filters placed on the equipment have a pore size of around 30 to 40 microns, which does not capture those smaller contaminants. Hence, it is important to use bypass filtration with the correct Beta Rating and Efficiency when transferring oil to the system during top-up or changing oil. The problem with these contaminants is that they are invisible to the human eye. This is because our naked eye can only see a range of 40 microns and above. Smaller than these would not be visible and we might be thinking that these invisible contaminants do not exist at all. The fact lies that they do exist and these contaminants are the ones causing more harm and damage to our equipment and assets.

According to lubrication experts and tribologists, the smaller the contaminant, the more destructive it can be inside our machines and equipment since these smaller contaminants are likely to be trapped in tight metal clearances which can cause abrasion and generate a new set of solid contaminants. One may wonder that since this is a metal contaminant, it will just settle at the bottom. The truth is, that these contaminants will suspend in the oil. The oil contains an additive called dispersant. The function of these additives is to suspend the contaminants in the oil so that it joins the flow, which means that they can have a better chance of being captured by the oil filter.

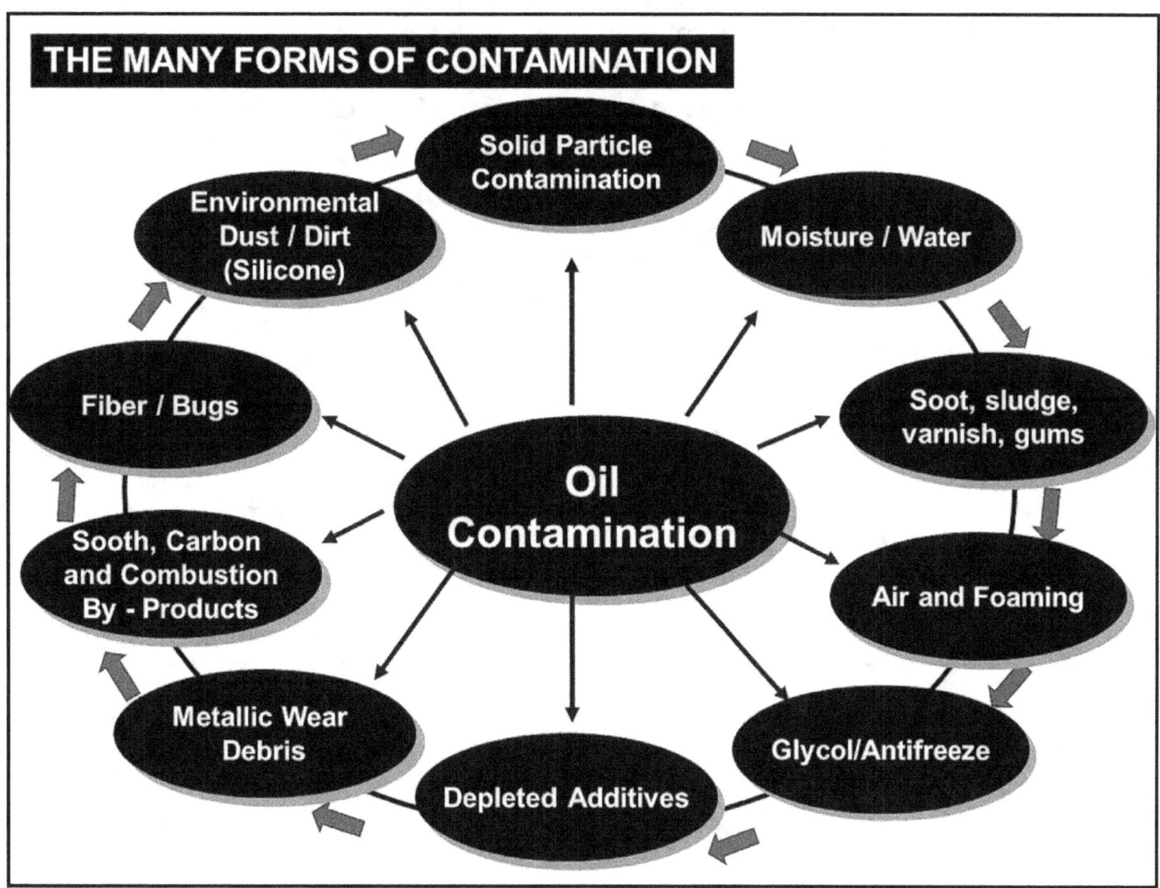

Figure 10.1: The Many Forms of Oil Contamination

Contaminants can also be ingested, which means that it is being ingressed into the equipment. Dust and moisture come from the environment. Usually, an oil analysis that reports an element called Silicon definitely means that there is dust and dirt ingression. Dust is a hard particle and can cause abrasion between clearances, especially when the contaminants' size is larger than the clearance. The more contaminants mean that the more the equipment is prone to failure. This means that the more contaminants present in the oil, then the more chances of failure the equipment can experience as there is a direct relationship between the number of contaminants and the operating age of the equipment. These contaminants present in the equipment and machines cannot be extracted or removed 100% by the existing oil filter since most oil filters placed on the equipment are nominal in rating, but the good news is that 99.9% of these solid contaminants can be removed through specialized absolute filtration with high efficiencies and beta rating. It is important for the maintenance people involved in the lubrication process to understand how contamination eventually leads to failure and how we can reduce the number of contaminants present in our system through several means, such as absolute filtration, centrifuge, and others.

The absolute rating of a filter refers to the smallest size particle that will be removed during filtration while nominal ratings refer to the average particle size that will remain in the fluid after filtration. A nominal five-micron filter will trap particles around 5 microns in size. An absolute 5-micron filter will guarantee to trap all particles 5-microns in size and larger. Tests have shown that particles as large as 200 microns will pass through a nominal rating of a 10-micron filter. An absolute rating gives the size of the largest particle that will pass through the filter or screen. This is the size of the largest opening in the filter.

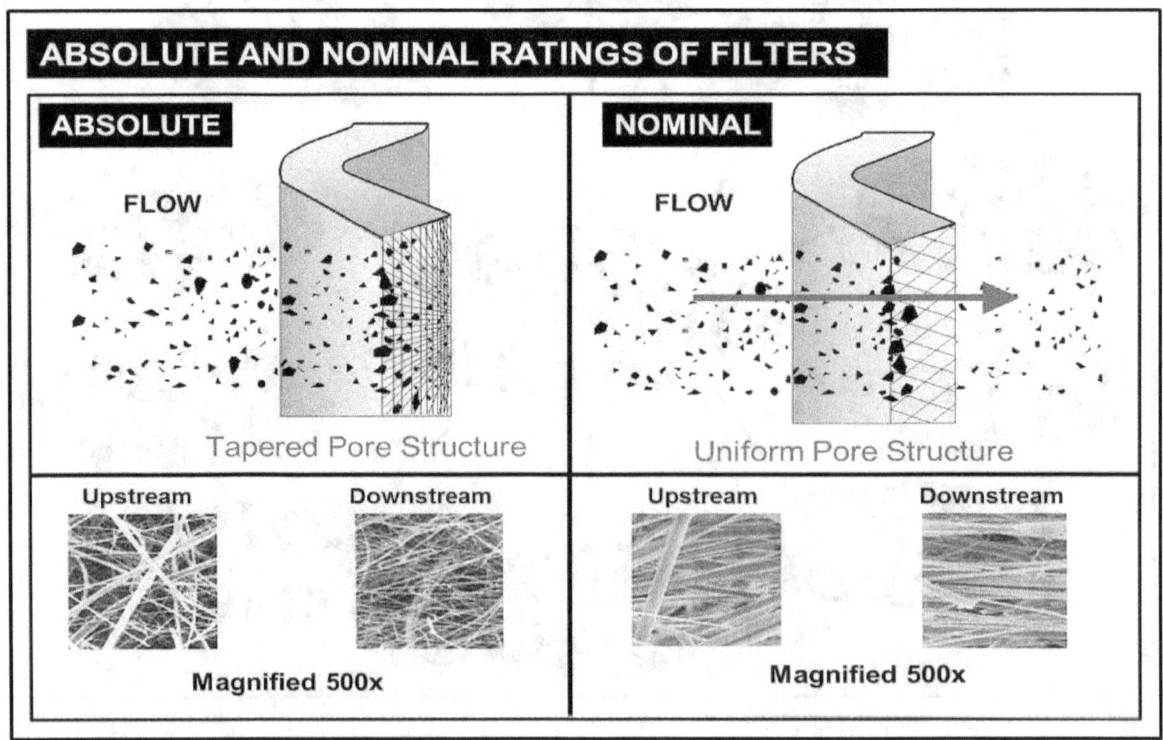

Figure 10.2: Comparing Absolute and Nominal Filtration

Figure 10.3 is a Table for Life Extension Factor if contaminants can be reduced from the lubricating oil. This means that in the given table, if the cleanliness of the oil can be reduced from 22/20/17 to 18/16/13, then the life of the equipment can be increased by a factor of 3. This means that if the life of the equipment is 10 years, then this can be extended to 30 years. For the reader's benefit, first, let us explain what these codes are all about.

LIFE EXTENTION TABLE FOR ISO CLEANLINESS LEVEL FOR HYDRAULIC AND DIESEL ENGINES										
ISO CODE	21/19/16	20/18/15	19/17/14	18/16/13	17/15/12	16/14/11	15/13/10	14/12/9	13/11/8	12/10/'7
24/22/19	2	3	4	5	6	7	>10	>10	>10	>10
23/21/18	1.5	2	3	4	5	7	9	>10	>10	>10
22/20/17	1.3	1.6	2	3	4	5	7	9	>10	>10
21/19/16		1.3	1.6	2	3	4	5	7	9	>10
20/18/15			1.3	1.6	2	3	4	5	7	>10
19/17/14				1.3	1.6	1	3	4	6	8
18/16/13					1.3	1.6	2	3	4	6
17/15/12						1.3	1.6	2	3	4
16/14/11							1.3	1.6	2	3
15/13/10								1.4	1.8	2.5

Source: From Noria Corporation

Figure 10.3: Life Extension Table for Fluid Cleanliness

ISO Code 4406:99	From	To
26	320,000	640,000
25	160,000	320,000
24	80,000	160,000
23	40,000	80,000
22	20,000	40,000
21	10,000	20,000
20	5,000	10,000
19	2,500	5,000
18	1,300	2,500
17	640	1,300
16	320	640
15	160	320
14	80	160
13	40	80
12	20	40
11	10	20
10	5	10
9	2.5	5
8	1.3	2.5
7	0.6	1.3

Figure 10.4: ISO 4406:99 Contamination Code Table

All lubricating systems have their corresponding ISO 4406 Cleanliness Code Standards, as reflected in figure 10.4. The aim of the ISO 4406 Cleanliness Standard is to get as near as possible to these standards, depending on the type of lubricating oil being used or even lower to be assured that our lubricating oil is clean at all times. There is a direct relationship between the number of failures and the number of contaminants present in the lubricating oil. This means that the lower the ISO Codes, the lesser will be the failures and the cleaner the oil. These codes can be measured using an oil analysis instrument called a particle counter.

The particle counter is the most recognized and widely used oil analysis instrument. This method aims to check the cleanliness by determining the total number of particles in several different size ranges. Service oils used must be within the target cleanliness level. Particle Counters can measure particles from 2 to 100 microns depending on the brand. This means that an increase in the number of particles can indicate a potential failure. These instruments also can quantify particles in different size ranges. This quantification is often referred to as the particle size distribution. Particle count is the procedure recommended by the ISO for establishing fluid cleanliness standards. It measures both metallic and non-metallic particles alike. The objective of having a particle counter is to control the fluid cleanliness levels and not trending dirt levels.

Most particle counter is equipped with a laser, and this laser is used to estimate the number of contaminants present in the oil analysis sample. Once the path of the laser beam is blocked by the contaminant, it will automatically count. This test aims to determine the number of contaminants present and control the cleanliness of the oil. The result of the particle counter will be provided in either NAS or ISO standards. Although I would prefer to use the ISO count as it provides a three-digit figure, which shall be explained later. There is, however, a conversion from NAS to ISO and vice versa. The ISO count is always done in three digits. The standard for contamination class use is ISO 4406-99.

The particle counter will provide a three-digit code such as 18/19/20. The first digit 18 refers to the small contaminants usually equal to or greater than 4 but less than 6 microns. The ISO 4406 Contamination Code in figure 10.4, estimates the amount of contaminant from the oil. The first digit 18, estimates that there are around 1,300 to 2,500 contaminants. The second digit 19 refers to the number of contaminants equal to 6 but less than 14 microns. In the table, an ISO Code of 19 will estimate the contaminants from around 2,500 to 5,000. The last digit 20 refers to all contaminants equal to or larger than 14 microns. Again, in the table, the number of contaminants will range from around 5,000 to 10,000. The ISO 4406 table simply implies that the higher the codes, then the dirtier or the more contaminants are present in the oil.

All lubricating systems have targeted cleanliness standards as indicated in figure 10.5. This means that the cleanliness of the oil should be within these standards or lower at all times. For example, in figure 10.5, the cleanliness standard for turbine oil is 17/15/12, which means that the cleanliness of the turbine oil must be within this level. However, if we can obtain a reading lower than 17/15/12 then the better and the cleaner the oil. The ISO cleanliness codes simply mean that the higher the ISO Code, then the more the oil is contaminated. Hence, placing a by-pass filtration permanently or cleaning the oil using an offline filtration whenever the reading of the particle counter exceeds and goes beyond its cleanliness standard can assure us that the oil is maintained clean at all times. Although in figure 10.5, the hydraulic oil should have a cleanliness standard of 16/15/12 or lower, however, if a servo valve is present inside the hydraulic system, then we need to aim at having a cleanliness standard of 13/12/10 for the hydraulic oil since the clearance of the servo valve is small at 2 microns. This means that particle sizes of 2 to 3 microns can be trapped between the clearances of the servo valve, which can cause more abrasion. Although during my employment days, more than a couple of decades, we only have these expensive particle counters, which cost around 18,049.00 European Dollars or more than 1 million in Philippine pesos. Today with these digitalization and smart sensors as in figure 10.6, we have these inline particle counter sensors which will be way less expensive that can monitor the cleanliness of the lubricating oil for 24 hours. However, if you have 20 hydraulic systems in the plant, then you need 20 of these portable inline particle counter sensors.

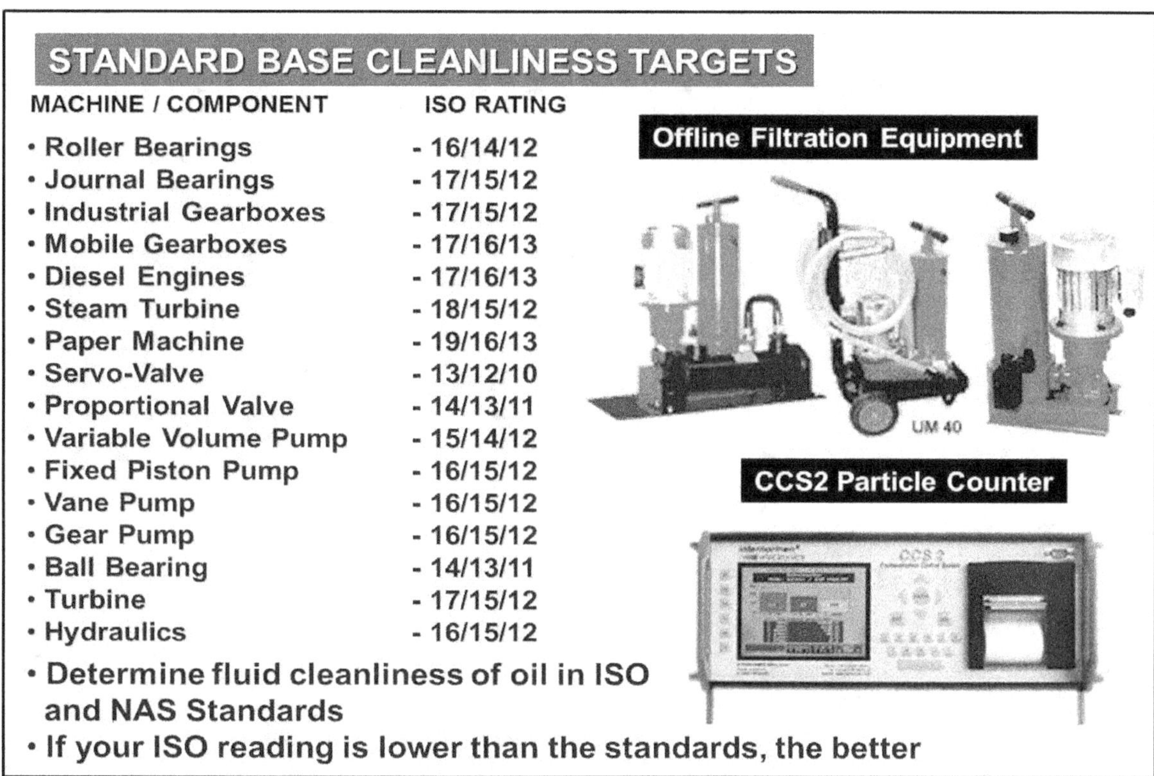

Figure 10.5: Standard Cleanliness Targets for Different Lubrication Applications

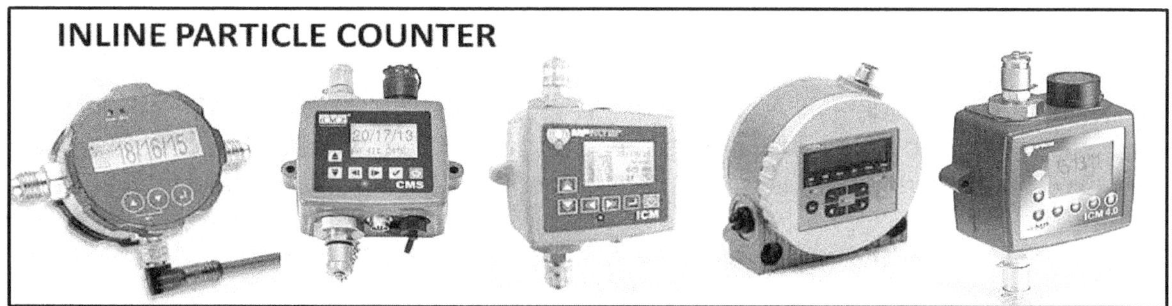

Figure 10.6: Inline Particle Counter

Let us give a case study of two oil filters as in figure 10.7. One oil filter has a nominal Beta Rating of 2. This means that the filter has an efficiency of 50%. The other filter has an absolute Beta Rating of 1000. What does this mean? This simply means that the efficiency of this absolute filtration is 99.9%. The efficiency of the filter refers to the number of solid particles or contaminants that the filter element can remove. Let us assume that the pore size of the oil filter is 5 microns, with an efficiency of 50% with a Beta rating of 2. This means that for every 100,000 contaminants in the size of 5 microns and larger that will pass through the upstream of the filter, the oil filter is only designed to capture 50% of the contaminants, which is around 50,0000. So what happens to the other 50,000 contaminants? The other 50,000 contaminants will escape downstream of the filter together with the oil and goes back once again into the system. On the other end, if we have an absolute filter of 5 microns in size with a rating of 99.9% absolute, this means that the filter is guaranteed to trap all five-micron in size contaminants and larger 99.9% of the time.

Figure 10.7: Which Oil Filter Will Save You on Costs?

Now we understand how both absolute and nominal filtration works, the next question is how much will the absolute filter cost. Both Oil filter A and filter B cost 490. When asked which filter you would prefer to use in the equipment, I'm pretty sure that the reader would prefer Oil Filter B. The only difference, in this case, is that the currency used in Oil Filter A is in Philippine Peso, while Oil Filter B is in US dollars. (Note: 1 USD is equal to 50.00 Peso as of this writing) Hence, the cost of Oil Filter A is roughly 8.00 USD, while Oil Filter B is priced at USD 490.00. Absolute filters will be more expensive when we speak about their initial cost, but the difference in the cost proves to be seen when operating the equipment for 6 months or even less using an absolute filtration, which will very much favor Oil Filter B. The savings that will be generated through the use of absolute filtration will very much far outweigh its initial cost in the long term.

Beta Rating and Efficiency				
		Number of Contaminants		
Beta Rating	Efficiency %	Filter Upstream	Trapped by Filter	Filter Downstream
2	50.0000%	100,000	50,000	50,000
4	75.0000%	100,000	75,000	25,000
10	90.0000%	100,000	90,000	10,000
20	95.0000%	100,000	95,000	5,000
30	97.5000%	100,000	97,500	2,500
60	98.3300%	100,000	98,333	1,667
75	98.6700%	100,000	98,667	1,333
100	99.0000%	100,000	99,000	1,000
125	99.2000%	100,000	99,200	800
150	99.3333%	100,000	99,333	667
200	99.5000%	100,000	99,500	500
300	99.6667%	100,000	99,667	333
500	99.8000%	100,000	99,800	200
1,000	99.9000%	100,000	99,900	100
2,000	99.9500%	100,000	99,950	50
4,000	99.9750%	100,000	99,975	25
5,000	99.9800%	100,000	99,980	20
10,000	99.9900%	100,000	99,990	10
20,000	99.9950%	100,000	99,995	5
50,000	99.9980%	100,000	99,998	2

Figure 10.8: Beta-Rating and Efficiency of Oil Filters

On the other side, if we speak about the Procurement or Purchasing Department, the decision to purchase will be more tempting for Oil Filter A since this is much cheaper in the initial cost, and tell me if I am wrong in this matter. What these people failed to understand is the chaos and havoc it will bring to the equipment. Nominal oil filtration is not designed to capture submicron particles that can get their way into tight clearances, which can cause abrasive wear of metal parts. The difference between these two filters is not only about the initial cost but, more importantly, about the filter's effectiveness and efficiency rating in removing these contaminants, which are designated by the Beta Rating of the oil filter itself. In this case, we are not only speaking about the comparison in terms of the initial cost of the two filters but we are also speaking about the lives of the people affected by this decision. Imagine the day-to-day failures, pressure, frustrations, and loss of revenue as a result of this on the part of the person involved. This is perhaps a non-measurable or intangible measurement, but the effects on the benefits that can be gained have something to do with the Life Cycle Costs of the filter itself, and that is what really matters most for industries.

Dynamic Clearances	Size in Microns
Servo Valves	1 to 4 microns
Servo Valve Flapper/Nozzle Spacing	40 to 80 microns
Servo Valve Spool to Sleeve, Radial	1 to 4 microns
Proportional Valves	1 to 6 microns
Directional Control Valves	2 to 8 microns
Journal Bearings	0.5 to 100 microns
Roller Element Bearings	0.1 to 3 microns
Hydrostatic Bearings	10 to 25 microns
Anti Friction Bearings Radial	0.5 to 1 micron
Gear Pump (Gear to Side Plate)	0.5 to 5 microns
Gear Pump (Tip to Case)	0.5 to 5 microns
Vane Pump (Tip to Vane, Estimate)	0.5 to 1 micron
Vane Pump (Sides of Vane)	5 to 13 microns
Piston Pump (Piston to Bore, Radial)	5 to 40 microns
Piston Pump (Valve to Plate Cylinder)	0.5 to 5 microns
Control Valve Orifice	130 to 10,000 microns
Control Valve Spool to Sleeve (Radial)	1 to 23 microns
Control Valve Poppet Type	13 to 40 microns
Control Valve Disc Type (Estimate)	0.5 to 1 microns

Figure 10.9: Clearances Between Metal Parts

Obviously, the filter that can trap more contaminants will provide better and smoother operations compared with a nominal rating oil filter. How is this so? Usually, the design of an absolute filter, let us say 5 microns absolute is that the pore size of the upstream or

242

where the oil and contaminants will enter the filter will be larger say at 6 to 7 microns, but the downstream which is where the oil will return back to the system will be smaller let us say at 4 microns making it certain that almost all 5 microns contaminants in size will be captured by the oil filter. However, this will not happen with a nominal 5-micron oil filter since it is only designed to capture 50% of all contaminants in the size of 5 microns and above. This means that the fewer the contaminants in the oil, the lesser the chances of abrasion to occur and the longevity of the parts.

According to tribologists and the people who specialized in the subject of lubrication, the smaller the contaminants, then the more destructive they can become inside the equipment since these are the contaminants that are possible to be trapped inside tight clearances between two mechanical parts that can result in abrasion which can generate more abrasive metal parts replicating the abrasion process. The problem is that these are the contaminants too small to be removed by the nominal filters since these filters are only designed to remove 50% of the contaminants in the size of 30 microns and larger.

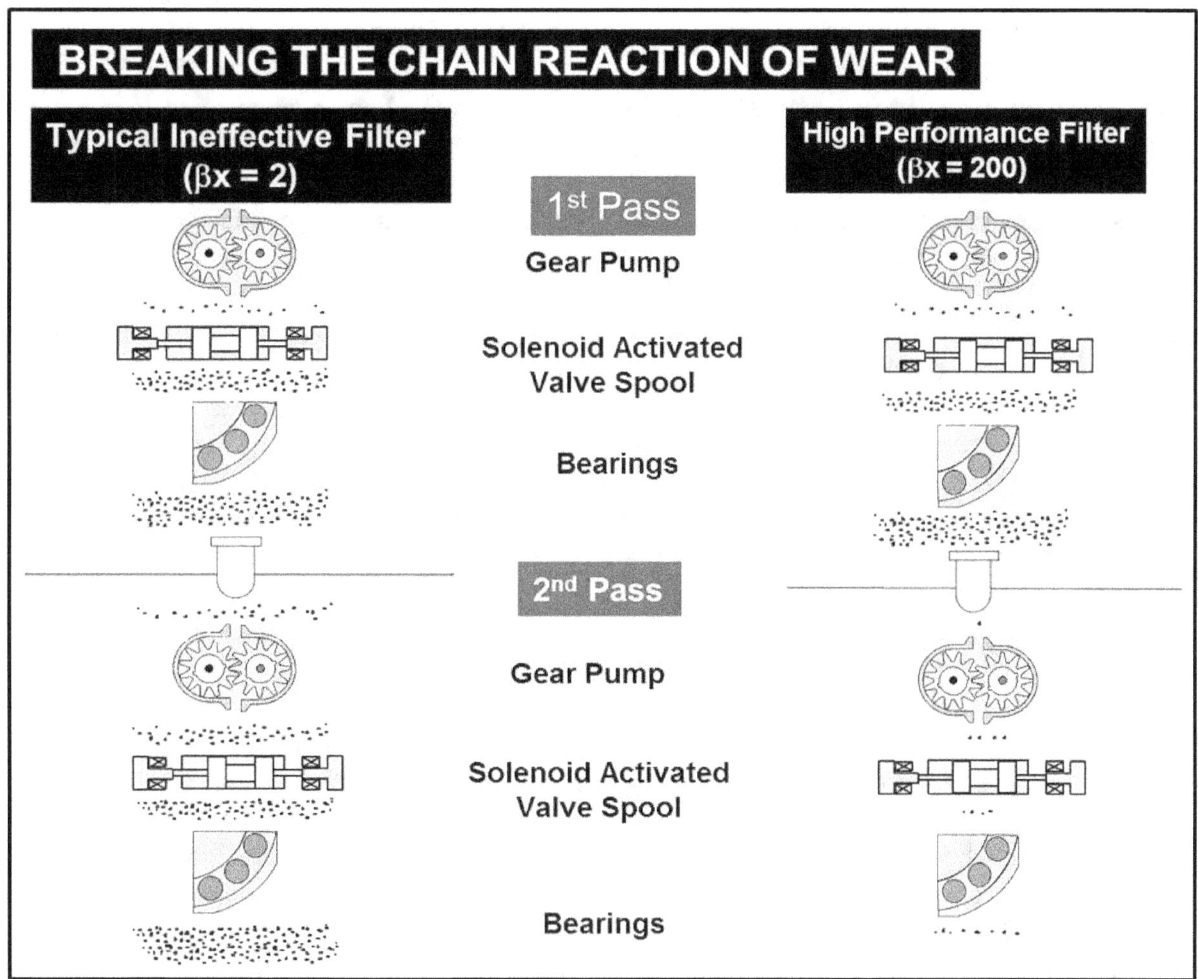

Figure 10.10: How Nominal and Absolute Filtration Works

Therefore, the higher the beta rating of an absolute filter, the better and more effective it can be in trapping these contaminants. Most OEMs will recommend using a nominal filtration so that both filters and oil will be changed. While using absolute filtration and complying with the ISO 4406 cleanliness codes will definitely reduce abrasion occurring on mechanical parts and extend the interval of changing the oil itself. The cleaner the oil, the fewer breakdowns will occur on the equipment. This means that initially, we may be paying for a more expensive oil filter compared to what we are currently using, but the benefits and cost savings of these oil filters generated will be felt in the long run. Although the OEM or vendor will not recommend replacing the oil filter with a low micron, rating by-pass filtration since this will create oil starvation. This is similar to when we take a shower and it floods since the sink is clogged with hair. Once we remove the hair then the flow becomes normal once again. That is why they called them bypass filtration since we are not replacing the original filter but only including a secondary filtration that is entirely separated from the main system. My question is, why do OEM and equipment manufacturers do not provide a secondary filtration that can remove these smaller contaminants. Well, the answer, in this case, is pretty much obvious. Here are just a few industries that have benefited from adopting oil contamination control and the use of absolute filtration in their plant.

• GM Corporation test shows that 82% of internal wear comes from contaminants less than 40 microns.
• SKF states that improper lubrication accounted for 54% of bearing failures.
• Ford states that 80% of hydraulic failures can be treated as particulate contamination.
• Cummins Technical Center indicates that wear can be reduced by 91% using a bypass filter combined with a full-flow filter.
• SAE states that contamination in the lubricant of engines, transmissions, and hydraulic systems causes 70% of equipment failures.
• Canadian National Research Council studies indicate that contamination was found to be the leading cause of wear in a variety of industries investigated. In fact, 82% of all wear was found to be particle-induced
• Nippon Steel reduced bearing failures by 50% through aggressive contamination control.
• International Paper's Pine Bluff Mill reported a 90% reduction in bearing failures again through an aggressive contamination control program.
• Caterpillar claims indicate that dirt and contamination are the number one cause of hydraulic system failures. This means that with regards to hydraulic systems, they must be kept clean at all times.
• British Hydromechanics Research Association, which covered a three-year study on 117 hydraulic machines composed mainly of injection molding, machine tools, mobile, construction, marine, and metals, concluded a dramatic relationship between fluid contamination and its service life.
• General Motors AC Delco Division tested their diesel engines and found an 8-fold improvement in wear rates and engine life with lower lube oil contaminant levels. GM reports that compared to a 40-micron filter, engine wear was reduced by 50% with 30-micron filtration, and wear was reduced by 70% with 15-micron filtration.
• Albertson's Inc, a supermarket chain, conducted a study on a series of over-the-road Cummins tractor engines that found reduced wear rates with greater lube oil cleanliness. After analyzing 6 engines having 600,000 miles, engine crankshaft journals showed only 0.0005 inches of wear. The

rod and main bearings had not even worn out through to the copper layer. The compression ring and oil ring shows no wear

- Naval Air Development Center in Warminster, Pennsylvania, performed on hydraulic pumps showed nearly a 4-fold wear life extension with 66% improvement in filtration covering a 3-year study on 117 hydraulic machines composed of injection molding, machine tools, mobile, construction, marine, and metal. They concluded a dramatic relationship between fluid contamination and service life. Improving fluid cleanliness can improve MTBF from 10 to 50 % of its lifespan or even more.
- U.S. Department of Defense states that approximately 30% of all engine failures are caused by metal contamination in lubricating oil systems.
- Alumax of South Carolina reported a per machine reduction in component replacement cost from $15,000 to $500 per year.
- Kawasaki Steel implemented a similar contamination control program and achieved an almost unbelievable 97% reduction in hydraulic component failures.
- Oklahoma State University reports that when a fluid is maintained 10 x cleaner hydraulic pump life can be extended by 50x.
- Machine Design Magazine reports that less than 10% of all rolling element bearings reached the fatigue limit since contamination usually causes wear or spalling failure earlier.
- MIT states that 6 to 7% of the gross national product ($ 240 billion) is required just to repair damage caused by mechanical wear, which is a result of contamination

Hydraulic equipment, turbines, engines, and other lubricating systems' primary defense against these contaminants will be their oil filtration system or what we call its oil filter. The typical function of an oil filter is to remove abrasive particles from the oil to achieve the desired fluid cleanliness level. Clearly, from comparing a system with just a nominal filtration with one that has a bypass or secondary filtration, we can see, therefore, where frequent failures can be experienced. We should also note that fluid pressure would play a critical role in selecting the correct micron size filters. These high-efficiency absolute by-pass filters' role is not to replace the original filters but rather to be part of a secondary filtration process, most commonly known as by-pass filters.

There are high-efficiency and beta-rating filters in the market that is being sold to remove contaminants in the oil 99% and higher. This means that for every 100,000 contaminants 99% of contaminants are guaranteed to be removed by the filter, and 1000 will go back to the system. A bypass filter will be permanently installed in the equipment while an offline filtration is a portable filtration cart that can be used to schedule the cleaning of oil when the ISO 4406 Codes increase based on the ISO 4406 cleanliness standards. These filters are considered absolute. The absolute rating of a filter refers to the smallest size particle that will be removed during filtration while nominal ratings refer to the average particle size that will remain in the fluid after filtration. Absolute filtration usually has a high beta rating. A beta rating is derived from the Multi-pass Method for Evaluating the Filtration Performance of Fine Filter Element (ISO 4572). During testing, Automatic particle counters measure the upstream particle size and quantity per unit volume of fluid, and the particle size and quantity downstream of the filter.

Lastly, most maintenance people assumed that when they purchased new oil from the vendor, and the oil is delivered to the plant and stored in their warehouse, the assumption is that since the oil is new, then it should be clean. The word new and the word clean are not synonymous. Just like a child, if a child is fat, it does not mean that the child is healthy. The word fat and the word healthy is not the same words. In fact, it may be the opposite. This means that new oil taken from the storage or warehouse should be filtered using an offline filtration before placing it on the equipment. Otherwise, if this is not done, then both oil and contaminants will be introduced to our equipment and assets.

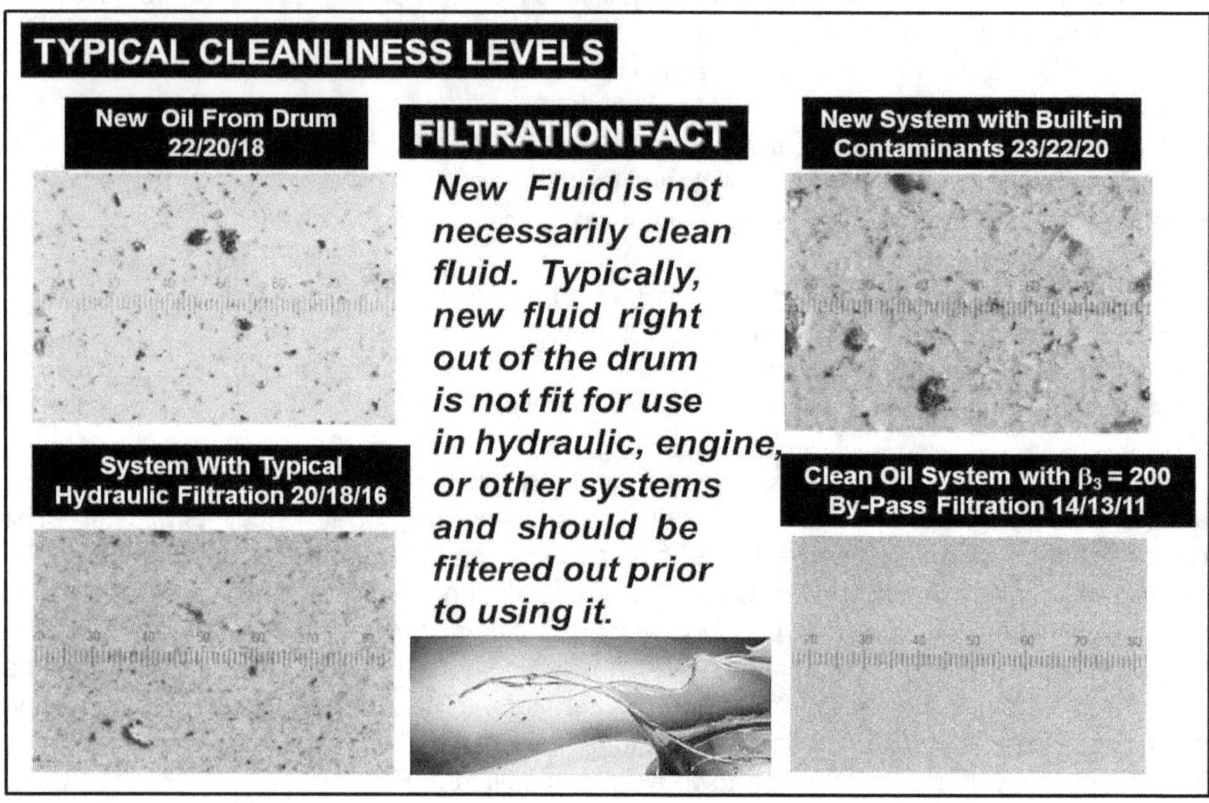

Figure 10.11: Do Not Assume that New Oil is Clean Oil

10.2: Effects of Excessive Moisture on Lubricating Oil

Another deadly contaminant in oil is moisture. The limit of moisture in a 100 ml sample of oil should be at least 100 to 300 ppm. This is considered the alert level. If the moisture content of around 300 to 800 ppm, this will be considered a danger zone while if the content of moisture is 800 ppm and above, this is considered extremely dangerous which means that the oil needs to be changed immediately. Moisture presence in lubricating oil can cause many oil problems and will affect the component or part it lubricates. It can cause premature breakdown of the lubricant itself through oxidation and deplete several additives making them useless. The effects of oxidation can increase the oil's viscosity and make the lubricating oil acidic. This can be traced when doing an oil analysis test on Total Acid Number (TAN). These contaminants can lead to the formation of sludge and varnishes. Excessive moisture can also promote rust and corrosion on metal surfaces by depleting rust

246

and corrosion inhibitor additives. Other problems from excessive moisture in the oil will accelerate surface fatigue, reduce the lubricating film thickness, and jamming components due to ice crystals forming. For transformer oil, this can cause a loss of dielectric strength in the oil.

Moisture Level Equipment Lifespan Extension									
Current Moisture Level in ppm	2	3	4	5	6	7	8	9	10
50,000	12,500	6,500	4,500	3,125	2,500	2,000	1,500	1,000	782
25,000	6,259	3,250	2,250	1,563	1,250	1,000	750	500	391
10,000	2,500	1,300	900	625	500	400	300	200	156
5,000	1,250	650	450	313	250	200	150	100	78
2,500	625	325	225	156	125	100	75	50	39
1,000	250	130	90	63	50	40	30	20	16
500	125	65	45	31	25	20	15	10	8
250	63	33	23	16	13	10	8	5	4
100	25	13	9	6	5	4	3	2	2

1% Water = 10,000 ppm, Estimated life extension factor for mechanical systems utilizing mineral based fluids

Figure 10.12: Life Extension Factor for Moisture in Oil

Sources of Moisture Contamination in Lubricating Oil

• Leaks from within the system such as the cooling system
• Leaks from damaged coolers and heat exchangers
• Leaks through worn out or damaged seals
• Condensation due to humid air
• Reservoirs without desiccant breathers
• Ingression from condensation
• Precipitation due to a fall in temperature
• Introduction of contaminated top-up fluid
• Rain in external reservoirs
• Rain leaking into incorrectly stowed barrels
• An oil barrel stored outside in a vertical position is likely to collect rainwater
• Through reservoir breather caps humid air. System fluid absorbs some of this moisture, while some are condensed on the reservoirs inside surfaces.

Different Types of Moisture Present in Lubricating Oil

Figure 10.12 is a life extension factor if moisture content can be reduced in lubricating oil. This means that if the moisture level can be decreased from 2000 to 156 ppm continuously, then the life of the equipment can be extended by a factor of 5. This means that if the life of the equipment is 10 years then it can reach up to 50 years if the moisture content can be reduced to 156 ppm consistently. Three types of moisture can be present in the oil, which can be free, emulsified, and dissolved, which are explained as follows.

Free Water: This can be any water or gas, which is not dissolved in the oil. Free water is when the water and the oil separate. Since the water is denser and heavier than the water, it will settle at the bottom. Once the water is settled at the bottom, we can easily untighten the drain plug and remove free water from the reservoir.

Emulsified Water: These can be microscopic droplets of water distributed in the form of an emulsion. Emulsified water is when the amount of dissolved water is greater than the saturation point. When the water is in an emulsified state in lubricating oil, its color turned into a white or milky state in layman's terms. It is best to address the problem's root cause in emulsified water and change the oil immediately. When the water is emulsified, it is recommended to change the oil immediately to avoid further damage and destruction.

Dissolved Water: While both free and emulsified water can be visible to the naked eye, dissolved water cannot be seen visibly, but it can be measured using a Karl Fisher Titration Test or Fourier Transform Infrared Thermography (FTIR). In this case, the water is dispersed in the oil in a homogeneous molecular solution. The gas or moisture cannot be removed from a conventional' nominal filtration; however, several absolute oil filtration on the market can remove a certain amount of moisture in the oil. Dissolved water is similar to humidity and cannot be visibly seen in the oil.

Excessive Moisture Can Cause the Following
• Fluid breakdown due to additive depletion
• Water can cause cavitation in hydraulic pumps
• Promotes oxidation, hydrolysis, and aeration, causing the oil to form acid
• Increase in viscosity that can cause varnish and sludge.
• Moisture can accelerate metal surface fatigue
• Promotes rusts and corrosion on metal surface parts
• Jamming of components due to ice crystal formation at low temperatures
• Loss of dielectric strength in insulating fluids such as transformer oil

There are several ways to check the moisture content in oil. The simplest test is the crackle test, which can determine both free and emulsified water, but for dissolved moisture, a Karl Fischer titration test is much more accurate and can measure as little as 1 ppm of water in the oil when used correctly. This method uses two types of titration to identify trace amounts of water in a given sample: coulometric or volumetric, measured either in ppm or percentage. The principles are the same for each, except volumetric uses a titrant solution. This testing method can be very beneficial because it allows you to analyze solids, liquids, or gases. The disadvantages of the Karl Fischer titration test include its cost and the time required when water concentrations are high. There are several ways to remove moisture from the oil, such as coalescence, centrifuge, adsorption, and vacuum dehydration. There are also absolute filters in the market that can remove a certain amount of moisture from the oil. However, the process of removing water in oil is not the solution to the problem, as the moisture will just return to the oil. Interpreting the oil analysis report can explain where the oil is coming from so we can finally address the root cause of the problem. If both metal debris and moisture are present in the lubricant, these metal fragments can act as a catalyst that can accelerate the oxidation of the oil. This means that if both iron and moisture are present in the oil, then less than 400 running hours, the Total Acid Number (TAN) can reach

7, while if both copper and moisture are present, then less than 100 running hours, the TAN can reach a value of 11 mg KOH/g in which the oil is in a degraded state and needs to be changed.

10.3: Increasing a Parts Lifespan Means Changing its Interval

One final note, increasing the lifespan of a part, spare, or item also means changing the maintenance interval. There are possibilities that redesign or modification can change the pattern of failure from random to age-related. Lengthening the lifespan of an item or spare also means that the period where the part or item will fail will be lengthened.

In this case, people responsible for the improvement or modification should advise both the MRO Storekeepers and Purchasing people regarding the replenishment period especially if the item or spare is stocked inside the storeroom. If the replenishment or reordering point will not be changed, then this will just create an overstock of parts in the storeroom. Another point that needs to be covered is what to do with the existing or original parts. If the item will no longer be used, then it should be removed from the storeroom, as these parts will tend to be considered obsolete parts. Reordering of these modified or redesigned parts will be based on the lead time to deliver these parts.

10.4: Detailed Steps in Monitoring Equipment Life Cycle

1. Conduct Machine Inventory and Ranking: A group of people can be assigned initially to conduct an inventory on all active assets, machines, equipment, systems, and sub-systems used currently in their plant. These will be their official machine inventory lists. Only equipment being used should be included in the list. Exclude assets that are planned to be decommissioned shortly. After completing the list of equipment and assets, conduct a machine ranking for all assets and equipment. Rank A will be considered the worst, Rank B will be in between, and Rank C should be tagged as a good machine. Priority for Planned Maintenance activities will be given only for equipment and assets categorized as Rank A or worst equipment then followed by Rank B machines. All Rank C or good machines will just be sustained. After the inventory has been completed, set up criteria for ranking equipment and rank them all accordingly. Perform this on all assets and equipment lists. It is highly recommended to include all facilities and utility equipment. The focus on Machine Ranking is to convert all Rank A to Rank B and convert all Rank B to Rank C, or good equipment.

Rank A: Very Severe Consequences: Loss of any one of these components will result to plant outage, total loss of production, severe environmental, or safety consequences, and critical, stand-alone equipment that does not have any redundancy, backup, or standby in the plant. Equipment failure takes 4 hours or more, fails 3 times or more per month, and the impact of the failure is definitely very high or unacceptable.

Rank B: Severe Consequences: Failures will limit production capacity by 30% or more. Also included in this critical classification are machines with chronic failure history or have high repair and replacement costs. Rank B can be classified as that equipment failure, which takes less than 4 hours of downtime, and frequency is less than 3 times per month.

Rank C: Less Severe Consequences: Failures that do not have a dramatic impact on production but still contribute to the maintenance costs. An example will be a component with a redundant system. Since the inline spare could maintain production, loss of any one component would not affect operations. Rank C can be classified as equipment failure that can be left unrepaired until a convenient time occurs. This will be less critical equipment and assets.

2. Determine the Vertical Start-Up Time: The vertical start-up time will begin from the time the equipment is decided to be purchased by the industry, which includes the time to complete both installation and commissioning. This will provide a basis regarding the length of time it needs for the equipment to be operational. The study and implementation of IFCA will determine the different processes involved and what particular process needs to be improved to reduce the vertical start-up time. An IFCA or EEM team will be responsible for these activities. The vertical start-up process should be documented as this will be the baseline to be used if there will be chances that the same equipment type will be reordered in the future. If this is a case of new equipment or machine that is purchased for the very first time, it is important to initially document the activities involved during the vertical set-up time, as this will be the basis for all repeat orders machines.

Average Monthly Life of Equipment

Division	PM Comm	Area	Equipment TYPES Rev 2	Equipment TOTAL Rev 2	Head Count Rev 2	Machine Ranking A	B	C	TOTAL	MOUNT	SAW	DA	WB	MOLD	MARK	DB/DJ	DTFS	Others	Remarks
												FRONT OF LINE			END OF LINE				
PLCC	RPAND	FOL	29	269	43	1	66	202	269	---	47	51	59	65	53	62	57	53	Include OS machines
	BBERN	EOL	26	117	44	4	46	67	117										
SOT	EBUST	FOL	3	18	5	0	0	18	18	---	----	23	13	24	24	24	24	---	
	CCRUZ	EOL	4	4	5	4	0	0	4										
PSOP	JJAVI	FOL	14	100	21	18	28	54	100	13	13	50	38	61	47	54	54	---	
	MAPAD	EOL	12	28	14	5	11	13	29										
MQFP	DGALA	FOL	6	202	35	117	91	0	208	---	55	48	55	63	46	48	48	---	
	DBALA	EOL	13	63	28	5	13	45	63										
SSOP	OPARP	FOL	9	150	24	5	119	26	150	---	48	32	54	62	46	55	55	---	
	GAVYB	EOL	14	33	18	0	33	0	33										
TQFP	RONAO	FOL	9	141	19	11	129	1	141	48	38	23	65	48	36	60	24	---	
	ROMES	EOL	15	32	20	1	23	8	32										
TSSOP	RFABR	FOL	9	218	28	8	53	157	218	---	59	49	33	48	48	61	25	---	
	CGONZ	EOL	17	53	32	8	14	30	52										
SOIC	RSUAR	FOL	20	423	118	125	92	206	423	72	13	39	33	66	49	37	37	---	
	NMOLE	EOL	22	82	28	48	29	5	82										
PDIP	JBAYB	FOL	14	158	38	33	30	95	158	---	---	70	53	45	30	105	105	---	
	ZBALO	EOL	19	77	28	5	15	57	77										
Hermetics	TLIM/GM	SEL / Cer	17	92	34	25	51	16	92	---	91	117	81	---	90	---	49	92	FOL average life
	ASOLO	Analog	9	40	17	24	11	5	40										Average life for test
CLF	CDELA	Plating	8	21	29	4	3	14	21	---	---	---	---	---	---	---	---	61	
Facilities	AJUSAY	---	30	199	17	35	84	80	199	---	---	---	---	---	---	---	---	78	
TOTAL			319	2,520	645	486	941	1,099	2,526	44	46	50	48	53	47	56	48	71	

Figure 10.13: Machine Ranking and Average Equipment Monthly Life

3. Determine the Average Life for All Machines: Once the total equipment inventory had been completed, we need to determine the operational or running life of each piece of equipment starting from the very first day it was put into operations. For example, in figure 10.13, PLCC Front Line Saw Station has a total of 259 machines. The average monthly life for all these 259 machines is 47 months. This figure provides us a summary of the average life in months, however, the CMMS should provide the real-time running life of each piece of these 259 machines. IT or CMMS vendor's assistance may be required in this matter. Each machine or asset should be regularly updated in real-time. CMMS or EAM software should provide real-time in determining the average life of each piece of equipment. The average life will include both uptime and downtime on the equipment. Calculating the total downtime whether Planned or Unplanned is also important to track in each of these pieces of equipment. Having software or a system in place can ease up the burden of doing everything manually. Hence, if I click a machine with a particular model or type in the system, it will show all the machines and their real-time running hours, which can be in days or months.

4. Identify Top 10 Machines with the Highest Running Costs: Once the system can track the individual running cost of each piece of equipment. We can sort our data to determine the Top 10 equipment that consumes the highest maintenance cost and determine the details of the cost. It is important from the start to determine what will be included in the running costs of each piece of equipment and this should be consistent for all equipment and assets included in the inventory lists. This means that if Preventive Maintenance is performed on Machine 001 and the annual Bill of Materials is 15,000 USD, then this should be reflected when looking for the running cost of Machine 001 in the system or CMMS. All costs involved should be coded for us to determine the details and breakdown of the running cost. If the estimated running costs are also given, we can compare the actual running costs with the estimated cost. It is important at the very beginning to provide a code for each maintenance cost as explained in Chapter 6, figure 6.3.

5. Determine the Maintenance Strategies to Adopt: There will be different strategies that will be adopted depending on the problem. Again this is clearly explained in Chapter 6, in figure 6.4 regarding the different strategies to adopt for maintenance. For parts with inherent design weaknesses, we can revert back to Chapter 9, Section 9.3 on how to proceed with Planned Maintenance Phase 2 redesign and modifications. If MRO Spare Parts Cost is the main contributor to the cost, then we can adopt an MRO Spare Parts Management Strategy depending on the problem. Whatever strategy to adopt, the ultimate goal is to eliminate or mitigate the problem and reduce the running cost of the equipment.

6. Determine the Improvement Plan: Once the corrective actions are finalized, assign the best person responsible for the design and target the dates of completing the corrective action, redesign, or modification. The person responsible may not be the one who will do the design changes but should be held responsible for providing updates regarding its completion. Priority of corrective actions should be based on the consequences of the failure, especially for those with safety and environmental consequences.

7. Implement the Improvement Plan: Improvement plans that require redesign or modification should be evaluated and observed on the equipment to determine whether the problem still exists or has been completely eliminated. It is also important to observe if there are abnormalities or new problems that emerged as a result of implementing the improvement plan. Once the improvement plan has been implemented successfully, operators and other maintenance people should be aware of the improvement and why it was done in the equipment.

8. Horizontal Replication: Replication of improvement changes and modifications will only be done on the same type of asset or equipment that poses similar symptoms of the problem. This means that the modified part or improvement will also be adapted to other same equipment in the plant that provides the same symptoms of the problem. Note that if there are similar pieces of equipment in the plant that does not possess this type of problem, it is not recommended to place the modified part on that equipment. Once these steps had been completed successfully move on to the next machine or equipment and repeat the steps.

10.5: Decommissioning – The Final Phase of Equipment Life Cycle

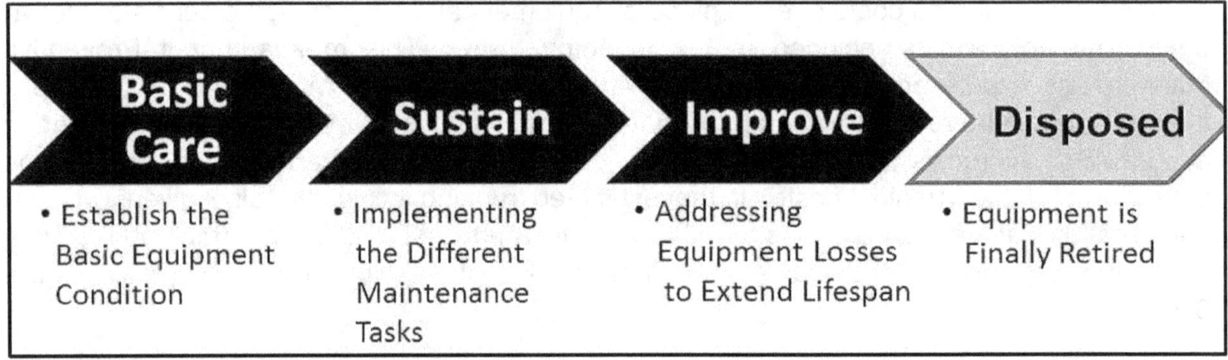

Figure 10.14: The Final Phase of Equipment Life Cycle

This will be the final phase of the equipment Life Cycle. Once the equipment will be retired the machine will be removed from operations. If the equipment does not have similar models in the plant, then the spare parts should be identified inside the storeroom since they will be considered as obsolete parts. Once a piece of equipment is identified for decommissioning, careful planning should identify the activities to be done and the resources needed. The safety of the people involved in decommissioning should be given priority. Risks and hazards should be identified so as not to pose any safety issues to the people performing the decommissioning such as live electrical connections, oil leaks, compressed gases, corrosion, hazardous chemicals, and others. Proper procedures should be observed and followed. Qualified people should perform the decommissioning process. If third-party contractors will be used, it is important to have a background check on the contractor to determine how long they have been doing this, and the possibility of reviewing their checklist on decommissioning. Discuss the plan with these contractors as well as the duration and expectations. Proper safety gear and apparel should be worn especially when hazardous or toxic chemicals are present. Safety signs should be visible during the decommissioning process.   Once removed, this decommissioned equipment can be auctioned or sold as scrapped by the purchasing and Finance Department.

There might also be cases where some parts of the equipment can be disassembled and serviceable parts are salvaged which can be used for other equipment and assets in the plant. Those parts that were salvaged will be tagged and can be used as spares. Unusable parts will be disposed of together with the equipment or machine itself. The machine will be placed in a box and can be sold to other industries or as scrap.

Chapter 11

ISO 55000 on Asset Management

> *Achieving ISO 55000 certification is not a guarantee nor a gateway to World Class Maintenance or World Class Reliability. A good Asset Management structure will not solve every problem industry faces, rather it should provide industries a structured framework to make better decisions in managing risks that they encounter.*

11.1: Definition of Risks Management

According to the dictionary, risk is something or someone that can cause problems, harm, injury, damage, danger, hazard, or loss. Risk is also the potential for uncontrolled loss that is of value to an organization. It has something to do with uncertainty. Uncertainty is a potentially unpredictable and uncontrollable outcome of an event. It is a key aspect of risks. **ISO 31000** provides guidelines on managing risks faced by different organizations. The guidelines can be tailor-made to different industries on how they can manage to operate their business. It provides a common term and understanding of how to manage, reduce, eliminate, or mitigate the risks involved. The guidelines can be applied to any instance, including decision-making at different levels. When we speak about the risk, it is the possibility that something bad or uncertain can happen. It means that the failure or problem had not yet happened but is likely to happen. This means that if we do something, the failure can be averted. It is the probability in case it happens. If we have data, statistics, or previous experience, it can allow our judgment to indicate if the risk is imminent and if there is a high probability that it will happen. If the risk is low, the failure will be less critical, and the consequences will be moderately low to none. If the risk is high, obviously, the failure will be critical, as well as the consequences of the failure. Here are some cases to indicate if the failure will be critical:

• If the equipment is in an unmanned location
• If the failure can result in environmental consequences
• If the failure can result in safety consequences or can kill or injure someone
• If the failure can permanently shut down the whole operations (loss of job)
• If the failure can result in industrial accidents
• If the failure can breach any known environmental regulations and laws
• If there is an absolute certainty that the failure cannot be detected

• If there is no redundancy or standby for the equipment
• If the cost of the failure will be extremely high and can result in a penalty
• If the company lawyers and insurance people will be involved in the failure
• The worst case will be if the media and those crooked old politicians will be involved

Bhopal India
December 2, 1984
Methyl Isocyanide Leaked
(6000 to 20000 killed)

Piper Alpha Disaster
July 6, 1988
167 killed, 3.4 billion loss USD,
World's worst offshore disaster

Chernobyl Disaster
April 26, 1986
Nuclear meltdown killing 50
people estimate cancer deaths
of 4000 to 100,000 overtime

Triangle Shirtwaist Factory fire,
New York, March 25, 1911
Killing 147 workers, inuring 71

Fire on Garment factory
in the Pakistan Sept. 11, 2012,
Killing 257 people, injured 600

Dona Paz Disaster
December 20, 1987,
Considered the worse Maritime
Tragedy of all times

Figure 11.1: World's Worst Industrial Disaster

ISO 31000 is an international standard published in 2009 that provides guidelines for an effective risk management structure for industries. It outlines a generic approach to risk management, which can be applied to different types of risks that can affect organizations. Although the standard does not provide detailed instructions, however, it remains on a generic level that can be applied by different industries. According to ISO 3100, risk management is the identification, evaluation, and prioritization of risks as the effect of uncertainty on objectives followed by coordinated and economical application of resources to minimize, monitor, and control the probability or impact of unfortunate events to maximize the realization of opportunities.

Risks can come from various sources, including uncertainty in financial markets, threats from project failures at any phase in design, development, production, or sustaining of life cycles, legal liabilities, credit risk, accidents, natural causes, disasters, deliberate attacks from an adversary, or events of uncertain or unpredictable causes. There are two types of events. The negative events can be classified as risks while positive events are classified as opportunities. Different types of risks can have different impacts and consequences on industries. What is important for industries is to identify the risks involved that would have a major impact on their business and find ways to eliminate, prevent, if not reduce, or mitigate their consequences and impact as well.

11.2: The Need for a Structured Asset Management System ISO 55001

Many industries around the world have suffered from major industrial accidents, disasters, environmental regulatory breaches, and recessions. In most cases, industries affected have halted their operations permanently. With this, the need to create a robust and structured asset management approach is indeed necessary to prevent, reduce, minimize, mitigate or even eliminate the risks involved to lower the consequences of the risks involved, that is, if industries still want to remain in business and stay tough with their competition. Industry's assets and equipment must not only be cared for and well maintained, but the need to develop an effective asset management structure will be of extreme importance. Although both PAS 55 and ISO 55001 can provide the industries with a universal framework and guidelines for managing their assets, it still is not a guarantee that passing ISO certification is a gateway to achieving a level of World Class Maintenance or World Class Reliability. The two are entirely different. ISO 55002 provides the guidelines for industries that would like to acquire the certification process. These will be the requirements needed to get the certification. Still, again, the ISO standard does not aim to eliminate the losses and failures experienced on the equipment. The thing is ISO certification will be focused on the documentation as a way of compliance and not the results.

ISO 55000 was launched in London in June 2010, followed by a series of meetings from experts from around the world. There were 31 participating countries including Argentina; Australia; Belgium; Brazil; Canada; Chile; China; Colombia; Czech Republic; Denmark; Finland; France; Germany; India; Ireland; Italy; Japan; Korea (Republic of); Mexico; Netherlands; Norway; Peru; Portugal; Russia; South Africa; Spain; Sweden; Switzerland; United Arab Emirates; United Kingdom; and the United States of America. There were 11 observing member countries, which include, Armenia; Austria; Hong Kong, China; Iraq; Israel; Malaysia; Morocco; New Zealand; Slovakia; and Thailand. Since then, many industries have been adopting this asset management system around the globe as a tool for integrating and improving their business practices, raising performance, mitigating risks, and assuring greater consistency, and transparency in their business and operations.

After several reviews and consultations with different industries, professionals, and consultants from around the globe, a guideline was provided on how to manage the different risks involved for different assets in industries. The final draft was concluded in a meeting in Calgary, Canada, in April 2013. After a few more deliberations, the standard was finalized. It was given to the International Standards Organization (ISO) as the basis for a new set of standards for asset management. The ISO Standard was finally approved in January 2014. ISO 55001 is an international standard that was founded on the belief that the worth of an asset can be defined as how much value it contributes to the achievement of the company's goals and objectives. This ISO standard is designed to optimize the value of assets and decrease organizational risks through an organized and comprehensive asset management strategy that is built around the premise of its business goals and objectives. According to this ISO procedure, asset management is defined as the activities of executing that set of processes to realize value as the organization defines it from those assets.

Good asset management entails the identification, assessment, management, and mitigation of the risks involved in a specific problem as well as the consequences associated with the risks themselves. Risk management is one of the important functions of any corporation that directly can have an impact on their asset management activities. It is also important to understand the relationship between asset management with the other corporate functions of an organization that provides substantial benefits. Good asset management can mitigate or reduce the risks involved, reduce costs, and improve the industry's productivity and revenue. This means that before we can manage our equipment and assets, we need to change the way we maintain our equipment.

Having a good asset management is not about just having the most expensive and the Top-of-the-Line CMMS or EAM software in place. Remember that a good asset management system will not solve every single problem or risk that the industry faces. Rather it should provide a structured framework and decision-making process, which will enable industries to make better decisions on every challenge they face. It allows industries to think more and explore things that will provide the most beneficial solution to the problem that they may encounter in the future instead of doing the traditional way of doing things in the plant. Good asset management is all about doing the right things, in the right way, at the right time.

Even though ISO 55001 is not all about maintenance, what I do believe is that before an asset or equipment can be managed correctly, it should be well preserved and maintained. ISO 9000 was spearheaded by quality, while both ISO14000 and ISO 45001 were championed by the Environment, Health, and Safety, or EHS function of the organization. What I do believe is that the best people to champion and spearhead this ISO 55001 will be the reliability and maintenance people from industries. Still, nevertheless, it should be a top-down approach. This means that top management and C-Level people should be the leaders in this venture. Although the scope of asset management in ISO 55000 will be much broader in scope compared to PASS, 55 since it not only includes the physical assets and equipment but all assets that are of value to the organization. Not only can the use of ISO 55001 in all sectors of industries improve the way we do maintenance, but most likely, the maintenance team will be responsible for fully grasping and implementing these standards to improve the way they both maintain and manage their equipment and assets in the plant.

ISO 55001 embraces the philosophy that any asset from the industry with income-generating value will deteriorate much faster if it is not maintained and managed well. TPM termed this as accelerated deterioration where the life was not reached because of the basics of the equipment were not taken care of. While actions taken to maintain the different types of assets may differ from one type of industry to another, a good maintenance management structure should be in place. A specific asset management system, such as ISO 55001, can add more value by allowing industries to get to know their equipment and assets better. It can also be used to maximize the life cycle of an asset, which can lead to better decision-making business performance and profitability. The goal of any asset management is to take care of the equipment's Total Life Cycle at the most reasonable cost by having a structured plan on how to maintain and manage the equipment, improving the asset's design flaws, and reducing the cost of operating and maintaining the asset itself.

11.3: Transitioning from PAS 55 to ISO 55000

PAS 55 started in 2004 by several organizations under the leadership of the Institute of Asset Management (IAM). It then underwent a substantial amount of revisions with 50 participating organizations from 15 industry sectors, in 10 different countries. PAS 55 was approved in December 2008, along with a toolkit for self-assessment required for complying with the requirements. PAS 55 provides guidance and a twenty-eight-point requirement checklist for good practices in physical asset management. This standard is applicable for gas, water utilities, electricity, road, aviation, rail transport systems, public facilities, manufacturing, process industries, and natural resource industries. It is also applicable to public and private sectors, regulated or non-regulated environments. PAS 55 standards were split into two parts:

• Part 1: Specification for the optimized management of physical infrastructure assets
• Part 2: Guidelines for the application of PAS 55-1

The scope and inclusion of PAS 55 include the management of physical assets primarily. As generally recognized, all equipment and asset types are highly interdependent. The optimal management of physical assets also includes managing the people, resources, information, finances, tangible and intangible assets such as performance and activities. Indeed, it is aimed to reduce barriers and silos among the different functions of the organization, and the consideration of assets in systems, along with the cross-functional optimization of their life cycle are the core principles of a good asset management plan. ISO 55000, on the other hand, is an international standard that is founded on the belief that the value of any asset can be defined by how much it contributes to the achievement of its company goals and objectives. ISO 55000 is composed of the following three main standards.

• BS ISO 55000, Asset Management: Overview, Principles, and Terminology
• BS ISO 55001, Asset Management: Management Systems, Requirements
• BS ISO 55002, Asset Management: Guidelines for the Application of ISO55001

ISO 55000 is a standard described as the parent document for ISO 55001. It provides a general overview of asset management as a discipline. It contains definitions of terms used in the ISO 55000 series of standards and terminologies used. This standard aims to establish a clear, precise, and consistent understanding of the principles and requirements, which must be applied when developing and implementing an asset management system in the entire organization.

ISO 55001 provides the requirements for achieving an asset management system. These requirements will enable ISO 55001 to be applied in the best possible way to any type of organization. It states what must be done to achieve the standards in ways that fit an organization and its operating environment. This is the actual certification process. ISO 55002 provides the guidelines for implementing ISO 55001. It provides the requirements taken from ISO 55001 with explanations and examples that will assist industries on how to proceed with the requirements. It also describes the benefits asset management can bring to the organization and define important clauses and sub-clauses that must be accomplished in acquiring ISO certification. In ISO 55000 asset refers to any item, thing, or entity that has potential or actual value to an organization, while in PAS 55, asset simply refers to the physical asset.

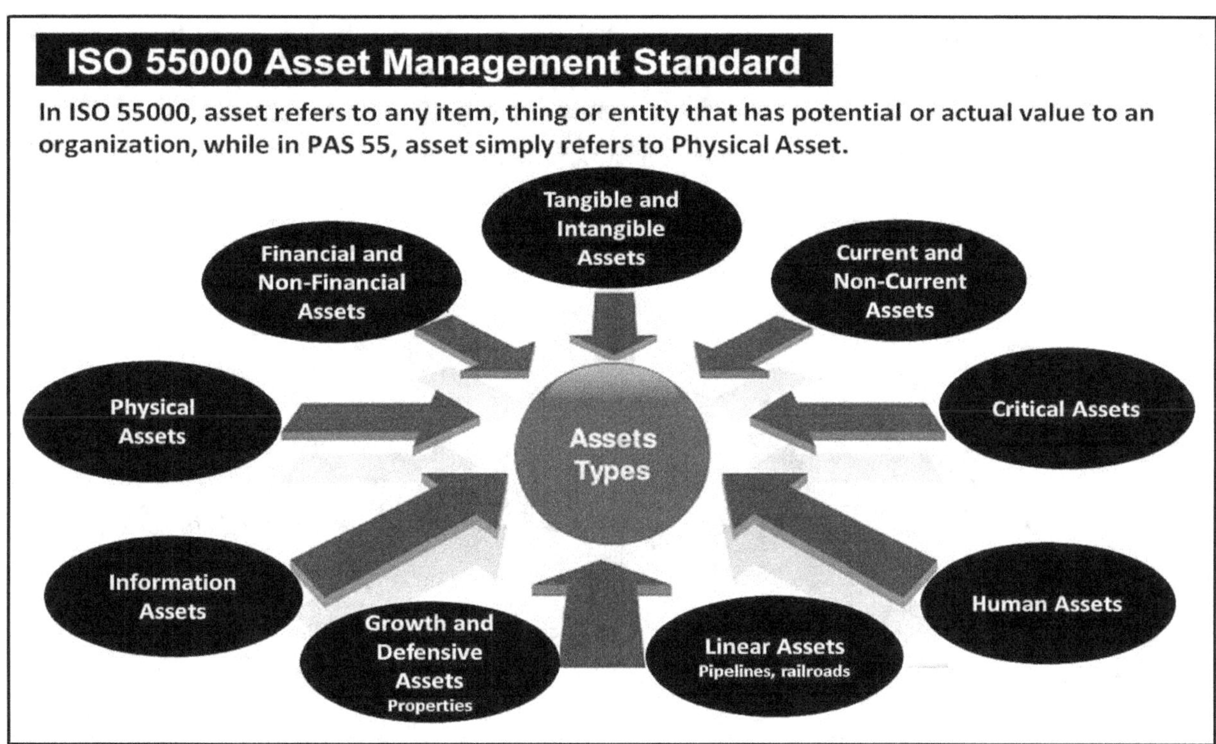

Figure 11.2: Definition of Asset in ISO 55000

Asset Management Definition on PAS 55 is the systematic and coordinated activities and practices through which an organization optimally and sustainably manages its assets and asset systems, their associated performance, risks, and expenditures over their life cycle for achieving its organizational strategic plan.

Asset Management Definition on ISO 55000 is the coordinated activities of an organization to realize value from its asset. In ISO 55000, the definition of an asset is much broader in scope as it refers to any item, thing, or entity that has a potential or actual value to an entire organization. In contrast, in PAS 55, asset simply refers to the industry's physical asset. An asset in ISO 55000 can include critical assets, human assets, linear assets, growth, and defensive assets, information assets, physical assets, financial and non-financial assets, tangible and intangible assets, and current and non-current assets. The scope of assets is much bigger.

Asset Management Policy is a written statement that dictates the guiding principles in which the organization intends to apply asset management to achieve its organizational objectives and goals. The asset management policy can be authorized and approved by the c-level people in the organization. It includes the purpose, scope, intent, policy statement, where the policy will be applied, commitment to continuous improvement, compliance with government laws, and the roles and responsibility of the stakeholders. The policy provides a template for decision-making so that people can achieve the best possible outcome for each task while meeting the organization's goals and objectives. It is concise and sets a strong direction and clear expectations.

Strategic Asset Management Plan (SAMP) is a document that specifies how the organizational objectives can be converted into asset management objectives. Included in the SAMP document is the relationship between organizational objectives and asset management objectives. It also includes structured details to achieve asset management objectives.

11.4: Elements of ISO 55002

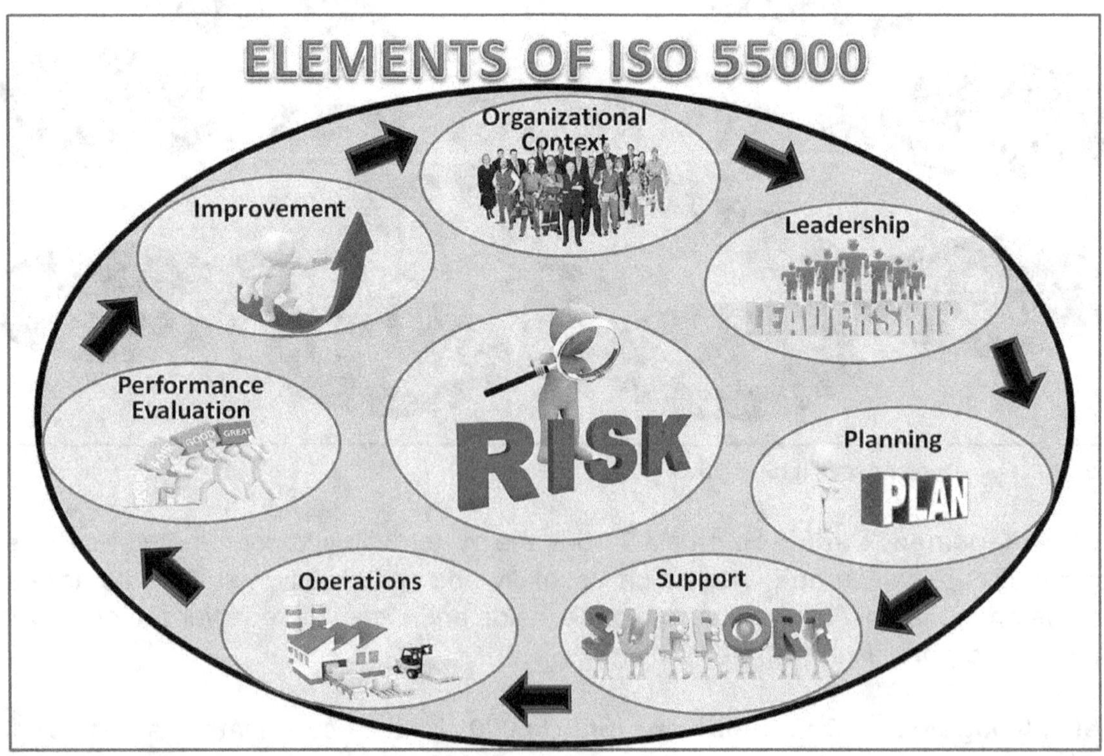

Figure 11.3: Elements of ISO 55000

(Clause 4) Context of the Organization: An organization should know its internal and external contexts. External context refers to the social, cultural, economic, and physical environments, as well as regulatory, financial, and other restrictions. An internal context, on the other hand, includes the organizational culture and environment, as well as the organization's mission, vision, values, or simply the overall ideology of the organization (refer to figure 11.4). The stakeholder inputs and expectations are also part of the organization's context. It is important to identify who are the stakeholders of the organization and those that will affect your asset management system. Determine their involvement and what they need to know about the assets being managed. Determine their roles, duties, needs, and what is expected from them. This should be well documented and included in the asset management policy.

The Strategic Asset Management Plan (SAMP) should document the alignment of both the organizational objectives and the asset management objectives of the organization. It must also define the details needed to achieve the asset management objectives. The objective of this organizational context is to ensure proper communication occurs between the people involved. They also need to understand their involvement as well as their boundaries in the asset management policy. The organization should clearly identify both the internal and external stakeholders' involvement, together with their needs and expectation, in asset management. External stakeholders must also be aware of the expectations of the organization from them. This may differ from one type of industry to another.

Internal stakeholders can include the following:
• All employees within the organization
• Different functions and departments within the organization
• C-level people, shareholders, top management, owners, the board of directors
• Company lawyers who are also employees

External stakeholders can include the following:
• Customers, OEM, vendors, suppliers, service providers, and 3rd party contractors
• Non-governmental organizations, civil society groups, consumer organizations
• Media with an interest in issues related to asset management
• Government organizations, government agencies, regulatory authorities, and politicians
• Investors or taxpayers
• Local communities
• Those interested in social, financial, environmental, or other forms of sustainability
• The Financial institutions, rating agencies, and insurance groups
• Employee representatives
• External auditors such as ISO, or any other government regulatory board
• Consultants, advisers, external trainers
• Company's security who are considered 3rd party contracts

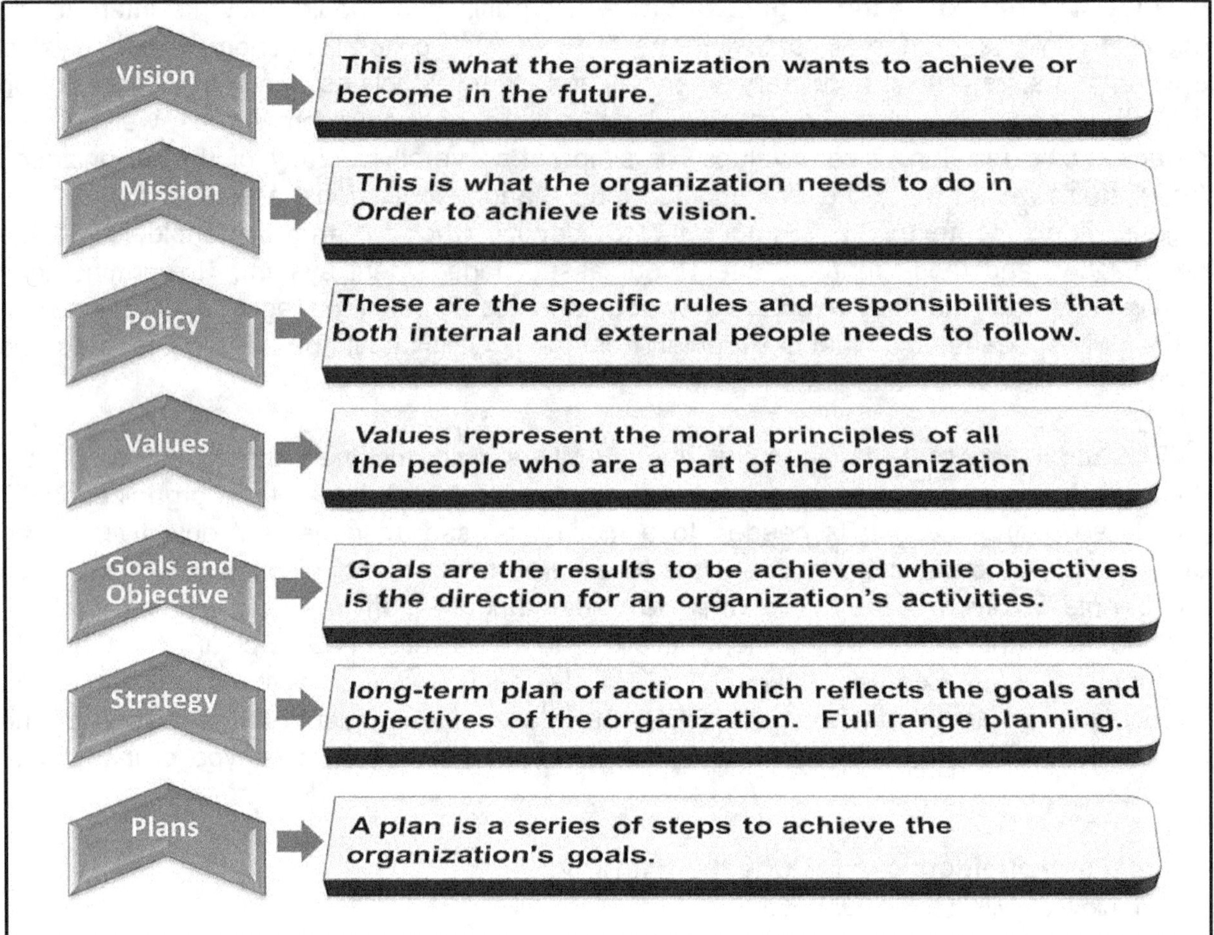

Figure 11.4: Company Credo

(Clause 5) Leadership: Although I have mentioned that maintenance might be the champion to spearhead ISO 55000, top management, and executive decision-makers in the organization must be the leaders of this initiative. They need to provide strong leadership and commitment not only to asset management but also to ensure the reliability and sustainability of their equipment and assets. Both commitment and ownership are very important to ensure the success of asset management. I would prefer to use the word commitment than support since management can support but never commit to the initiative. Top management should establish an asset management policy aligned with their organizational strategic plan for the company. One of the key highlights of this clause is that top management must provide equal priority for safety, quality, and asset management. For industries aiming for ISO 55001 certification, maintenance, and reliability, leaders must take this opportunity to provide their top management with how reliability can affect their overall business objective. Management should be the leader as well as the drivers of reliability within their organization. An asset management policy is a written expression that sets out the principles in which the organization intends to apply asset management to achieve its organizational objectives. The policy can be delegated to others with their full consent. Management should adhere to this policy.   The asset management policy must be disseminated to all the people in the organization. It should be kept updated for any changes in the context of the organization.

(Clause 6) Planning: This section deals with addressing the risks involved as well as the opportunities that need to be considered when planning for an asset management system to achieve its intended objectives and outcome. The purpose of planning is to understand the cause and effect of these adverse events occurring so that organizations can provide an appropriate plan to manage the risks involved to an acceptable level. We also need to provide an evaluation to ensure that the risks have been prevented or mitigated. The intent of this clause is for the organization to ensure that the asset management system achieves its objectives, prevents any undesirable effects, identifies opportunities, and achieves continual improvement. The asset management plan must require a set of defined activities that must be done on the asset including specific and measurable objectives, such as timeframes, resources to use, costs, and the activities to be done. The asset management objectives should be specific, measurable, achievable, realistic, and time-bounded (SMART). An organization should identify the required actions for managing the risks for its asset management system. An organization should have a risk assessment criteria or a matrix to assess the risks involved. For physical assets, an FMEA or FMECA equipment criticality analysis table can be used for assessing the risks and providing the corrective measures to address the risks in line with the organization's asset management objectives. The activities, once implemented, must be evaluated to check the effectiveness of managing or mitigating the risks being addressed. Some risks have a low probability of occurrence but with extremely high consequences. This should, likewise, be included in the planning process. There must be an additional dimension in the risk analysis to include the capability of an asset system to monitor and continually assess the probability of these rare, but potentially catastrophic events. While the probability of such events can be very low, the organization should establish a system of monitoring most, especially if the risks cannot be detected or are considered hidden.

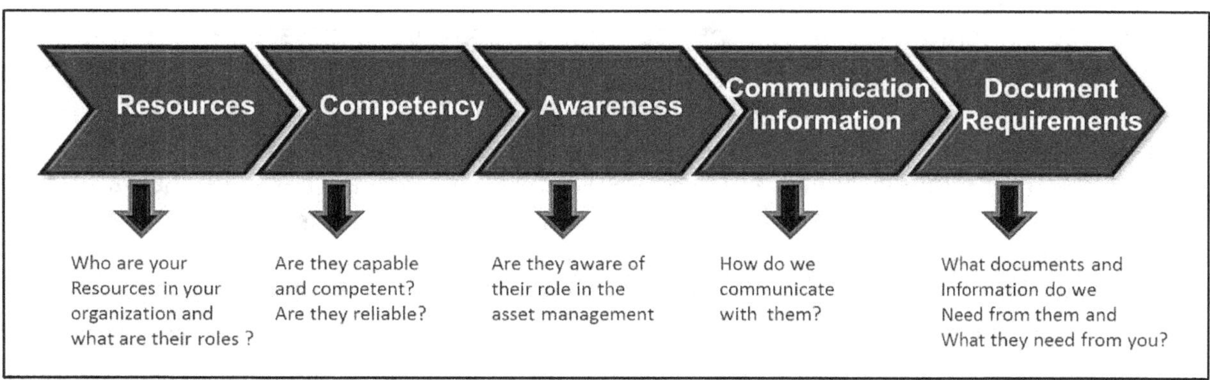

Figure 11.5: Support Criteria for ISO 55000

(Clause 7) Support: During the development and implementation of this clause on ISO 55000, which is supported, the following must be crystal clear to the organization.

• Resources, which can be either internal or external

• Competence of the resources
• Awareness of the resources on the asset management they are involved with
• Communication between the resources and people directly involved in asset management
• Information requirements needed by the resources
• Documented Information

In clause 7 on ISO 55002, the organization should know who their resources are for both internal and external, as indicated in clause number 4 on organizational context. The organization should map out the capabilities and current competencies of its resources and must have a way to determine any gaps. When undertaking a competency gap analysis, the organization needs to provide the business process a way to develop necessary plans for improvements whenever needed. All people involved that can have an impact on the asset management system should have their roles, responsibilities, and accountabilities communicated to them. They should be provided with the correct training, education, development, and other support needed to perform their roles and be able to demonstrate the competencies required from them. Asset management activities carried out by the organization must be communicated well to involve stakeholders in a consistent and coordinated manner under the organization's asset management activities and asset management system. These stakeholders must be updated regularly for any changes in the document information. In establishing its documented information needs, the organization should consider important and relevant information that will be managed, maintained, and regularly updated over the entire life cycle of the asset, taking into account its period of responsibility for the assets. This clause in ISO 55001 deals with the entire life cycle of an asset and the commitments needed for both internal and external stakeholders to fully maximize and optimize the performance of the asset used by the organization.

(Clause 8) Operations: This clause on ISO 55002 covers the actual cycle of operations based on what has been planned in the Strategic Asset Management Plan (SAMP). The operations should be as close as possible to what they have planned so that they can produce the outcome and results expected. In the sub-clause on Management of Change (MOC), any changes, improvements, or modifications made in the process must have to be clearly evaluated and checked before approving and implementing such outcomes. A different group of bodies should be responsible for evaluating any changes from the originator to make certain that the change is feasible with no additional problems or risks involved in the process. Once the changes have been approved, this must be disseminated to all concerned people in the entire organization, both internal and those likewise who are externally involved. For works and projects that are being outsourced, ISO 55001 requires a formal, documented process for outsourcing should be provided if the work or project is to be done by an external contractor. According to ISO 50001, outsourcing is a common method for an organization that prefers to perform certain asset management activities not by itself, but by an external or internal service provider. Any asset management objectives, processes, and activities that are outsourced should be controlled by the organization to provide assurance that performance is as planned and expected. Contractors should be competent enough to provide the scope of work required from them. They must assure compliance with the organization's requirements. It would be highly recommended that a qualified person do a thorough checking for the completeness of work from the contractor after the completion of the contracted work to assure compliance with the requirements is satisfied.

(Clause 9) Performance Evaluation: This clause on ISO 55002 contains three parts. The first part is monitoring, measurement, analysis, and evaluation; the second is having an internal audit, and finally, a management review. The purpose of this clause is to ensure the desired outcome based on the plan is achieved effectively. Measurements indicated can either be either quantitative (measurable) or qualitative (non-measurable), financial or non-financial. These indicators and measurements will determine the success and areas that require room for improvement. The information collected from monitoring, measurement, analysis, and evaluation is needed to guarantee the effectiveness of the operations based on the plan itself. The organization should conduct a consistent and regular evaluation of its process to ensure its effectiveness.

Figure 11.6: Performance Evaluation Criteria for ISO 55000

The evaluation can be done either internally or externally with a third-party consultant knowledgeable about the process. One advantage of having a third-party consultant is that since he or she is from the outside, the results of the evaluation can be considered unbiased. The evaluation process can be either an audit or an assessment. The frequency of conducting the audits will be based on the organization's requirement, as there is no mention of the clause about its frequency, or can be based on government laws and regulations, depending on the size, nature, and legal status of the organization and type of operations. The organization should conduct internal audits at planned intervals to ensure the asset management system conforms to its requirements on ISO 55001. It is important to have these internal audits conducted particularly on critical assets and systems to ensure that the asset management system fulfills its expectation from the users. The results of an internal audit of an asset management system can be used to correct or prevent non-conformities identified as input for continual improvement, and to provide input for management review. The persons conducting the audit should be competent and in a position to do so impartially and objectively. Finally, top management should review the organization's assets management system. Audits are conducted to test the conformance of the organization to its asset management system and a management review is done to assure the effectiveness of the asset management system. Although there is no mention of the clause as to the frequency of the review and who will be part of the review team. I strongly recommend that the review be conducted once a quarter and that the CEO and all c-level people should be part of the review team.

(Clause 10) Improvement: This final clause on ISO 55002 on improvement refers to any non-conformity, corrective, preventive actions, and continual improvements performed on any risks involved, deviation, problems, or any incident by taking action to control and correct it. Although this clause on ISO 55002 does not include what type of analysis or investigative methods are used to address its non-conformity. It should likewise be clear that tools such as RCFA or RCA must be used to deal with the root cause of the problem. The organization should establish plans and processes to control non-conformities that occurred together with their consequences to minimize any adverse effects on the organization and on the stakeholder needs and expectations. Corrective actions are actions taken to address the root cause of identified non-conformances, or incidents, to prevent, manage, reduce, or mitigate their consequences, and to prevent the likelihood of recurrence. The identification and execution of these corrective measures should be both done for the short and long-term process. Preventive actions, which may include predictive actions, are those taken to address the root cause of potential failures or incidents, as a proactive measure, before such incidents are likely to occur. Preventive action deals with the entire process of proactively identifying problems and providing improvements to prevent their recurrence. Continual improvement should be regarded as an ongoing activity to deliver organizational objectives. The asset management system and its operation must be continually improved by identifying opportunities to make things simpler, and more cost-effective than can provide better results for the organization.

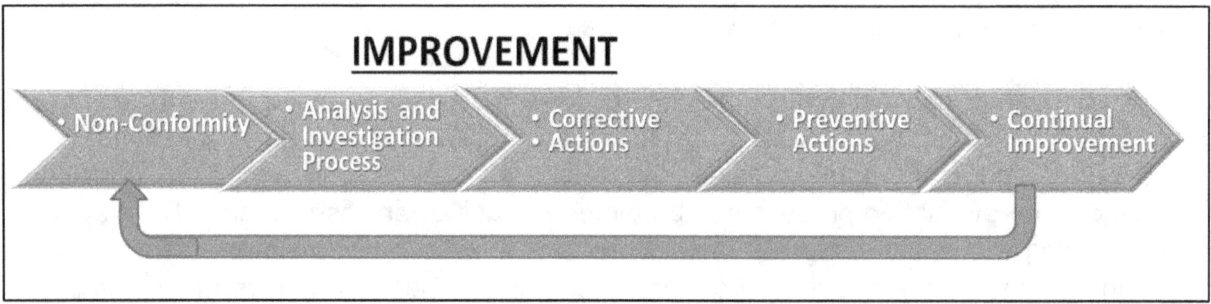

Figure 11.7: Improvement Criteria for ISO 55000

11.5: Benefits of ISO 55000

If by any chance, you visit an industry, you will be staying in the lobby area of the plant for visitors where the vision, mission, policy, and objectives are all hung on the wall for everyone to see and read. It's nice to have these things around in the plant, but a few people who may no longer be connected with the plant itself drafted most of them. Try to ask any employee in the plant about their corporate values or even their mission and vision. Only a handful of these people can give what you have read on the wall. ISO 55000 must not be done just to satisfy the visitors of the plant or to have these credos hanging up on the wall for aesthetic purposes. The strategy for ISO 55000 must be disseminated to all employees within the organization. Providing all the documents needed for ISO 55000 will give you the certification. Still, if the purpose of certification is merely to satisfy all the required documentation needed and not for its intended purpose, which is to manage risks successfully, then everything is just a big joke, and I believe that industries are fooling no one but themselves.

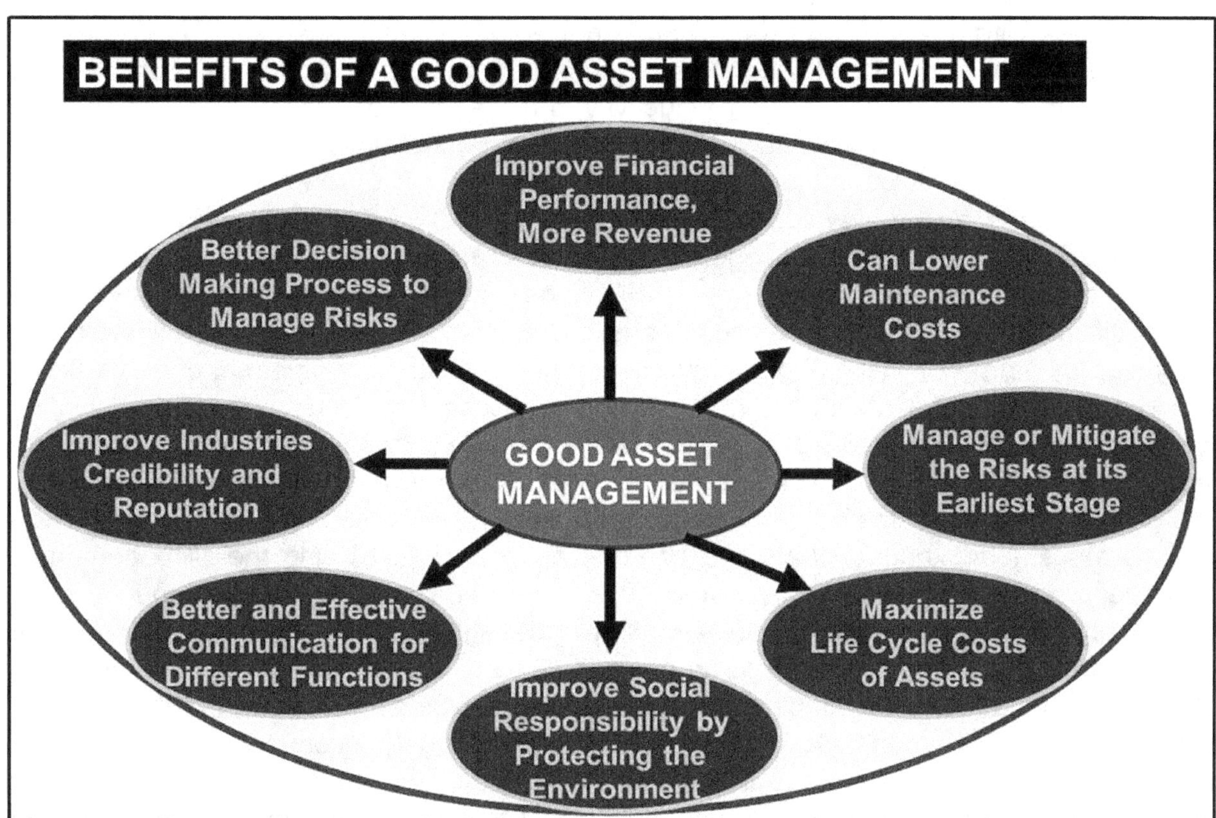

Figure 11.8: Benefits of ISO 55000

ISO 55000 is a top-down approach strategy where the decision-making process begins in the boardroom and aligns its asset management strategy with the overall corporate strategy of the organization. An organization certified on ISO 55001 must provide a framework for cascading the asset management plan throughout the entire organization. It provides top management and C-Level people with regular feedback on risk assessment, opportunities, and asset performance to inform the ongoing decision-making process. Many benefits can be derived if ISO 55000 is implemented religiously, wholeheartedly, and not merely for documentation purposes.

Although having an ISO 55000 in place in the organization will not totally remove barriers and silos among the different functions in the organization, it would provide a good start on building effective communication with the other functions within the organization. These functions need to understand that they cannot co-exists without each other and constant communication, collaboration, cooperation, and coordination should exist among the different functions of the entire organization.

Good asset management in place will definitely not solve all problems. Still, it can help industries by enabling organizations to make better decisions based on the challenges they face as they come. A good asset management system entails the identification, assessment, management, and mitigation of these risks as well as their consequences. Risk management is one of the important corporate functions that directly can impact asset management activities. It is also an excellent example of where the integration of asset management with other corporate functions provides substantial benefits to industries that adopt these practices. Good asset management can lower a broad spectrum of risks, reduce maintenance costs, improve productivity, provide decision-making, strengthen social responsibility, especially to communities, and revenue-generating capacity.

One more precaution I would like to reiterate in the ISO 55002 guidelines is on the sub-clause on 8.2 Management of Change and 10.3 on Continual Improvement. Typically, in an industry that implements a Management of Change process, any changes, improvements, or modifications made in the process must be fully evaluated, tested, validated, and approved by a group of people for the purpose of not generating another set of new risks for the organization. The Management of Change process (MOC) usually slows down the process of continuous improvement, as everything needs to be evaluated, tested, and approved by a group of people with other priorities. Usually, those assigned to handle the MOC process are part-time people, mostly managers with other priorities in the plant. There are no industries I know of that have full-time people for just doing the Management of Change process. They also have their main duties and obligations to attend to in the organization. If we do everything by the book, then it might take several months for a kaizen improvement done by an operator to be approved and take effect. This must be discussed thoroughly in the ISO 55002 implementation process to avoid these kinds of problems. I am not against the process of MOC, but perhaps it would be highly recommended that a simple decision diagram could be made if a change or improvement will undergo a MOC or be allowed to proceed. I use to consult with a plant that has a MOC system. Even incremental or small improvements recommended by operators take months or even longer only to be rejected in the end, which demotivates people to further performance improvements. MOC should have a timeline for big and incremental improvements. For example, incremental improvements suggested by operators should be processed in less than 48 hours if they will truly benefit the process while innovations and modifications will take more time in the process, as it needs to be evaluated, and tested for their effectiveness to ensure that it will not generate new problems within the organization.

11.6: New ISO 55010 Alignment of Financial and Non-Financial Functions

In many industries, there seems to be a grudge between finance and maintenance, and other technical functions. I have seen this happen in many industries where finance people interrogate maintenance for every single cent of purchase maintenance people did and why they needed them. I have also experienced several industries where the two have opposite, contradicting, and conflicting views thinking that their paths are the correct ones to follow. Although I have mentioned in my other books that I resigned in one industry because finance cut back my training expenses. My thinking then was, why on this planet did you hire me if I will have no budget for training. I told the finance manager, are you stupid or something? The following day I resigned. In one of the plants I served, one person from maintenance was requesting a tire for his service pick-up truck. He told me that finance called him, asking him why he needs four sets of tires instead of only one, his reply was that all four tires are in contact with the road. Lol ☺.

I have a client in the Philippines. They have three power plants in different provinces. Although the three power plants have different capacities, they also have similar equipment. What the finance people did was to have a separate stock item (others called this part number, stock number, item code, or codification) for each spare part for the three power plants. This means that if I have a common oil filter for a diesel engine, the stock item or part number of this will be different for the three power plants. My thinking was that if plant A has an excess of these filters, and plants B and C need them, then they could just borrow from plant A and return it at a later date. Now that would not be possible in this case because the three plants have a different stock item or part number for this filter even if it is the same part. Although both departments are keen on controlling the cost and budget of their company, they are the ones contributing to the cost itself. People from Finance must also understand that there are better and more proven ways to reduce costs by doing maintenance correctly. Just a piece of advice to industries, do not ever try to restrict or cut off the training budget of your industry, especially for maintenance.

Organizations are typically structured along hierarchical lines for different organizations or departments within a company, which in most cases leads to barriers, red tape, bureaucracy, and functional silos. In September 2019, a new ISO standard was published. This standard provides guidance on the alignment between Asset Management functions such as Finance, Accounting, and other technical functions within an organization. These non-technical and technical people are commonly the key people responsible for delivering asset management within their organization. However, in most cases, the two functions developed their own way to manage and control risks creating conflict with one another. A new ISO was established to address this problem.

Figure 11.9: Prestigious International Awards for Industries

ISO 55010 specification was written to help all functions within an organization to align both their financial and non-financial, or what we consistently called technical and non-technical issues to maximize value from their assets. ISO 55010 provides guidance to bridge the gap between the financial and non-financial functions of an organization to help them resolve their conflicts with one another. However, we all know that the finance voice is stronger in an organization compared to the technical people such as maintenance since they are near to God, I mean the CEO. Over several years of discussions involving several hundreds of experts from around the globe, ISO 55010 specifications were now approved for publication. By having this alignment, it can enable the organization's different functions to share and utilize their information better and improve the achievement of their organizational objectives.

Although none of these awards specified in figure 11.9 is a guarantee that the industry will remain in business and exists in the future, but rather these awards are meant to help industries leap forward within their competition. I know of a couple of industries that were (past tense) certified in TPM and are no longer in business today. As we reiterated consistently, businesses are always in a survival mode as the law of the jungle applies. Look at what happened during this **covid 19** pandemic, it shut down so many industries globally, and nobody was even prepared or saw it was coming.

The industry's brand and reputation are what we are paying for when industries aim for these certifications and awards. Once a brand or a company's reputation had been damaged, it is very difficult or impossible to bring it back once more. Perhaps that is why the industry hired lawyers, to tell everyone that the matter is taken care of. Likewise, it is the responsibility of the organization to take care of and value both its internal and external resources. While industries become more aware of their value to their customers, these awards will help them win more business and clients. That is why top management and executive decision-makers must fully understand their role in reliability and how it affects their business because the reputation of the organization is at stake. Reliability should be given the same priority as safety and quality. Industries must understand that reliability is not only the responsibility of the reliability and maintenance people, but it is everyone's responsibility in the organization as each function is connected like a chain and that the strength of the entire organization lies on its weakest link.

Chapter **12**

The Conclusion

Implementing Equipment Life Cycle Costing simply means what it takes for the entire maintenance function to operate and maintain the equipment with the least amount to own over its entire lifespan, but the entire challenge of Life Cycle can be realized if we can extend the lifespan of our equipment and assets. This is where the least amount to own can be experienced.

12.1: Focus on the People First, Before Technology

The role of maintenance is not just about sustaining and preserving the equipment but also understanding that there are parts, spares, and items that have design weaknesses and challenging themselves to further lengthen their lifespan through improvements, redesign, or modification. It is wrong to assume that the mere role of maintenance is just to sustain their equipment and assets, just as what other consultants remain to preach.

As time flies by too quickly, so with technology. As we move towards digitalization, automation, Maintenance 4.0, cloud, IIoT, smart sensors, and all these stuff, many industries are jumping into the bandwagon thinking that finally a Silver Bullet solution now exists in maintenance, but the truth of the matter is that as equipment becomes more complex and automated, majority of industries are still using what the prehistoric caveman used and that is firefighting. Many industries are still reactive and perhaps trapped indefinitely. I have nothing against technology, but what we also need is not just upgrading equipment's technology, but also upgrading the skills of our entire workforce, especially the maintenance workforce.

I recalled my last work in the mining industry, where I was present in one of the operations meetings and the resident manager raised a concern and asked what can be done to increase the utilization of their mobile equipment. The maintenance manager replied that we could increase the utilization from 40 to 60% if you can purchase four additional loader trucks. In a couple of months, they purchase four brand new fully automated loading trucks with a lot of additional new features. After three months of using these new loaders, one loader was severely damaged beyond repair, and after a couple of months, two of the new loaders were no longer operational. What seems to be the problem, they invest in the equipment, but not with the operators and maintenance on the assumption that the plant has many loaders, thinking that it will have the same functions and features as the existing loaders.

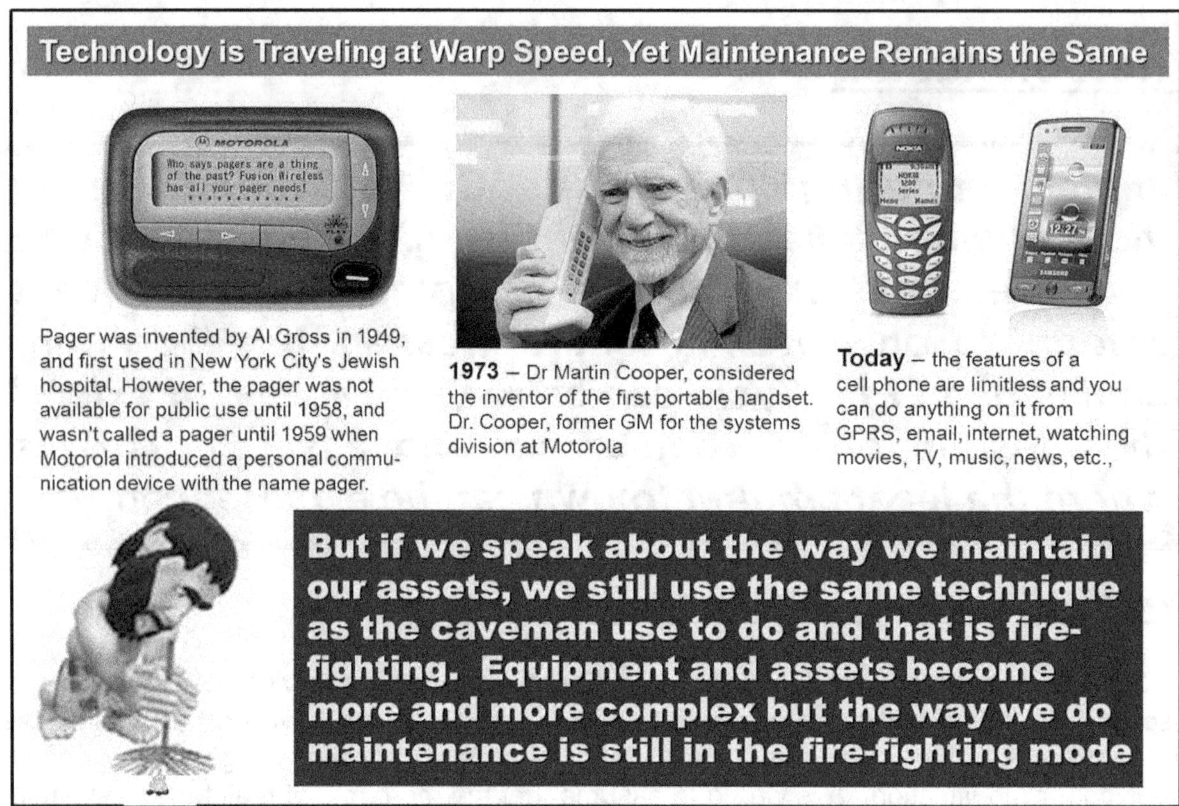

Figure 12.1: Many Industries Still Used the Caveman's Technique of Fire Fighting

Before we can even accomplish our objective on maintenance, our people should possess the correct knowledge on how to maintain their equipment and assets. Unluckily, the majority of the knowledge needed for maintenance to perform their work correctly was not taught in college. Few western countries offered courses on reliability and tribology, but the majority of the countries, including my country, The Philippines, only have the University of Hard Knocks, where everything is learned from mistakes and experimentation, which later on we called experience. Few people can provide this knowledge-based training to maintenance. Without this correct knowledge of maintenance, then we can be writing the wrong maintenance procedures or executing maintenance tasks that may be detrimental to the equipment without knowing it. Traditionally, what happens in industries is we just follow what the previous generation on maintenance does since this is how we do things here in the plant and this had been passed on for generations and will continue to be passed on for generations to come. Perhaps other industries are too dependent on their OEM that everything will be followed to the letter. Many industries I know are reluctant to send their people to maintenance and reliability training because their management thinks that training maintenance is just an added cost to them not knowing that the reason for the majority of their equipment failures, breakdowns, and problems can be attributed to their maintenance people's ignorance. I have experienced more than a dozen times during my training where maintenance people will approach me during the coffee breaks and tell me how they wish they had known me 10 or 15 years ago. To my surprise, I asked why is that? They simply told me that they are scheduled to retire soon. In my frame of mind, these people have worked their butts in their plant for many years and have not even attended a single training on maintenance. The sad thing is that we cannot bring back the time. Hence, my question is how on the planet can these people perform their jobs correctly. They simply can't and problems will continue endlessly. In a survey, I conducted on the Top Ten Problems on Preventive Maintenance, lack of training on the maintenance function garnered the majority of votes which indicates that this is the number one problem for industries.

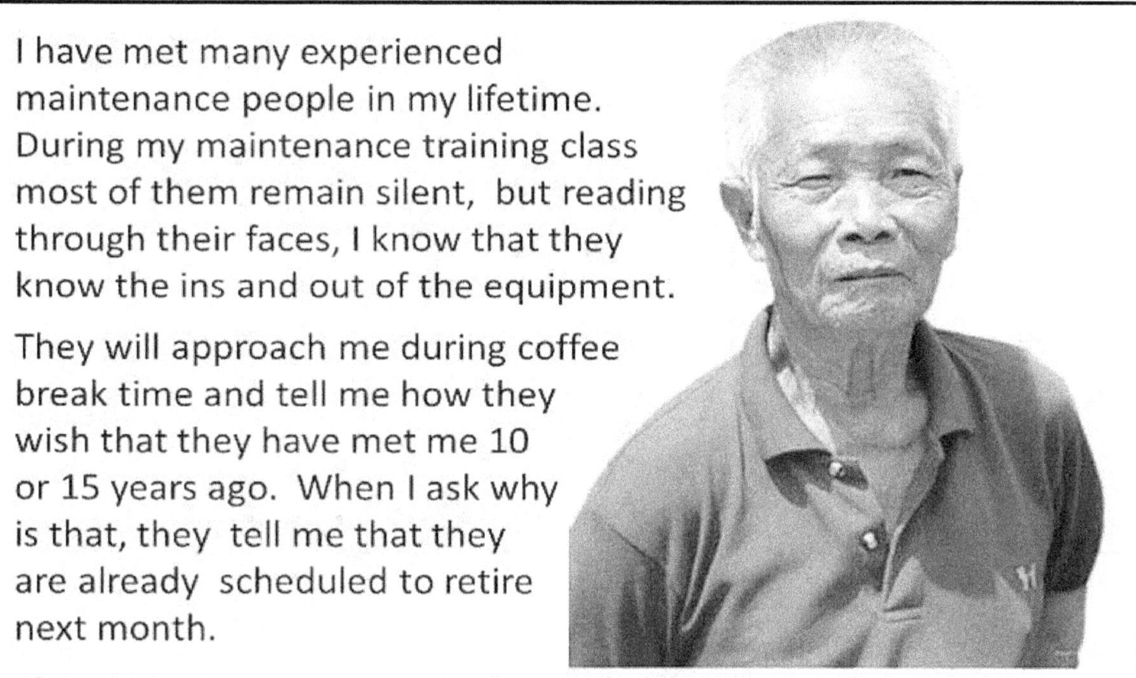

Figure 12.2: Let Us Not Deprived Our People with Knowledge

However, to be fair and have a balanced perspective, training is still not a guarantee that what has been learned will be applied in their industry. Honestly, there are maintenance people who attended my training seminars whose purpose is to update and upgrade their CVs and Resumes so that it looks cool when they apply for greener pastures. What is important in this case is that before sending people to maintenance training, human resources or their respective managers should talk to their people and set expectations on what will be done after attending the training.

Before this **Covid 19 Pandemic**, I used to go twice or thrice a year to India to deliver maintenance and reliability training event, which is sponsored by one of the leading training providers, more than 60% of my class are management people, deputy managers, general managers, engineering head and people with authority. I even have classes where the CEO or a Senior Vice President of a large industry is sitting right in front of me. What tickles my mind is how come in India, management people have the time to attend these training, while in my country, the Philippines, or other Asian countries I teach, they send the people directly involved in the maintenance and if the discussion becomes hot and sensitive, I often hear people say that Rolly, I think my manager should be here, or I think you need to talk to my management. When I called their managers, they will reply that is the reason I send this guy to your training and everything is just a Merry-Go-Round which makes me dizzy.

Figure 12.3: Wih Mr. Rajendran

I remembered my first training experience in India where I was in Chennai from July 6 to 8, 2008, teaching World Class Maintenance Management, the 12 Disciplines, during the last day of our afternoon tea break, a delegate introduced himself, Mr. Rajendran, and told me that he is the Maintenance and Engineering Head at his plant and told me that he learned many things from my training as he handed me his calling card. I said I was happy to hear that and then we go each other's way and I went straight to the airport back in my country. After a couple of months, they invited me once again to teach a different course, which is Reliability-Centered Maintenance in September 2008. Once again, Mr.Rajendran was in my class and I immediately recognized him. In fact, it is very easy to distinguish Mr. Rajendran from the rest since he is always the tallest person in the class. He shook my hand, gave me a bag of peanuts, and said that it came from his hometown, hence, it is special. He told me that there are two reasons why he joined the training, the first is to understand what RCM is all about and the second is to personally thank me for the previous training on World Class Maintenance Management as he shook my hand. He said that he applied one of the disciplines on lubrication and realized the benefits immediately. I was indeed very happy for him. A month later, he emailed me and requested an in-house training on WCM, which I went and taught their key people on maintenance. What makes me ponder is how come maintenance managers and those with authority have the time to attend training in India, but how come in other Asian countries where I teach including my very own country, the Philippines people with authority simply have no time for these kinds of reliability and maintenance training. It just makes me wonder why. I just hope that if this is not the case for this generation, I do hope that things change in the next generations to come.

12.2: Cost Cutting the Wrong Way to Save on Maintenance Costs

Cost cutting has always been the focus for many Finance and Top Management in several industries, and perhaps we need to learn from the lessons of history that there are many cases where this can be detrimental most of the time, especially when done on maintenance. Costs must be studied thoroughly, not just based on their initial cost but on the entire life cycle cost of the equipment since this is where real savings can be generated. Considering the initial cost is just like seeing the tip of the iceberg, yet underneath the iceberg lays the true cost far greater than the initial cost of the equipment. If one has seen an iceberg, multiply the area 3 to 10x, and that will be the estimated area or size of what is underneath the iceberg.

If we conduct a survey and ask each maintenance manager in every industry which of these two items is more important, reducing costs or preserving equipment's reliability, almost everyone will agree that preserving equipment's reliability is more important than reducing costs. If failures and breakdowns are reduced, then the cost of maintenance will definitely go down. However, this is not reversible. In fact, many cases state that cutting costs on maintenance will create more problems with the equipment than you can ever imagine. If we conduct a survey on what most industries are doing, I think it is more on cost-cutting, and that is the irony of it all.

Cost cutting is the name of the game most industries play, which is dangerous. How many times has your industry requested training on maintenance that their people need only to be rejected by their top management? Maintenance often realized a cut or slash in their maintenance budget, and their budget decreases every year. Headcounts are reduced as industries think they are truly saving money by removing the person and their benefits. Most of these good people will be laid off permanently, sad to say. There are cases when old maintenance people were forced to retire as they reach their retirement age. At the same time, industries replace them with fresh new graduates with absolutely no experience in the field whatsoever thinking that they are saving more. Worst if someone retires, the majority of the time, the industry has no program on what to do to capture these people's experiences so that they can be taught to the new generation. When the person retires, their experience goes with them to the grave and everything will start with the beginning. Here is a classic case of management cost cutting that led to the worst industrial accident of all time you can ever imagine.

The Case of Bhopal India: On December 2, 1984, 40 tons of highly poisonous methyl isocyanate gas leaked out of the pesticide factory of Union Carbide in Bhopal, India. A cloud of methyl isocyanate gas escaped from the Union Carbide plant in Bhopal. More than

6,400 died the following day, many from the effects several years after the disaster. There were at least 30,000-40,000 who were seriously injured in the world's worst chemical disaster. It was not the first incident as there was the first fatal accident at the plant that happened in 1982, where one plant worker died after being exposed to phosgene gas, which was an intermediate chemical used in making Sevin. More than 500,000 men, women, and children have been exposed to the poison clouds, and at least six thousand people died within the first week of the disaster. The current death toll is well over 16,000 and rising. Pregnant women were miscarried, and babies who survived were stillborn. The poison affected some people's lungs and nervous systems. Therefore, the question is what caused the methyl isocyanate gas to leak, leading to the deaths of thousands of people on December 2 to 3, 1984, at Union Carbide Plant in Bhopal, India.

Figure 12.4: Horrifying Scenes in Bhopal India

The physical cause of the failure is that MIC (Methyl Isocyanate) mixed with water, created a very violent reaction that caused the MIC to heat up, boil, turned into gas, and escape through the stack at the Union Carbide factory. MIC gas is dense and heavier than air; which mostly stays in the ground. There were too many mistakes and human errors to mention if we speak about the human causes of the failure. The plant is located in Bhopal where many people were residing near the plant. Most of them were slums and lived in shanties. There was no orientation done by the plant to the people living near the plant just in case some emergency cases, happened, as management deprived it thinking that it may cause some panic among the community. Since MIC gas is denser or heavier than air, then people can inhale it. If you are in your home, you just need to close down the windows and door. Place a wet towel on the bottom of the window or door, and in your face, the MIC gas that escaped will be absorbed by the towel and will not be inhaled by the people inside the house. This was actually what one of the survivors did when she called the person who knew what to do (a nurse who resigned at Union Carbide) in case of the MIC gas escaped. That person survived the tragedy. There were already massive amounts of errors committed before the disaster, which include the following:

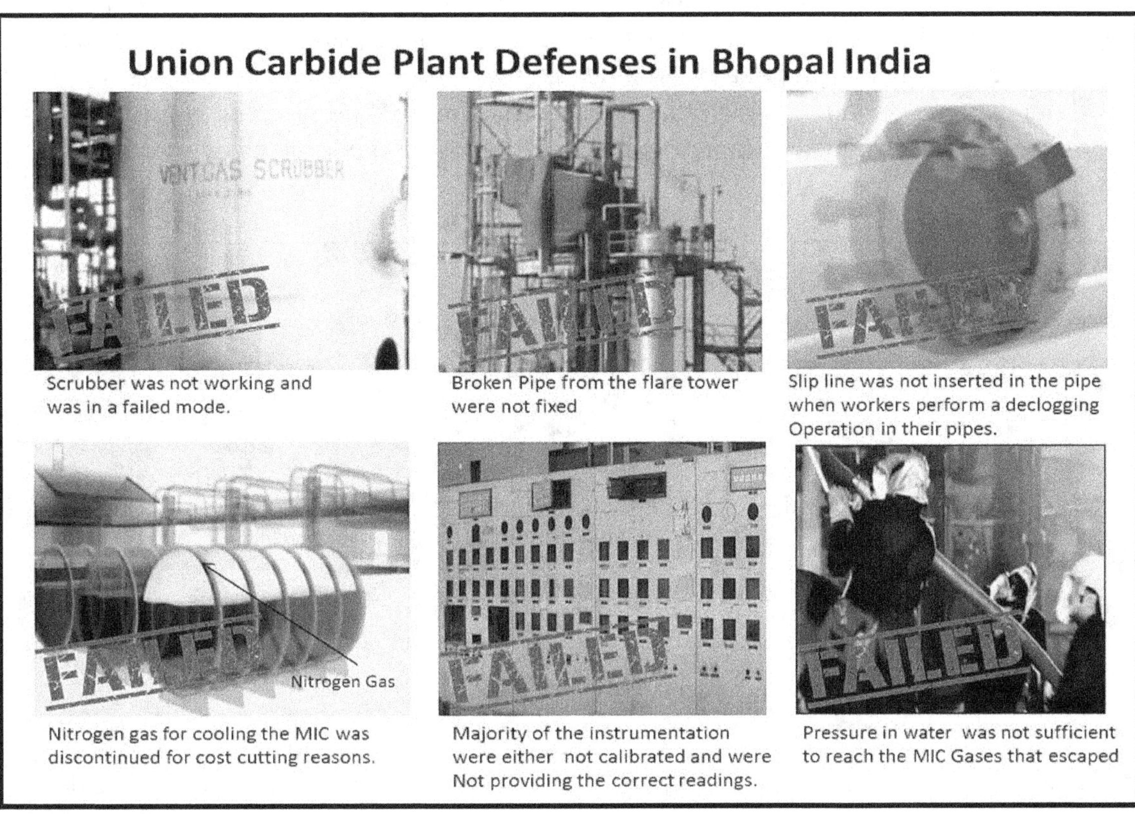

Figure 12.5: Union Carbide Defenses

- The maintenance manager was laid off from his work months before the tragedy.
- Instrumentations were not calibrated due to a lack of resources and budget.
- The scrubber was not working during the time of the tragedy for several weeks. Although, the scrubber can only contain a fraction of the leaked MIC gas that escaped.
- Gauges and instrumentation were uncalibrated and provided false readings. There was no one to check since the maintenance manager and 1/3 of the employees were laid off.
- The corroded pipe was not replaced on the flare tower. Should the flare tower have been operational, it might have averted the tragedy.
- Emergency water hose spray insufficient pressure to reach the vapor exiting the exhaust stack.
- Nitrogen gas used for cooling the MIC tank was removed. It serves as cooling for the MIC. If the Nitrogen gas was not removed, then the temperature would not have increased abruptly, or the leak would never happen as it cools the MIC. It was removed due to a cost-cutting scheme by management to save $ 74.00 per day.

What Happened in Union Carbide: Union Carbide is an American Factory that manufactured pesticides for farmers. These pesticides contain a chemical called MIC or Methylisocyanate. MIC is a toxic chemical that when inhaled can be fatal to humans. To the people of Union Carbide, they called this as liquid dynamite. This MIC should be cooled at all times. The most dangerous way of heating MIC is it will be mixed with water, which tends to create a violent reaction and can turn MIC to vapor or gas.

Kumkum Saxena a health and medical officer at Union Carbide knows the danger of MIC and proposed to his supervisor to have an emergency evacuation drill for the people living near Bhopal on what to do in case of an emergency in the plant. However, senior managers ignored her recommendation, which can cause panic among the people. Kumkum even told his supervisors that American plants have these drills. His supervisor even quoted that we are Indians and not Americans.

Figure 12.6: Aftermath of Bhopal Tragedy

By 1984, a drought-hit Bhopal agricultural sector and farmers were not buying the pesticides manufactured by Union Carbide as hundreds to thousands of pesticide boxes were returned to Union Carbide. Losses of revenues accumulate at the Union Carbide plant forcing its management to cut costs on all corners. One-third of its employees were laid off and production of their pesticides slowed down dramatically.

During the day before the tragedy, a declogging operation was to be done on the pipes. A fatal flaw in the work procedures did not include that before declogging a slip blind should be paced between pipes so water would not go directly to the MIC tank. With the absence of the slip blind, water went directly to the MIC tank, which immediately created a violent reaction, and the temperature of the MIC began to increase which started the beginning of a horrible ordeal.   One safety measure if the MIC temperature increase is that the tank contains a refrigerant but it was shut off for cost-cutting purposes. As the temperature of the MIC liquid reached 200 ° Centigrade, it starts to become a lethal gas. The scrubber, which was another defense in case of a leak, was not functional. The alarms suddenly sound off but were switched off to avoid a wide city panic. Since the MIC toxic gas was dense and heavier than air, it was inhaled by the people living near the Union Carbide Plant. People who inhaled the toxic gas find difficulty breathing and people rush through the hospitals, however, the doctors and nurses do not know immediately the cause of the problems, and the medication they provided to the victims was not working and people started to die. Those who survived realized a slow death as there were people who became blind, pregnant women were miscarried and those babies who survived suffered birth defects, and heart problems, Neurological system: Impairment of memory, mental depression, and uncontrolled health problems seem to affect those who survived that fateful evening.

The CEO of Bhopal, Mr. Anderson reached Bhopal four days after the accident, where he was immediately arrested, but after paying bail, he went back to the United States and never returned to face trial in India. If Saxena's proposal to create an emergency drill was followed, people can be aware of what to do and the tragedy might be mitigated.

Root Cause is due to Management Cost Cutting: A compilation of human errors led to the worst industrial disaster of all times. Although the CEO of Union Carbide, Warren Martin Anderson insists that the incident was caused by sabotage; however, with the pieces of evidence presented, this was a clear case of top management negligence. MIC (Methyl Isocyanate) was a lethal and toxic liquid that, when turned into gas and inhaled, can cause many problems to a human being. Small leaks were already experienced before the incident, yet nothing was done, even if one employee had already died. After the Bhopal Tragedy and litigation in court in 1989, the Supreme Court of India approved a settlement of the civil claims against Union Carbide for $470 million to pay damages to the people affected by the tragedy. This is approximately 400 US dollars for every individual who died and has been affected by the disaster.

This is the worst industrial disaster of all time that created a lasting horrifying impact on the survivors of the tragedy, which was caused clearly by the Management Cost Cutting Schemes. The devastation inflicted on the people was for a lifetime, and Top Management and C-Level people from different industries still never learned from the history of these disasters and still continue to play this dangerous game. All I can hope and pray for is that this tragedy never happens again in a lifetime.

12.3: The Challenge with Maintenance

Although in Chapter 7, we discussed that maintenance, as tasks cannot further improve the reliability of the equipment and assets. However, considering maintenance as people and human beings can definitely improve the reliability of their equipment through improvements, redesign, and modifications, which can lead to the possibility of extending the equipment's life cycle.

Every equipment, machine, and asset has different problems experienced by both operators and maintenance as they operate them in their plant. In fact, industries may have different equipment that poses different problems, especially in manufacturing plants. There are machines and equipment that experienced many failures, others experience minor stoppages, while there are also machines and equipment that have a long set-up or changeover. Still, there are several equipment and machines in the plant that experiences many product defects and quality problems. Each of these problems will require a different approach and even a different team. What is important for the maintenance function is to understand what the problem of the equipment is and what specific problem-solving tools they need to counteract the problem. These problems contribute to delays, downtime, increase in maintenance costs, and the possibility of having unacceptable consequences.

However, one very important thing is to identify what particular analytical problem-solving skill should be used to address the problem. We also have to understand how to manage human error as it will happen and this is given as part of being human. Even the smartest person in the plant can generate the worst error or mistake.

As we define the subject of Life Cycle Cost in Chapter 1, which is the sum of the Initial and Running Costs. This refers to the least or minimum amount of cost that we need to spend in the entire lifespan of the equipment or machine. However, the question is how can we expect to have the least amount to own if the equipment is suffering from losses and our industry is reactive? To do this, C-Level people need to invest in their human resources so that maintenance can understand how to provide a sustainable process for each and every piece of equipment and asset they have in the plant. Repairing the machine every time it fails will just improve the efficiency of the person repairing the failure since the failure keeps on recurring. What we need today are people who understand how to analyze their equipment problems. But what matters most is even before we can sustain the equipment, we need to understand and take care of the basic equipment condition. These are the small things that have been neglected on the equipment and assets. In fact, the majority if not all of the equipment failures and breakdowns experienced are merely an accumulation of small problems such as leaks, missing bolts, dirty equipment, and incorrect lubrication practices.

Once the basic equipment condition and a sustaining process are in place, the next challenge with maintenance people is to learn and understand the possibility of extending the life span of their equipment, which is detailed in Chapter 9 of this book. The machine or equipment does not fail by itself, it is a part, spare, or item that actually failed. What we need is for our maintenance to understand the cause of the failure, challenge these parts with inherent design weaknesses, and the possibility of addressing them, which in the end can further lengthen its lifespan. This is how we can extend the lifespan of our equipment and assets. The role of Top Management and C-Level people is to provide their people with the knowledge and resources they need to perform their work. What I believe is that if Top Management and Decision Makers in industries take care of their people, then these people will take care of their equipment, and assets.

12.4: When Maintenance Can Move Mountains

The case study provided in Chapter 9 on Planned Maintenance is a realization of the fact that there is simply a lot of room for improvement that can be done on maintenance if there is a clear direction and roadmap on how to achieve such a feat. However, the journey to reliability will always be a challenge in every industry as they battle the pressures of being reactive in their day-to-day activities. There will always be resistance at the very beginning. Expect resistance at the beginning.

I recalled my time in 1996 when my boss told me that I will be handling Planned Maintenance and gave me the lists of members I need to coordinate with. Hence, I organized our very first meeting on a Friday, and everyone was present. The Maintenance Department Head was also present in the meeting (JGRAN the one in the middle). This guy was the shortest in height but he was a very tough head. Hence, when I started my presentation on the reason we are here, 5 minutes into my presentation, he interrupted and asked me what was my position. Calmly I said, that I was just an Engineer 2. The people in the room were composed of Senior Engineers, Section Managers, and himself who was the Maintenance Department Head and Manager at that time. He said, why should they follow me if I am the head of maintenance. I was stunned and it silenced me as I do not know how to respond. Another person, Gabby, said that as far as he is recalled regarding his job description, he never read anything about following Rolly Angeles on Planned Maintenance. Another guy said that they are already too loaded with their work and will the company increase their salary. Soon everyone was talking, and complaining at the same time and the Maintenance Head walked out of the room. One by one, the people walked out of the room and I was left all alone stunned just like in boxing when you are knocked out face down on the floor. I went home still thinking about the incident and embarrassment and thought about writing my resignation letter. I did not report for work for a week thinking and telling myself that I was better than this. Finally, I come up with the idea of teaching them. I recall that the first slide I thought them was about the difference between MTBF and MTTF, which they appreciate. Every time I met weekly with these guys, I spent just one topic I taught them which took me 30 minutes or less until everyone was on board.

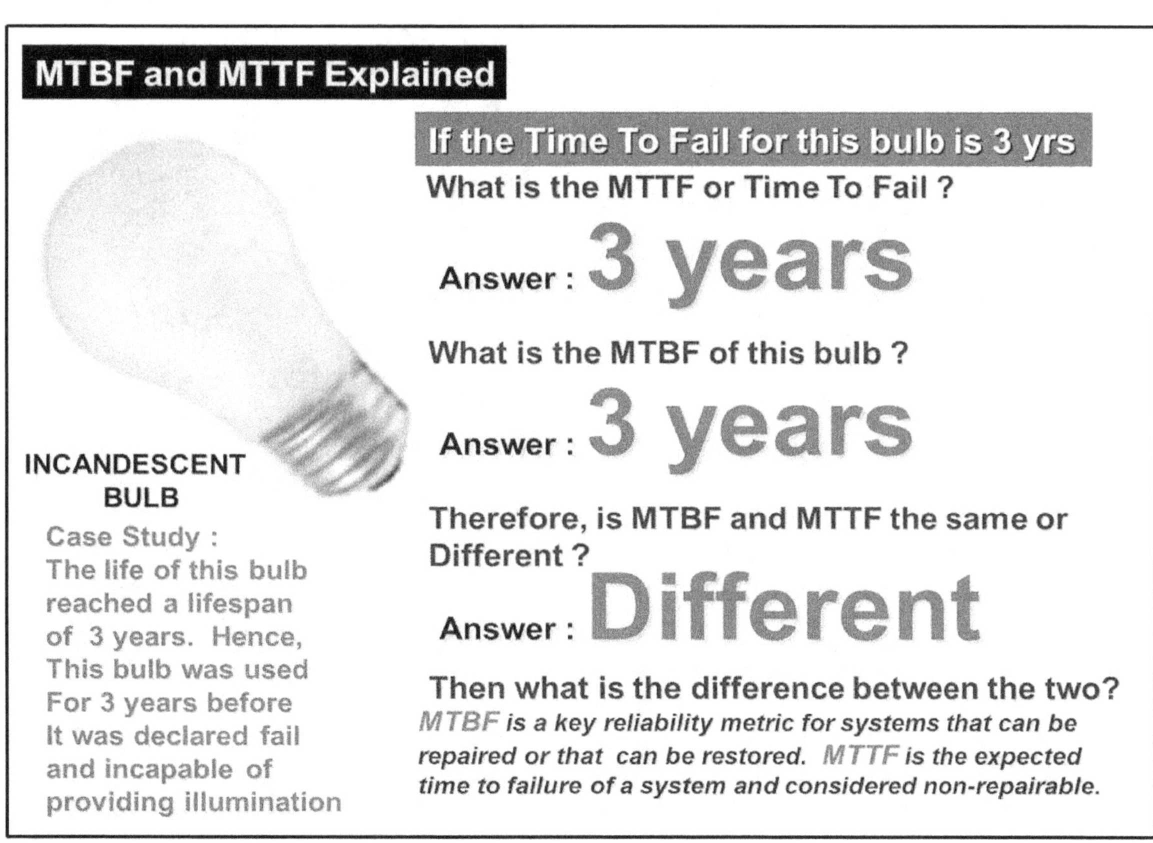

Figure 12.7: Difference Between MTBF and MTTF

I drafted the roadmap on how to proceed with our journey on Planned Maintenance of course with the input of the team during our meetings. Our meeting became regular every Friday from 3 to 5 pm. Once the roadmap was finalized, each of the members deploys their teams, monitor the results, and finally, realized a reduction in their maintenance costs, machine breakdowns, downtime, and an increase in their equipment's Availability, and MTBF. One Friday afternoon, during the meeting with the different Planned Maintenance Committee members, the maintenance head was present and asked me again, Rolly did you remember the question I asked you during the first meeting? I said, yes sir, you were asking for my position and I said that I was just an Engineer II. He handed me a letter and said that he together with all the members of our Planned Maintenance petitioned you to be a Section Manager to the Human Resources, but unluckily it was rejected since you will be leaping a lot of positions from Engineer 3, 4, Senior Engineer and finally Section Manager, but the Human Resources agreed to promote you to a Senior Engineer. Hence, effective today you are one and the members applaud and I was closed to tears, once again keeping silent and doing my best to compose myself. I went home early that afternoon after the meeting and told my wife about it and we were very happy about that.

Figure 12.8: Official 2001 Planned Maintenance Committee (Myself Extreme Right)

Remember that there is always a better way how we can maintain our equipment and assets. Real savings can be realized on maintenance if we improve the reliability of our equipment and assets. Industries must understand the basic needs of maintenance. Everything will start with its basic foundation, and that is acquiring the right knowledge since this is where they will build the skills in doing their job the right way. Training will always play an essential part in building the skills of our maintenance people. It will play an important role in achieving a clear and concise roadmap on reliability. There is always a better way of taking care of our equipment and assets. Industries don't have to be reactive all the time. World Class industries were not born that way. They started reactive themselves. I think this is the message I want to convey to the reader.

My experience in our journey on Planned Maintenance is tough but worth sharing. We were successful, as proven in our records that we have achieved dramatic improvements in our breakdown reduction, increase in MTBF and a dramatic reduction in maintenance costs. I see people change because of the results they achieve, and we recognize them for their efforts. Those who were pessimistic during the beginning became the frontline preaching, motivating, and driving the journey with their people. Not only does the equipment improve, but there is also some peace of mind for maintenance function. We are all proud of our accomplishments. It was a collaborative effort for the whole maintenance in that plant, and each one of us has contributed to its success. In short, despite the odds, our Planned Maintenance teams have moved mountains. What I believe is the key to improving our equipment and assets lies within our people. If we take care of the people, then the people will take care of their equipment and assets. Our Planned Maintenance journey lives on and will always be remembered in my books. And I think that is all I have to say about that.

Appendix A: Answer to Maintenance IQ Quiz

Answer to LCC IQ Quiz Part 1

1) c	11) c	21) d
2) d	12) a	22) b
3) c	13) a	23) a
4) a	14) c	24) a
5) b	15) b	25) c
6) b	16) a	
7) c	17) d	
8) c	18) b	
9) a	19) b	
10) c	20) b	

Answer to LCC IQ Quiz Part 2

1) b	11) b
2) a	12) b
3) b	13) a
4) a	14) b
5) b	15) a
6) b	16) a
7) a	17) b
8) b	18) b
9) b	19) a
10) b	20) a

Appendix B: RSA Maintenance Courses

RSA Reliability and Maintenance Consultancy Firm have been around for 17 years, and through these years, we have developed more courses suited for our reliability and maintenance people in industries. Here is a complete list of maintenance courses and services that we offer that can be catered in-house in your plant.

RSA Courses on Total Productive Maintenance
1. Total Productive Maintenance (3 days)
2. Planned Maintenance 4 Phases to Zero Unplanned Breakdown (3 days)
3. Understanding Autonomous Maintenance, Operators 7 Steps to Empowerment (3 days)
4. Understanding Focused Improvement-Kobetsu Kaizen (1 day)
5. Relationship between OEE and Equipment Losses (1 day)
6. Advance Maintenance Strategies on Planned and Autonomous Maintenance (2 days)

RSA Courses on Reliability and Maintenance Strategies
7. Lubrication Strategy-Understanding Tribology, Importance of Oil Contamination Control (2 days)
8. Reliability-Centered Maintenance for Industries (3 days)
9. Condition-Based Maintenance, Total Approach to Failure Prediction and Analysis (2 days)
10. Root Cause Failure Analysis-Understanding Equipment Failure (3 days)
11. Optimizing Equipment Reliability-Streamline RCM Approach (2 days)
12. Optimizing Preventive Maintenance Strategy (2 days)
13. World Class Maintenance Management-The 12 Disciplines (3 days)
14. Understanding MRO Spare Parts and Storeroom Management (3 days)
15. Failure Mode and Effects Criticality Analysis (FMEA/FMECA) (1 day)
16. Practical Best Maintenance Practices (3 days)
17. Advance Maintenance Leadership in TPM, RCM, LUB, and RCFA (5 days)
18. Advance Maintenance Strategies on RCM and RCFA (2 days)
19. Advance Maintenance Strategies on Lubrication and CBM (2 days)
20. Cutting-Edge Maintenance Management Strategies (3 days)

RSA Courses on Reliability and Maintenance Concepts
21. Meaningful Measures of Equipment Performance - Understanding MTBF, MTTF, MTBA, MTTR, \ MTTS, Failure Rate, OEE, and Weibull Overview (2 days)
22. Basic Maintenance Concept, Understanding Reactive, Preventive, Predictive and Proactive Maintenance (1 day)
23. Understanding Proactive Maintenance (1 day)
24. Maintenance Best Practices on LUB and CBM (2 days)
25. Advance Maintenance Strategies on RCFA and RCM (2 days)
26. Preventive and Predictive Maintenance Strategies

RSA Facilitation, Guidance, and Consultation Services Includes
• Facilitation and consultation on Total Productive Maintenance Implementation
• Facilitation and consultation on Root Cause Failure Analysis
• Provide guidance on starting up a Predictive Maintenance Strategy in the plant
• Facilitation and consultation on Reliability-Centered Maintenance

• Conduct initial assessment on maintenance Technical Training Needs Analysis
• Facilitation of Strategic Planning for Maintenance and Reliability Professionals
• Provides assessment on World Class Maintenance-The 12 Disciplines
• Consultation on setting up an Oil Contamination Control in your plant
• Implementation of TPM Planned Maintenance Pillar
• Implementation of TPM Autonomous Maintenance Pillar
• Implementation of World Class Maintenance Management-The 12 Disciplines
• Maintenance Assessment to determine where your industry currently stands

RSA Reliability and Maintenance Consultancy Firm accept in-house training services for industries for both local and international overseas countries. Special arrangements can be offered for overseas countries on any of our maintenance courses selected. These courses are what industries need to improve the way we maintain our equipment and assets in the plant.

The following are lists of training I offer to plants and industries concerning reliability and maintenance courses. My mission is to uplift the technical competence of our maintenance human resources in industries in search of seeking ways to achieve a level of maintenance excellence by capturing the industry's reliability and maintenance best practices. These courses provide in-depth details and a wealth of information regarding maintenance best practices from the most basic to the most advanced strategies. These powerful courses have been proven by industries that the best way to reduce their maintenance cost is to sustain the reliability of their equipment. With consistent focus on output and productivity and secondary to maintenance and reliability, the latter results in frequent failures, costly unscheduled repairs, unexpected downtime, and an inevitable high cost of maintenance that calls for the adoption of a more rigorous and more effective maintenance strategy that is truly world-class.

Finally, it is also undisputed that among the challenges every maintenance manager faces is maximizing the equipment reliability through a traditional and often self-designed Preventive Maintenance system. This practice is the very reason why the approach to maintenance management seems to remain reactive rather than proactive. Truly, these courses are designed for all maintenance managers, engineers, and professionals whose mandate is to optimize their equipment capacity and reliability at the lowest possible cost. It is in the preceding light that we designed these courses that can be tailored fit and made available to be conducted in-house in your plant for wider participation of your people. Should you be interested in any of these courses, you may reach me through my website at https://www.rsareliability.com or email me at rollyangeles@rsareliability.com.

Serving Maintenance Mankind Worldwide

This book is dedicated to all "Maintenance" out there in industries. My mission is to reach out to industries searching for ways to improve their maintenance human resources. I believe that the key to improving reliability is to provide our maintenance people the knowledge and education they need to maintain their equipment and assets. At my small firm, *We Serve Maintenance Mankind Worldwide.*

WCM in Thailand
October 2013

WCM in Indonesia
June 2013

WCM in South Africa
June 2012

WCM in Malaysia
August 2010

WCM New Delhi, India
December 2010

WCM at Meralco, Philippines
August 2016

WCM in Bangladesh
November 2009

WCM in India, Chennai
July 2008

WCM at Bharat Petroleum, India
September 2008

WCM Kolkata, India
February 2018

WCM in Mumbai, India
December 2016

WCM in Malaysia
July 2013

We're Serving Maintenance Mankind Worldwide

Bibliography

- Angeles, Rolly, **Implementing Preventive Maintenance for Industries the Right Way**, Central Books Printing, Philippines 2022
- Angeles, Rolly, **RSA Reliability and Maintenance Newsletter Vault Collection, Subscriber's Edition,** Central books Printing, Philippines 2021
- Angeles, Rolly, **Investigating Equipment Failures through Root Cause Failure Analysis,** Central books Printing, Philippines 2021
- Angeles, Rolly, **Maintenance Indices, Meaningful Measures of Equipment Performance,** Central books Printing, Philippines 2022
- Angeles, Rolly, **Cutting Edge Maintenance Management Strategies,** Central books Printing, Philippines 2020
- Angeles, Rolly, **Decoding Reliability Centered Maintenance Process for Manufacturing Industries,** Central books Printing, Philippines 2021
- Angeles, Rolly, **Lubrication Tactics for Industries Made Simple,** Central books Printing, Philippines 2020
- Angeles, Rolly, **Maintenance Roadmap to Reliability**, Central books Printing, Philippines, 2016.
- Angeles, Rolly, **Problems, and Solutions on MRO Spare Parts and Storeroom,** Central books, Printing, Philippines 2020
- Angeles, Rolly, **Reliability, A Shared Responsibility for Operators and Maintenance,** Central books Printing, Philippines, 2018.
- Angeles, Rolly, **World Class Maintenance Management – The 12 Disciplines**, Central books Printing, Philippines, 2009
- Bazovky, Igor, **Reliability Theory and Practice,** Dover Publications Incorporated, 2004
- Campbell, John Dixon, **Uptime, Strategies for Excellence in Maintenance Management**, Productivity Press, 1995
- Gotoh, Fumio, **Equipment Planning for TPM, Maintenance Prevention Design,** Productivity Press, Portland, Oregon, 1991
- Lapierre, Dominique, and Moro, Javier, **Five Past Midnight in Bhopal**, Warner Books Inc., 2002
- Masaji Tajiri and Fumio Gotoh, **Autonomous Maintenance in Seven Steps – Implementing TPM on the Shop Floor**, Productivity Press, 1999
- Moubray, John, **Reliability Centred Maintenance II, Second Edition**, Great Britain: Butterworth, Heinemann, 1997.
- Mobley, R. Keith, **An Introduction to Predictive Maintenance**, Butterworth-Heineman Elsevier, Heinemann, 2002.
- Nakajima, Seiichi, **Introduction to TPM, Total Productive Maintenance**, Portland Oregon, Productivity Press, 1984
- Nowlan, Stanley, and Heap, Howard, **Reliability-Centered Maintenance**. Springfield, US Department of Commerce National Technical Information Service, 1978.
- Pall Industries Hydraulics Company, **Contamination Control, and Fundamentals**: New York, 1994.
- Reason, James and Hobbs Alan, **Managing Maintenance Error, A Practical Guide**, Ashgate Publishing Limited, 2003
- Robinson, Charles, and Ginder, Andrew, **Implementing TPM, The North American Experience**, Portland Oregon: Productivity Press, 1995.
- Schein, Edgar. **Coming To a New Awareness of Organizational Culture**, Sloan Management

Review, Winter, 1984.

• Suzuki, Tokutaro, **TPM in Process Industries,** Productivity Press, Portland, Originally published by Japan Institute of Plant Maintenance, 1992
• Whittingham R.B**., The Blame Machine, Why Human Errors Causes Accidents**, Elsevier Butterworth Heinemann Publications, 2004

Glossary on Maintenance

This glossary is a collection and compilation of all my books on reliability and maintenance, which includes the following;

• Volume 1: World Class Maintenance Management – The 12 Disciplines
• Volume 2: Maintenance – Roadmap to Reliability
• Volume 3: Reliability – A Shared Responsibility for Both Operators and Maintenance
• Volume 4: Cutting–Edge Maintenance Management Strategies
• Volume 5: Problems and Solutions on MRO Spare Parts and Storeroom
• Volume 6: Lubrication Tactics for Industries Made Simple
• Volume 7: Decoding Reliability-Centered Maintenance Process for Manufacturing Industries
• Volume 8: RSA Reliability and Maintenance Newsletter Vault Collection, Subscriber's Edition
• Volume 9: Investigating Equipment Failures through Root Cause Failure Analysis
• Volume 10: Maintenance Indices – Meaningful Measures of Equipment Performance
• Volume 11: Implementing Preventive Maintenance for Industries the Right Way
• Volume 12: Extending Equipment Life Cycle - The Next Challenge for Maintenance

8-Disciplines is an analytical problem-solving tool designed to determine the probable or most likely cause of the problem. This method can be applied to defects and equipment-related failures. This method establishes a permanent corrective action based on data and provides the probable cause of the problem.

ABC Analysis also called inventory stratification is a technique used to classify and optimize inventory levels. Inventory is based on the value and usage of an item. Class C refers to standard parts, consumables, or commodity items that can be delivered by the vendor on a regular schedule or can be easily made available from a few hours to a couple of days where the cost is less than $100/unit. Class B are standard parts and can be made available in a few days to weeks. Parts are mid to high costing around $100 or more. Class A: are one-of-a-kind parts with long lead time, high cost, and low demand. The cost can range from $500 to $100,000 or more. Items classified in this category are critical and expensive.

Abnormality can be defined as any deficiency, disorder, slight irregularity, defect, bug, flaw, or any unwanted condition that could lead to other equipment problems. This is addressed during Autonomous Maintenance Step 1 activities.

Abrasive Wear is a type of wear that can be categorized by a single keyword, cutting. Abrasive wear occurs when hard particles are suspended in a fluid or projections from one surface roll or slide under pressure against another surface, thereby cutting the other surface. Abrasive wear can be either two-body or three-body abrasion.

Absolute Filtration refers to the smallest size of particle that will be removed during filtration. This means that if the filter size is 5 microns and the rating is absolute, almost all the contaminants in the range of five microns and above should be removed by the oil filter. The filtration efficiency for absolute rating usually is high at 99% and above, which is the capability of the filter to remove contaminants.

Absolute Viscosity measures the resistance to flow when an external force is applied like a spindle driven by a motor. For example, when we buy water-based paint and open it, it takes some force to stir the paint. However, if we add water and mix it with the paint, it is much easier to stir. This means that the viscosity decreases.

Accelerated deterioration means that the part, item, or spare had failed prematurely and has not reached its natural lifespan. We can state that we have a case of premature wear.

Additive is any blended chemical added to a lubricating oil or grease to improve its performance and properties to protect the base. It is also mixed with the oil to counter any negative effects of the oil from heat and contamination. Additives usually represent around 1 to 30 % of the lubricant with engine oil having the most amount of additives.

Adhesive Wear occurs when a peak or asperity from one surface comes in contact with the other surface's peak or asperity. There may be instantaneous micro-welding caused by the friction or heat involved due to the lubricant's loss of film.

Advance Discipline refers to the last couple of disciplines on World Class Maintenance that includes the state-of-the-art non-destructive tools and diagnostics instruments for Predictive Maintenance and software such as CMMS or EAM for the maintenance requirements.

Age is the measure of a unit, item, or spare's total exposure to stress which can be expressed as the number of operating hours or other stress units since the asset started operating until its retirement or decommission.

Aging is the process where certain parts of the equipment deteriorate or wear out because of stress-induced over a given period. It is also termed as wear and tear or deterioration. TPM refers to this as natural deterioration.

Age Exploration Method is a method for determining the interval for age-related items by increasing the interval for Preventive Maintenance overhauling and replacement by 10 % if aging or wear-out is not yet evident when it will be overhauled or replaced.

Air Traffic Fatalities is an incident where a person is fatally injured due to an occurrence associated with the operation of the aircraft. This definition starts at the time from when the first person boards until the last person disembark the plane.

Analytical Ferrography (ASTM D7690) is an oil analysis test where solid debris suspended in a lubricant is separated and systematically deposited onto a glass slide called a ferrogram. The slide is examined under a microscope to distinguish the particle size, concentration, composition, morphology, and surface condition of the ferrous and non-ferrous wear particles

Ancillary refers to providing something additional to a main part or function. The word ancillary implies that there is a primary function or activity.

Anthropometric Errors are a type of human error that occurs because the person simply cannot fit into the space provided, cannot reach something, or is not strong enough to move or lift something. Industries aiming for multiskilling should also consider anthropometric factors to avoid human errors.

Antifoaming or Defoamer is the process of removing entrapped air or bubbles in the lubricating oil. Almost all lubricating oil systems contain some form of air. Air is found in four phases: free air, dissolved air, entrained air and foaming. A defoamer or an anti-foaming agent is a chemical additive in oil that reduces and hinders foam formation in industrial process liquids.

Anti-Wear additives (AW) provide the protective layer on moving parts that minimize metal contact effects. Anti-wear additives are used in many lubricating oils to reduce friction, wear, and scuffing under boundary lubrication conditions, where full-film lubrication cannot be fully maintained. Anti-wear additives are also called boundary lubrication additives.

Asperity, in tribology, on a micro-scale are topographical irregularities that are similar to peaks and valleys of a solid surface. This can only be seen if the surface is enhanced through a Scanning Electron Microscope (SEM). Once the two surfaces' asperities come into contact, lubrication is displaced, and there is metal-to-metal contact and fracture later occurs.

Asset Hierarchy is an index of all your maintenance equipment, machines, and components, and how they work and are linked together. This explains the parent-child relationship of assets in the plant.

Asset Management is defined as executing that set of processes to realize value as the organization defines it from those assets.

Attrition Rate also referred to as a churn rate, is the rate at which people leave the industry. This refers to the number of people who have left the company, divided by the average number of employees over a given period.

Autonomous Maintenance is one of the main pillars of TPM which refer to the activities in which operators perform their daily inspection, lubrication, parts replacement, minor repair and troubleshooting, accuracy checks, and so forth on their own equipment, aiming to keep the equipment in good running condition.

Auxiliary equipment is defined as peripheral equipment that may be an integral part of the extrusion process to improve or optimize the extrusion process efficiency and ease of operation.

Availability is the proportion of time the equipment is available for use for its intended purpose, whether it is being utilized or not. The trend should be consistent and should not be lower than 98%. Calculation of availability can be reported on a daily, weekly, or monthly basis.

Available Time is given as the total time in a given period. 24 hours for one day, 168 hours for one week, 720 hours for 30 days, 744 hours for 31 days, or 8760 hours in a year.

Barcode is a machine-readable form of information on a surface that can be scanned by a barcode scanner. They are often known as UPC codes. The barcode is read by using a special scanner that reads the information directly. This information is transmitted into a database where it can be logged and tracked in real-time.

Basic Maintenance Discipline is the fundamental activity that should be performed on the equipment before proceeding with any other advanced or specialized disciplines. It also serves as the foundation of any maintenance management strategy. This includes training, KPI, Basic Equipment Condition, Autonomous Maintenance, and Preventive Maintenance.

Bathtub Curve is a conditional probability curve representing the age-reliability relationship of certain items or characterized by an early infant mortality region, a region of relatively constant reliability, and an identifiable wear-out age.

Bending Stress is the force an object encounters when subject to a load at a given point which causes the object to bend or warp. This type of stress usually occurs in objects when subjected to a tensile load.

Beta Filtration Rating is derived from the Multi-pass Test Method for Evaluating Filtration Performance of a Fine Filter Element (ISO 4572). The automatic particle counter measures the upstream particles and quantity per unit volume of fluid, the particle size, and the quantity downstream of the filter during the test.

Bill of Materials is a list of materials, items, spares, and consumables needed either for repair or Preventive Maintenance. Having a PM Bill of Materials will allow the people involved to identify easily what items and materials will be needed to conduct a repair or a scheduled routine Preventive Maintenance for a particular asset and the cost of PM.

Blending is a refinery operation that blends different component streams into various grades of gasoline. For gasoline, the fuel is blended to achieve a higher octane rating standard, thereby creating different gasoline types. Most gasoline stations offer three octane levels, including regular, about 87, mid-grade, about 89, and premium, with an octane rating from 91 to 94.

Book Value of equipment or asset refers to the book from the Accounting of Finance Department and is reflected on the industry's financial statement.

Bottleneck is a constraint or congestion in a production operation that occurs when the products' inventory is built up quicker than the equipment can process. In manufacturing, the bottleneck equipment includes those assets that frequently fail in the process, or have a lower UPH (units per hour) where inventory is built than other equipment in the process.

Boundary lubrication is affiliated with metal-to-metal contact between two sliding surfaces as the asperities come in direct contact with one another. This mostly happens during start-up and shutdown.

Breakdown Occurrences refer to the frequency of failure or breakdown of the system or component. The trend should be the lower, the better. This can be reported on a weekly or monthly basis. Care should be taken on what to include and not to include as a breakdown.

Breakdown Loss sometimes referred to as equipment failure loss, is a loss when the machine stops since its function completely failed. The Japanese term for breakdown is kosho. Breakdowns can be either function loss or a function-reduction breakdown.

Brittle Fracture is a fracture that involves little or no permanent deformation of metals.

Many non-metals lack ductility and are subject to brittle fractures. A brittle fracture occurs when a part is overloaded and breaks with no visible distortion and deformation.

C-Level Executives are high-ranking executives of a company or any organization in charge of making company-wide decision-making. C stands for chief. This will include the Chief Executive Officer (CEO), Chief Operating Officer (COO), and Chief Information Officer (CIO).

Calibration is the process in which instrumentation and equipment used in industries are monitored and maintained to ensure they continue to give accurate and reliable results and ensure that deviations in measurements are corrected.

Capital Spares are the items within the inventory that are purchased as spare parts for depreciable assets (e.g., capital equipment, backup engines, and redundancies). As such, these capital spares within the inventory can depreciate. In most cases, capital spares are not included as inventories since they are part of a company's property, plant, and equipment (PPE).

Carrying or Holding Costs is the accumulated cost of parts held in stock inside the storeroom. The holding or carrying cost begins to accumulate the day the items are put inside the storeroom. Usually, this will be around 10 % to 30% of the overall costs per year.

Causal Factor in root cause investigation is any unplanned, unintended contributor to an incident or failure. It can also be stated as the contributing cause to the incident. It can also be said that the causal factor is just one of the reasons that contributed to the incident.

Chronic Losses are usually made up of a wide variety of causes and frequently occur over time. The word chronic means repeating or recurring, which can be seen in defects and failures. Chronic Losses are always a combination of different causes.

Circadian Rhythm, which means around the day. It comes from the Latin word circa which means about and dies which means a day. These circadian variations are governed by a biological clock located in the human brain.

Classification is used in FMEA/FMECA to highlight those failure modes with a severity rating of 9 to 10 or failures with the highest impact or consequences on the equipment or system being analyzed in the FMEA analysis.

Cleaning Materials Inventories are items and materials needed to sustain and maintain the plant's cleanliness and facilities, comfort rooms, toilets, and office spaces, used by the janitorial services of the plant.

CMMS or Computerized Maintenance Management Software is a maintenance software used by the maintenance department to streamline and automate the maintenance process. CMMS should support these functions by capturing and automating administrative maintenance tasks and gathering relevant information to perform this process.

Combination of Tasks is the use of two or more types of maintenance tasks to address a particular failure mode such as Preventive and Predictive Maintenance, which can be done at different intervals and with different people. For example, if the failure mode is a bearing failure, operators may be performing lubrication, while the Predictive Maintenance group is performing Vibration Monitoring, while other operators are carrying out a daily reading of temperature using a thermal gun.

Commissioning a piece of equipment means carrying out all the necessary tests needed to be required by industry standards. It is the process of ensuring that all systems and components of a building are designed, installed, tested, operated, and maintained according to the operational requirements of the owner or user. A checklist is usually provided to check the process.

Complex Item refers to an item, equipment, or asset whose functional failure can result from numerous failure modes.

Compression is when we applied a downward force on a vertical cylinder, the object will have an equal upward force, which is equivalent to the normal force. Once we increase the downward force, then we also increase the upward force.

Conditional Probability of Failure refers to the probability that an item will fail during a particular age interval, given that it survives to enter that interval.

Condition-Based Maintenance or Predictive Maintenance checks the equipment's actual condition using sophisticated measuring non-destructive instruments with precision accuracy. Predictive Maintenance instruments are just a higher form of the human senses. Performing inspection through the use of human senses is also included in On-Condition tasks for RCM.

Consequences of Failure result from a given functional failure at the equipment level and for the operating organization classified in the RCM analysis: which may be a hidden failure consequence, safety consequences, environmental consequences, operational consequences, or non-operational consequences.

Consistency depends on the type and amount of thickener used and the viscosity of the base oil of grease. Consistency is also known as the resistance to deform caused by an outside force. The measure of consistency is called penetration.

Containment is a short-term action being initiated to keep the defects or failures from further damage. Containment is just a temporary solution and will not fix the problem.

Contamination in oil is anything that should not be present in the oil besides its base and additives. Contaminants may be in the form of solids, liquids, or even gases in bubbles that cause foaming.

Consumable spares are regularly consumed and used parts and items such as fasteners, seals, belts, oil filters, grease, lubricants, gloves, WD-40, rags, face masks, cleaning materials, chemicals, and so on. These items are stored in the storeroom. They are used by maintenance regularly. They are considered fast-moving items in the storeroom.

Contract Maintenance is defined as the contract between two parties that creates the agreement that one party will maintain an asset owned by another party. This is also defined as the contract agreement signed between the industry and a third-party contractor who will perform a maintenance activity based on the terms of the agreement on industries equipment, and assets. Usually, the scope of maintenance work will be dictated by the maintenance head, and performed by a third party and not by the plant's regular maintenance crew.

Corrective actions are actions taken to address the probable cause of identified non-conformances or incidents, manage their consequences, and prevent or reduce the likelihood of recurrence of failure. Identification and execution of corrective measures must be done for both the short term and the long term.

Corrective Maintenance is a maintenance task that has different meanings. In the majority of industries, this refers to repairing a failure after it happened. Other industries also refer to this as Predictive Maintenance tasks, which are the activities that are done when the P-F interval is nearing its functional failure. For those industries implementing TPM or Total Productive Maintenance, corrective maintenance means performing an improvement on the equipment.

Corrosion is a wear process of materials caused by chemical, electrochemical, or other reactions. The main difference between corrosion and rust is that corrosion occurs due to the chemical reaction on metal surfaces, while rust only affects iron metals exposed to air or moisture.

Corrosion inhibitor is a chemical compound substance that decreases the metal or alloy's corrosion rate when it comes in direct contact with the oil when added to the lubricating oil. Corrosion inhibitors are additives to the fluids that surround the metal or related object. The corrosive inhibitor's nature depends on the material being protected, which are commonly metal objects, and on the corrosive agent that needs to be neutralized.

Corrosive wear is the deterioration of metal due to chemical or electrochemical reactions with its environment. Corrosion can occur when the substance is exposed to air or some chemicals while rust mainly occurs when a metal is exposed to air or moisture.

Cortisol is a stress hormone that triggers your heart to work harder. Cortisol works with certain parts of our brain to control our mood, motivation, feelings, anxiety, and fear.

Cost avoidance focuses on taking actions that avoid incurring costs. For industries, these are measures to reduce their expenses. Any actions to avoid an increase in expenses are considered cost avoidance.

Cost cutting refers to measures implemented by a company to reduce its expenses and improve profitability. Cost-cutting measures are typically implemented during times of financial distress for a company or during economic downturns that can have repercussions.

Cost savings is the reduction from last year's spending on the same item. They will appear on the company's budget and in financial statements as a decrease in spending.

Cracking Process is used to maximize the effectiveness of heavier oils. Heavier oil contains large strings of hydrogen and carbon molecules. Using a catalyst, these long strings of molecules are broken into small chains that transform heavier oil into lighter fluids such as gasoline and diesel fuels.

Criticality is a measure of how important an asset is to your process. The more critical the asset, the more impact it will have when failure happens. Criticality is based on what could happen if the failure occurs.

Critical Failure involves the loss of function or damage that could directly affect safety and environmental consequences. It can also have operational consequences with devastating impacts or aftermath.

Culture means how people perceived and do these things around their plants. It refers to their common values and beliefs, while others refer to them as shared thoughts and feelings. According to Schein, culture is the pattern of basic assumptions that a given group had invented, discovered, or developed in learning to cope with its current problems of external adaptation and internal integration that had worked well enough to be considered valid which will be taught to new members as the correct way to perceive, think, and act concerning their day-to-day activities.

Decommissioning refers to the removal or retirement of a piece of equipment or asset from active use. At this point, the equipment will now stop operating and decision-makers finally decide that it is time to retire the asset for good. This is the last phase of an equipment's life cycle.

Dedicated Machines refer to manufacturing machines that produce only one type of product and there is no set-up or conversion involved.

Defect and Rework Loss is a type of equipment loss caused when defects on products are found in which the product has to be reworked or scrapped. Defects on the products occur for a variety of reasons such as machine, method, raw materials, human error, or the environment.

Deicing fluid is a mixture of a chemical called glycol and water, which is heated and sprayed under pressure to remove ice and snow on the aircraft's wings and body. While it removes ice and snow, deicing fluid has a limited ability to prevent further ice from forming.

Detection in FMEA/FMECA is an assessment of the design and machinery controls' ability to detect or capture that a failure mode is occurring independently. This is also termed as confidence.

Detergents are oil additives that keep the hot metal components free from deposits and neutralize acids, forming sludge and residues. The detergents used in lubricating oil can be organic soaps and salts of alkaline earth metals such as barium, calcium, and magnesium.

Design Speed Loss is a type of equipment loss on production caused by the difference between the design or theoretical speed and its actual operating speed. This is usually given in the PM manual document provided by the OEM.

Die, Tool, and Jig Losses are losses that include the cost of the physical consumption of spare parts or refurbishments of items that are used on the line. Examples of this include the cost of spares, cost of replacement, maintenance tooling, dies, and jigs.

Diffusional Interception is a filtration mechanism where the particles that strike through randomly moving gas molecules impact the medium and are held by adsorptive forces. The probability of removal is further increased by the effect of Direct Interception, Inertial Impaction, and Diffusional Interception combined together.

Direct Interception is where the contaminants are trapped in the pores between the medium fibers. It can also catch contaminants that are smaller than the pore size through the process of bridging. In direct interception, the contaminants come in physical contact and are attached to the filter media.

Dispersants are added to the lubricant to prevent the accumulation or particle attraction of sludge and dirt in the oil. The dispersants' function is to suspend the contaminant in the oil rather than for the contaminant to settle at the bottom of the crankcase and form deposits so that it can be removed by the oil filter easily when the oil flows and circulates inside the engine system.

Dissolved water is when the water is dispersed in the oil in a homogeneous molecular solution. The gas or moisture cannot be removed from a conventional' nominal filtration; however, several absolute oil filtration on the market can remove a certain amount of moisture in the oil.

Distillation Process is a refinery process where crude oil will undergo a process called distilling or the distillation process. The oil is heated in an atmospheric distillation vessel until the crude oil reaches its boiling point, which will turn oil into vapor.

Distribution Loss is a manpower loss that includes losses that occur due to incorrect or inefficient delivery of raw materials, packaging, or products to and from the factory or the production line. Example: Incorrect delivery of materials from the supplier to the store, late deliveries, excessive handling of deliveries (double handling).

Downtime is the period when equipment or asset is not operating. This is the time that the equipment is unavailable for use. The equipment's unavailability may result from equipment failure, malfunction, or non-machine-related downtime such as PM shutdown. Downtime is classified as machine-related (unplanned downtime) and non-machine-related (planned downtime). It refers to the amount of time spent on doing corrective maintenance on the equipment and assets.

Dropping Point is a test on grease that determines the temperature at which the grease drips off the testing unit in a non-decomposed condition. It indicates the grease's heat resistance or up to what temperature the grease can remain firm.

Duplicated Tasks are similar maintenance activities that are done by different people. This will create redundant work and a waste of manpower and other resources.

Economic Consequences refer to the consequences of the failure mode, which is evident and does not affect safety, and the environment and has no operational consequences. The only consequence, in this case, is the direct cost of the repair.

Economic Order Quantity (EOQ) is one technique that can be used to optimize inventory levels by ordering the right quantity at a specific time interval to minimize inventory cost but still meet the users' demands. EOQ will answer how many orders need to be placed per year and how many items to place per order.

Electronic Data Interchange (EDI) can be best used by Purchasers in automating their transactions with their vendors and suppliers. It provides a direct link between the vendor and the buyer by automating the Purchase Request and Purchase Order system directly linked with an internal system.

Empowerment means power, control, authority, or dominion. The prefix "em" means to put on to or to cover with. Empowering is the passing of authority and responsibility to an individual. Empowerment occurs when the power goes to the operators, who then experience a sense of ownership and control over their jobs.

Emulsified Water can be in the form of microscopic droplets of water distributed in the form of an emulsion. This is when the amount of dissolved water is greater than the saturation point. When the water is in an emulsified state in lubricating oil, its color turned into a white or milk state in layman's terms.

Energy Losses are losses in the input energy, which cannot be used effectively for processing. Examples can include start-up losses and insufficient compressed air for pneumatic machines. This usually occurs in the facilities/utilities in industries.

Enterprise Asset Management (EAM) software provides a holistic view of an organization's physical assets and infrastructure throughout its entire life cycle, starting from the design, procurement, installation, commission, operations, and finally, its retirement or disposal.

Environmental Consequences mean that a failure mode has environmental consequences if it causes a loss of function or other damage, leading to the breach of any known environmental standards or regulations. This means that the consequences of the failure go beyond the industry and affect society, people, or any government laws. This is considered the worse consequence of all.

Equipment failure refers to an event in which the equipment cannot accomplish its intended function and objective. It may also mean that the equipment stopped working, is not performing as desired, or is not meeting its target expectations.

Ergonomics originated from the Greek word ergon, meaning work, and nomoi, meaning natural laws. It is the science of refining the design of products to optimize them for human use. It is also sometimes known as human factors engineering. Ergonomics is a science that deals with designing and arranging things so that people can use them easily and safely.

Erosive Wear is a type of wear due to mechanical interaction between the surface and a fluid, a multi-component fluid, impinging liquid, or solid particles. Erosive wear occurs by the impact of particles upon a surface with a resultant material loss due to fracture.

Evidence is a piece of information that supports a conclusion. Evidence in its broadest sense includes anything used to determine or demonstrate the truth of an assertion. It is the lifeblood of any Root Cause Failure Analysis investigation. The three types of evidence include people, paper, and physical evidence.

Evident failures are failures that will become evident to both the maintenance and operators when they occur independently.

External set-up includes the activities, which can be performed while the machine is still running or is running the last production lot before the actual conversion will take place.

Extreme Pressure additives (EP) are oil additive that is usually used in gearboxes to provide a layer of film on the metal's asperities so that when the two opposite asperities meet, instead of breaking up, they will simply slip or slide with each other. It is not recommended to use EP additives on yellow metals.

Fact-Finding Group is a group of people composed of the Principal Investigator and Evidence Gathering Team. Its mission is to find out the root cause of the failure, learn from it, and do something to prevent, mitigate, or eliminate its occurrence.

Fail-Safe System is a system whose function is replicated or duplicated so that the function will still be available to the equipment after the failure of one of its sources, making failure tolerable and allowed to happen.

Failure is the inability of the equipment to perform its required function. The failure of a component is viewed as terminating its life. In general, failure refers to the state or condition of not meeting a desirable or intended objective. In life, failure may be viewed as the opposite of success.

Failure Analysis is an engineering approach to determining how and why a particular part, equipment, or component fails. The goal of failure analysis is to understand the physical cause of the failure.

Failure Development Period is the time when the potential failure is spotted by the instrument until the time the equipment finally reached its functional failure or this is when the equipment is actually declared fail. This is also known as the P-F Interval.

Failure Finding Tasks are maintenance tasks designed for hidden or unrevealed failures such as protective devices and redundant functions. These tasks are being performed to reduce or mitigate the risks of multiple failures wherein both the protective device and protected function are both in a failed state. It is also termed as functionality inspection or Detective Maintenance.

Failure Effects describe, most likely, what happens when each failure mode occurs on its own. Writing the failure effect should compose of 20 to 60 words.

Failure Modes are the most likely, or probable cause of the failure. These are not the root cause but are considered the possible or most likely cause of why the failure happened.

Failure Mode and Effects Analysis (FMEA) discipline were developed in the United States Military. The procedure for MIL-P-1629, titled Procedures for Performing a Failure Mode, Effects and Criticality Analysis, was written on November 9, 1949. It was used as a reliability evaluation technique to determine the effects of system and equipment failures.

Failure Rate is the expected rate of failure or the number of failures in a specified period. It is expressed in failures per million or billions of hours. It is also the probability of a failure in a stated unit of time and is the reciprocal of MTBF.

False Brinelling is caused by the continuous vibration of the duty equipment, thereby causing fretting damage on the bearing for standby equipment. The best way to address false brinelling is to rotate the shaft manually around 360 to 720 degrees + 60 degrees so that the ball's position changes from its position on the outer raceway of the bearing.

Fatigue Wear is a type of wear that results from a continuous cyclic slip under repetitive load applications for many thousands or millions of load cycles. Fatigue is the phenomenon leading to fracture under repeated or fluctuating stress having a maximum value less than the material's tensile strength. Fatigue fractures are progressive, which can start as micro cracks and propagate into larger ones that cause complete destruction of the components.

Fault-Tree Analysis was developed by Boeing Aerospace in the 1950s for use in the development stages of the design process. This mathematical tool yields probabilities. Its primary intent is to predict the probability of a specific failure.

Federal Aviation Administration (FAA) is an agency of the Department of Transportation that regulates all aspects of aviation in the United States. This group is responsible for all Air Traffic Control and sets standards for the maintenance and manufacturing of planes as well as pilot.

Filtration can be defined as the process of removing contaminants from a fluid, liquid, or gas stream through some means of a porous medium. An oil filter's function is to remove contamination from a fluid, liquid, or gas medium through some porous medium to achieve a required fluid cleanliness level.

Finished Goods Inventory refers to the finished products of the plant available for customer purchase. These products will either be shipped to their clients, picked up, or ready to sell.

Fire Point is defined as the lowest temperatures at which vapor of the material will catch fire and continue burning even after the ignition source is removed. The fire point is usually higher than the flashpoint, typically plus 10 to 24 °C or 50 to 75 °F beyond the flashpoint. The vapors produced at the flashpoint are not sufficient to ignite the fuel.

Flashpoint is the temperature at which the oil gives off vapors that can be ignited with a flame over the oil. The lower the flashpoint, the greater the oil's tendency to suffer vaporization loss at high temperatures and burn.

Flow rate is how much volume of a liquid can pass a certain point in a given time, measured in gallons per minute or liters per second.

Foaming is the formation of air and bubbles in a petroleum product, lubricant, or fuel oil that can reduce the product's effectiveness. Foaming can cause sluggish operation, air binding of oil pumps, and overflow of tanks or sumps. It can also result in excessive turbulence, improper fluid levels, air leaks, cavitation, or contamination with water or other foreign material.

Four-ball EP (Extreme Pressure) is a grease test called the Four-Ball EP, Four-Ball Extreme Pressure, Four-Ball Weld, or Load Wear Index. The Four-Ball EP test (ASTM D2596) measures the grease's ability to prevent wear during sliding contact under extreme pressure caused by heavy loads.

Fractography refers to the science and art of evaluating broken things. Fractography is often thought of as a means of getting information about how and why the part cracked by examining the fracture surface and the forces involved.

Fracture refers to the separation of a solid body into two or more pieces imposed upon by stress, which is usually static and at temperatures relative to the material's melting temperature. The applied stress can be considered tensile, compressive, shear, or torsional.

Free Water can be any water or gas which is not dissolved in the lubricating oil. Free water is when the water and the oil separate. Since the water is denser and heavier than the oil, it will settle at the bottom, which can easily be removed from the drain plug.

Friction is a force that is created whenever two surfaces move relatively across each other. As long as there is friction, heat is generated, and wear takes place. The area of contact will always be the point of wear. The main function of lubrication is to reduce friction.

Friction Modifier additives affect the frictional properties between two rubbing surfaces. These additives prevent scoring, reduce wear, and noise, and prevent micro-pitting in industrial gear lubricants. Friction modifiers are commonly used in gasoline engine oils.

FSNO Analysis stands for Fast-moving, Slow-moving, Non-moving, and Obsolete items. This form of classification identifies the items frequently issued, less frequently issued for use, items that are not issued for a longer period, say, two years, and those that are no longer used.

Function is the normal or characteristic actions of an item, sometimes defined in terms of performance capabilities. This is also what the users want the equipment or asset to do.

Function Loss Breakdown or failure of the primary function is when a failure occurs where the machine or equipment will totally stop and will definitely halt operations completely.

Functional Failure is defined as the inability of an item to meet its specific performance standard. This was when the asset had failed completely. This can also be called a function-loss breakdown.

Grease is defined as a solid to a semi-fluid product of dispersion of a thickening agent in a lubricant. It comes from the Latin word Crassus, which means fat. Grease is a thick, oily lubricant consisting of inedible lard, the fat of waste animal parts, mineral or petroleum-derived or synthetic oil, plus a thickening agent.

Grease Consistency depends on the type and amount of thickener used and the viscosity of the base oil. Consistency is also known as the resistance to deform caused by an outside force. The measure of consistency is called penetration.

Hard skills are technical knowledge gained from training needed to develop the skills to perform people's jobs correctly.

Hershey Number is the dynamic viscosity (η) multiplied by the speed (N) divided by the normal load (P) per length of the tribological contact. The load also means the average pressure.

Heterodyning is a process in ultrasonic analysis that translates these frequencies into the audible range. Heterodyning is the mixing of two waves, which produces both the sum and difference of their original waves, which allows the shifting of a high-frequency sound to the audible or sonic range. It converts the ultrasonic range to an audible or sonic range through this process, which is audible to the human ear.

Hidden Failures are failures that will not become evident to the operator or user of the equipment when the failure occurs independently. To consider a hidden failure, a secondary failure should occur. Hidden failures will only apply to protective devices and redundant components.

Hidden Failure Consequences refer to the risks of multiple failures due to an undetected earlier failure of a hidden function item. This mostly refers to the aftermath when protective devices and redundancies fail.

Horizontal Replication, also called fan-out, is the process of repeating or doing the proposed tasks derived from the RCM analysis to similar equipment with the same operating context or condition. This may also apply to modification, redesign, or improvement as long as the equipment replicated also possessed the same problems.

Human Cause is the second level of Root Cause Failure Analysis, which refers to human errors, omissions, or commissions resulting from the physical roots. Either someone did something wrong, or the person did the wrong thing.

Human error can be defined as an action planned but not executed according to the plan. Research by Dr. James Reasons of the University of Manchester in England found that humans commit an average of six errors per week. It is also an inappropriate action, or intention to act, given a goal and the context in which one is trying to reach that goal. Human error is the by-product of poor planned tasking or execution of a task resulting in the failure of a goal.

Human Nature is a concept that denotes the characteristics of human beings such as the way they think, their feelings, and the way they feel and react naturally. Humans react based on their instinct.

Human Sensory Perception means humans are gifted with five senses, which include the sense of smell, touch, hear, feel, and taste. For industries, forget about the sense of taste. Some failure modes will give some sort of warning or symptoms that they are on the verge of failing, such as smell, excessive vibration, noise, heat, or anything that can be detected by the human senses.

Hydraulics is the transmission and control of forces and motions through the medium of fluids. It is also the science of transmitting force or motion through the medium of a confined fluid. In a hydraulic system, the power is transmitted by compressing a confined fluid. This transfer of energy takes place because a quantity of liquid is subject to pressure.

Hydrocracking is a flexible catalytic refining process that can upgrade a large variety of petroleum products. Hydrocracking is commonly applied to upgrade the heavier fractions obtained from crude oils' distillation, including their excess.

Hydrodynamic Lubrication (HL) is also termed or called full-film or fluid film lubrication. Hydrodynamic lubrication is where the film is thick enough to separate interacting surfaces to minimize friction.

Hypothesis is a term used in root cause failure analysis, which means the probable or most likely causes that need to be verified. In RCM, they are termed failure modes, while in a crime scene, they are referred to as the suspects.

Industrial Internet of Things (IIoT) uses wireless smart sensors and actuators to enhance manufacturing and industrial processes by hooking them on the equipment and providing data directly to the computer.

Inertia Impaction is a process that helps remove contaminants smaller than the pore size of the filter medium. The contaminants are retained mechanically or through adsorption. This type of removing contaminants is more effective in gases than in fluids. This process is efficient for particles greater than 0.5 to 1 micron.

Infant Mortality Failures are failures, which occur at the beginning of life. Others refer to this as commissioning failures, start-up failures, or debugging failures that occur after conducting major Preventive Maintenance activities, which include overhauls and replacements.

Infrared is a type of light in the electromagnetic spectrum that is not visible to our naked eyes. Our eyes can only see a small portion of the electromagnetic spectrum, the visible light spectrum, or the rainbow's colors in layman's terms. Infrared radiation lies between the visible and microwave portions of the entire electromagnetic spectrum.

Infrared Thermography is the science of actually allowing us to see the heat. An infrared inspection can detect heat that would normally be invisible to the naked eyes and represent them as an image, which we can see.

Inherent Design Reliability means that the equipment or asset is only designed to function which it is intended to do. A pump with a rated capacity of 500 gallons per minute cannot discharge beyond that capacity since this is what it was designed for.

Inherent Reliability Level is the level of reliability of an item, equipment, or asset that is attainable by utilizing all the available maintenance tasks on RCM.

Initial Cleaning in Autonomous Maintenance Step 1 removes any form of unwanted abnormalities from the equipment. It consists of removing dirt, contaminants, grime, excess oil, grease, and other foreign objects that affect equipment parts and components that have accumulated over time in the equipment.

Initial Flow Control Activities (IFCA) also called Early Equipment Management (EEM) has something to do about purchasing new equipment, improving the vertical set-up time in the shortest time possible, developing easy-to-manufactured products, and extending the life cycle of equipment through MP (Maintenance Prevention) modifications and improvements. What is important is to have a detailed and structured process so we can identify the flaws and problems at the very beginning.

Initial task intervals also called proposed tasks are maintenance task intervals assigned before the service maintenance program, subject to adjustments based on the actual operating experiences.

Inspection Tasks refer to scheduled tasks requiring testing, measurement, visual inspection, or human senses for detecting failure evidence done by both operators and maintenance. For inspecting hidden failures, this will default to Failure Finding tasks, while for evident failure inspection, it will default to On-Condition Tasks.

Installation is the process of placing, positioning, building, and connecting equipment in its proposed and designed nominated location following design specifications.

Instincts can be defined as an inborn impulse or motivation to action typically performed in response to specific external stimuli. It can be considered intuition, feeling, impulse, or gut feeling. This is how we react to certain situations.

Insurance Spares are spare parts that will replace a failed part in a piece of equipment whose penalty cost for downtime is very high. These parts do not become obsolete until the equipment is retired from service. In most cases, these parts are big and classified as non-moving parts.

Intangible Measurements cannot be quantified, measured, or contribute to the plant's bottom-line results. They can also be termed as qualitative measurements.

Intermediate Discipline is a discipline on World-Class Maintenance that refers to the different reliability and maintenance strategies that can be adopted once basic equipment conditions have been well established in the asset. This includes Lubrication Management, MRO Spare Parts Management, Life Cycle Management, Root Cause Failure Analysis, Reliability, and Maintenance Strategies.

Internal setup are those activities that can be performed during conversion only when the machine had been totally shut down for conversion since a different product will be required to operate on the equipment.

Inventory Turnover is a ratio that measures the number of times an inventory is withdrawn or consumed in a given time by the end-user. Inventory turnover is sometimes called stock turns or stock turnover.

ISO 55000 is a standard described as the parent document of ISO 55001. It provides a general overview of asset management as a discipline and contains definitions of terms used in the ISO 55000 series of standards. This ISO standard aims to establish a clear and consistent understanding of the principles and requirements applied when developing and implementing an Asset Management System.

Karl Fisher Titration Test is an oil analysis instrument that measures the amount of dissolved moisture content in the oil. In this method, water reacts quantitatively with a Karl Fischer reagent. This reagent is a mixture of iodine, sulfur dioxide, pyridine, and methanol.

Kauro Ishikawa developed the Ishikawa or Fishbone Diagram in 1969. A fishbone is constructed by assigning the 4M's and 1E. 4M refers to man, machine, method, and materials, while 1E refers to the environment. The most common probable causes that relate to the problem are listed and grouped accordingly. The team brainstorms and focused on the most likely or probable causes of the failure.

Key Performance Indicators (KPIs) are measurements that are tracked and monitored regularly to indicate if an organization is on the right track to hit its goals, targets, and objectives.

Kinematic Viscosity is a viscosity test, which takes the time for the oil to travel through a glass orifice of a capillary tube under the force of gravity.

Kosho is the Japanese term for breakdown. It is a lost time due to equipment failure that may or may not cause downtime.

Knowledge-Based Mistakes are types of mistakes that occur when someone is, confronted with a situation that has not yet occurred before and had not been anticipated. In other words, there are no rules or procedures to follow. In situations like this, the person has to decide quickly about an appropriate course of action, and a mistake occurs due to the wrong decision.

Lagging Indicators are indicators and measurements that indicate the bottom-line results. Examples of this include productivity, repair, revenue, and maintenance costs. This will also depend on who is reading the KPIs, meaning that the lagging indicator for a maintenance manager will be the availability but not for the CEO of the plant.

Lapse occurs when someone misses out on a key step in a sequence of events or activities. For example, a mechanic leaves a tool behind after working on a machine or simply forgets to fit a key component while reassembling it. If a person will do steps 1, step 2, step 3, step 4, and step 5, and the person missed out on step 4 in the process, then a lapse occurred. As humans age, our memory will not be as sharp compared to when we were young, just like in our high school days. Lapse is when people tend to forget things as they age.

Latent Cause is the last level of Root Cause Failure Analysis, which refers to hidden or concealed causes that need to be exposed so that industries can learn from the things that go wrong. This is about looking at the man in the mirror and admitting that each of us is also part of the problem. The latent cause is composed of two things, which are organizational latency and personal latency.

Lead time is the time the storekeeper noticed that the part is below the reordering point and makes a request for the part until the part is finally delivered and received in the storeroom.

Leading Indicators are the process of how the results were achieved. These indicators influence the final outcome of the results. These are also the indicators that lead to the lagging results.

Life Data Analysis is when a product, part, or component has operated successfully, or the time it operated before it failed, measured in hours, miles, cycles, strokes, minutes, or other measures.

Life Cycle Cost is the sum of the initial costs and the running costs of the equipment. It is the sum of the overall cost of equipment throughout its entire lifespan, including the costs incurred initially during the design stage up to the stage the equipment will be put out of service, disposed of, or decommissioned.

Line Organization Loss is a loss that results from a shortage of operators on the production floor where the operator needs to operate other machines that were originally planned. This often results in a shorter break time for the operators.

Lubricating oil also called a lubricant or lube oil, is a class of oil used in equipment, engines, and machinery types to reduce friction, heat, and wear between mechanical components to avoid metal-to-metal contact.

Lubrication Tasks refer to the scheduled tasks to assure the existence of completeness of lubrication films on the equipment. These tasks can either be scheduled routine tasks or condition-based maintenance tasks.

Machine-Related Downtime, also known as unplanned downtime, is when the equipment is not operating due to machine-related downtimes such as breakdowns, set-up, and conversion, cutting tool change, start-up losses, minor stoppages, and quality defects.

Maintenance are the activities that are done before an asset or equipment will fail. These are the activities of preserving or sustaining the assets. This means ensuring that the assets continue to do what the users want them to do.

Maintenance backlog is a time indicator that consists of delays in performing the scheduled maintenance works. This consists of pending scheduled planned activities allotted on the maintenance that has already passed its due date.

Maintenance cost is a universal and common measurement for all types of industries, unlike other KPIs. By definition, maintenance costs are any expenses incurred on the equipment, machines, people, or asset by the maintenance function to sustain and preserve their equipment and assets.

Maintenance Induced Failures are direct failures caused by the maintenance itself. Examples of maintenance-induced failures include infant mortality failure or doing intrusive maintenance, incorrect overhauling practices, and inducing early failures right after endorsing the equipment back to operators. In this case, the maintenance reassembled the equipment incorrectly.

Maintenance Management is the art or science of managing maintenance resources. It is also the manner of managing to keep our physical assets in an existing state of condition.

Maintenance Prevention is defined as the use of the latest maintenance data and technology when planning or building new equipment to promote greater reliability, maintainability, economy, operability, and safety while minimizing maintenance cost and deterioration.

Management Loss includes losses caused by the waiting time that is lost due to management problems and delays such as no available MRO spare parts, lack of manpower resources, lack of insufficient utilities, and work instructions.

Marking is the process in the semiconductor industry of identifying, traceability, and distinguishing marks on the integrated circuit package at the end of line process.

Market Value will be the value or cost of the asset according to the stock market. This will be the cost of the asset when sold to the market.

Mastery is the last level of skill development where the person is capable of transferring both his skill and knowledge to another person. This is what we want our maintenance to become.

Material Hardness is the measure of the material's resistance to plastic deformation. The most common method to determine the hardness of the material is the Rockwell Hardness Test, which can test the hardness of metals and alloys. Usually, a compressive force will be applied.

Maxi-Event or large-scale RCFA is performed on maxi events or failures with a large impact or consequences in the industry. It is recommended that an outsider do this event or an independent third party person act as the principal investigator to avoid bias. There will be three persons assigned to the evidence-gathering team for the maxi event, which will require a stakeholder meeting.

Mean Time between Assists (MTBA) is the average time the equipment performs its intended function between assists. It is also the productive or the operating time divided by the number of assists, errors, or minor stoppages. The common unit used is in minutes.

Mean Time Between Failure (MTBF) is defined as the average time between failures and therefore is a measure of the trouble-free time. The common unit used is in hours. MTBF is derived from the US MIL-STD 217 for testing electronic parts. It is a reliability engineering term that means the average amount of operating time between the occurrences of breakdowns that requires repair divided by the frequency of failures. The frequency of reporting MTBF should be either on a weekly or monthly basis. MTBF trend should be the higher, the better.

Mean Time to Fail (MTTF) is the expected time to fail for a system, part, or component. It is a basic measure of reliability for non-repairable systems. It is the meantime expected until the first failure of a piece of equipment. This is also equal to the MTBF minus MTTR.

Mean Time to Repair (MTTR) is the average time required to repair a piece of equipment, machine, or component. Other terms used are Mean Time to Restore, Mean Time to Recover, or Mean Time to React. It is also the average time required to perform corrective maintenance or repair on all removable items in a piece of equipment, product, or system.

Mean Units Before Assists (MUBA) indicates how many units were produced before an assist or minor stoppage is encountered on the equipment.

Measurement and Adjustment Loss are losses caused by frequent measurement and adjustment to prevent the recurrence of problems. Examples include excessive inspection integrated into the process as a result of poor quality and failure to find the root cause. Adjustment loss is experienced when adjusting equipment back to the standard after routine cleaning and periodic consumable changes.

Mercaptan was introduced by William Christopher Zeise in 1832. The Latin word mercurium captan means capturing mercury because the thiolate group bonds strongly with mercury compounds. These mercaptans are used as an odorant in sour gasoline into disulfides.

Micron is also known as a micrometer and is exhibited by the Greek symbol Mμ. A unit of one-micron length is equivalent to 39 millionths of an inch or 0.000039 or 0.0009906 millimeters. The human eye can see around 40 microns and beyond.

Midi-Event or Medium Scale RCFA is performed on midi or medium-scale events. A Principal Investigator leads the investigation with one or two persons assigned as the evidence-gathering group. This will require a stakeholder meeting.

Mini-Event or Small Scale RCFA event is usually performed on small-scale failures where the Principal Investigator will be the ones collecting the evidence. This type of RCFA will not require a stakeholder meeting.

Molding is the process of modern molding used to encapsulate plastic material components to protect IC or passive devices at the end-of-line process in semiconductor industries.

Moisture Vapor Transfer Rate (MVTR), sometimes called Water Vapor Transmission Rate (WVTR), is a measure of water vapor passage through a substance. It is a measure of the permeability of vapor barriers. There are many industries where moisture control is critical such as semiconductor industries.

Moment represents the effectiveness of either a rotating, bending or twisting force upon an object. All materials have their respective hardness. The hardness of the material is the measure of the material's resistance to plastic deformation.

Motor oil, also called engine oil, is a lubricant used in internal combustion engines, power cars, motorcycles, diesel engines, engine-generators, mobile, mining equipment, etc. Inside the main engine are metal parts, which move relatively against each other, and the friction between these moving parts consumes more power by converting kinetic energy into heat. This type of oil consumes the most amount of additives.

MRO Spare Part is defined as a part of a machine ready to replace an identical part if it becomes faulty. It is also defined as those parts of the machine which are kept on standby to be substituted when a part of equipment fails, a repair is required, or the part simply becomes worn out and needs to be replaced

MRO Wear and Tear Spares are spare parts that must be replaced every time the equipment undergoes Preventive Maintenance or Shutdown, and the equipment is disassembled and re-assembled for overhauls and parts replacement.

MSG-1 Document is a working paper prepared by the Boeing 747 Maintenance Steering Group published in July 1968 under the title: Handbook Maintenance Evaluation and Program Development (MSG1), which includes the first use of the decision diagram techniques to develop an initially scheduled maintenance program.

MSG-2 Document is a refinement of the decision diagram procedures in MSG1 published in March 1970 under the title MSG-2 Airline/Manufacturer Maintenance Program Planning Document, which is the immediate precursor of the RCM document. This document was used to develop a scheduled maintenance program for Lockheed 1011 and Douglas DC10, military aircrafts such as Lockheed S3 and P3, and McDonnell F4J.

MSG-3 Document refers to the Operator/Manufacturer Scheduled Maintenance Development developed by the Airlines for America (A4A), formerly the Air Transport Association or ATA. This document was developed in 1980, revised in 1988 and 1993, and to this day process is used for maintaining all types of civil aircraft.

Multipass Test is a controlled laboratory test where the effluent fluid is recirculated through the filter element while a new contaminant is continuously added. This test is used to determine the efficiency and the beta rating of the oil filter.

Multiple Failure is where the protected function had miserably failed because the protective device is said to be in a failed state.

Multi-Purpose Grease can be defined as grease combining the properties of two or more specialized greases that can be applied in more than one application. This permits the use of a single type of grease for a wide variety of applications.

National Transportation Safety Board (NTSB) is an independent US Government body whose responsibility is to investigate every civil and commercial aviation incident in the United States.

Natural deterioration is a TPM terminology that means the part or item has reached its lifespan and eventually gradually wears out. This is similar to age-related failures.

Near misses can be considered minor accidents or close calls that have the potential for an injury, accident, property loss, or even death. A person was standing in front of a Wrecking Ball Crane. After a few minutes, the ball fell which nearly smashed the head of the person just a few feet away.

Needed Non-Productive Work includes those that cannot be totally eliminated in maintenance but are still needed. Examples of this include lunch breaks, attending training, mentoring operators about their equipment, conducting improvements, RCFA investigations, and the like. Although in this situation maintenance is not doing Planned Scheduled Work, these activities should also be part of the maintenance function

Net Operating Time is used to identify whether the equipment is operated at the stabilized speed within the unit time.

Noack Volatility test was named after Kurt Noack. It is measured by the principal European test called NOACK. It is the amount of oil lost (light molecules) over time at a given temperature and pressure. It directly impacts high engine temperature and oil effectiveness, especially on viscosity, emissions, and oil consumption. Today's oil has a NOACK volatility limit of 15 %. Oil's Volatility rate should be less than 15%. The test is done through a NOACK Volatility instrument for 1 hour at 250° Centigrade.

Nominal Filtration refers to the average particle size of contaminants that will remain in the fluid after filtration. The efficiency of nominal filtration is 50%, and the beta rating is 2. This means that this filter is only designed to remove 50% of the contaminants based on its pore size.

Non-Dedicated Machines are those machines in manufacturing that are designed to manufacture different products using the same machine. For example, in manufacturing, some machines are used to manufacture different products, hence, one measurement of importance will be the set-up or conversion time.

Non-Machine Related Downtime, also known as planned downtime. This is when the equipment is not operating due to non-machine-related downtimes such as PM activities, operators' break time, meetings, or any other downtime, which is not caused by the machine.

Non-Maintenance Induced Failures are failures on the equipment, which are not caused by the maintenance function, but by other factors such as how the equipment was operated, design errors, commissioning errors, purchasing going for the lowest bidder, or management cutting costs. This constitutes 83 % of equipment problems and defects.

Non-Operational Consequences is one of the consequences of a failure mode, which does not directly affect safety, environmental, or operational consequences. The only consequences of this failure mode are the direct cost of repair and the chances of secondary damages to the equipment.

No-Scheduled Maintenance is a default task on RCM, which means that there is no feasible scheduled maintenance. Failure is allowed to occur since the consequences are minimal and acceptable. Also similar to the terms run to fail, breakdown maintenance, reactive maintenance, unplanned maintenance, or corrective maintenance. When the failure occurs, it will be subject to repair.

Not-Needed Non-Productive Work includes all waste on maintenance. These are those added activities and resources which are wasted as a result of poor or inadequate planning. This includes wasted time for the wrong part number supplied, lack of manpower, wrong tools, MRO spare part not available, or going back once more to the storeroom since you needed this and that.

Obsolescence refers to one of the reasons why a piece of equipment or product can no longer be used. This means that the equipment or the product being manufactured on the equipment will no longer be used. 15% of why equipment can no longer be used is due to obsolescence

Occurrence in FMEA/FMECA is a rating according to the likelihood that a particular failure mode will occur within a specific period. Occurrence is an assessment of the likelihood that a particular failure mode will occur. This is also termed as the probability of failure to occur.

Octane Rating measures the fuel's ability to resist engine knocking caused by the air-fuel mixture. The higher the octane, the greater resistance the fuel has to knock during the combustion process. The octane ratings are a measure of the fuel's stability.

Office Supplies Inventory are inventory supplies used by the different plant offices ranging from printer ink, pencil, ball pen, bond paper, folders, envelopes, scissors, paper clips, memo pads, and so on in industries.

Oil Analysis is a maintenance management tool that allows users to monitor the oil condition for maximum equipment life and maximum lubricant drain interval. It tells us the actual condition of the oil in the equipment. There are hundreds of several oil analysis tests that can be done both on used and new oil.

Oil Contamination is anything that should not be present in the oil besides its base and additives. Oil contaminants may be solids, liquids, or even gases in bubbles that cause foaming.

Oligomerization is a polymerization process in which a few, usually three to ten, of the basic building block of molecules are combined to form the finished product. Therefore, the product is formed with varying molecular weights and viscosities to meet a broad range of requirements.

On-Condition Tasks are maintenance tasks that entail checking the equipment's actual condition or using human senses, pressure, temperature inspections, gauges, the use of SPC charts, or Predictive Maintenance instruments.

One-Point Lesson is a learning tool for communicating standards, problems, and improvements in work processes and equipment. Workers and supervisors use one-point lessons to provide key information about everyday work and improvement opportunities.

Operational Consequence is a type of consequence where operations will be affected. The primary function of most equipment in any industry is connected to the need to earn revenue or support revenue-earning activities. A failure mode has operational consequences if it has a direct adverse effect on operational capability.

Operator Motion Loss includes losses generated due to unnecessary or excessive movement by the operator, as a result of poor layout, and work for the organization due to excessive walking, wasted motion, unnecessary reaching, or equipment being far apart where the operator will take a longer time to transport the product to the next process in the manufacturing.

Outsourcing is a common method for an organization that prefers to perform certain asset management activities not by itself, but by an external or internal service provider such as 3rd party contractors.

Overall Equipment Effectiveness (OEE) is the primary measurement used for plants and industries initiating Total Productive Maintenance. It is calculated by multiplying the equipment availability by its performance rate and quality rate expressed in percentage.

Oxidation is the breakdown of oil due to the extreme heat in the engine or equipment. This is the reaction between the oil and oxygen, causing acidic gases, sludge, and varnishes to form in the crankcase.

Oxidation inhibitors are added to the oil to extend its operating life. These additives are said to be sacrificial (perhaps like a suicide bomber), consumed while performing their duty of delaying oil oxidation, thus protecting the base oil. They are present in almost every lubricating oil and grease. This inhibitor aims to prevent oxygen from reacting with the oil, thus slowing its aging rate.

P-F interval is the interval between the emanation of potential failure and its decomposition to its final stage, which is a functional failure. At this stage, the equipment has already failed. The P-F interval is actually a warning period or the lead-time to failure better known as the failure development period

Pareto's 80/20 Rule. Dr. Joseph Juran, a Quality Management pioneer, developed the use of the Pareto Principle for problem-solving. He worked in the US from 1930 to 1940. According to Vilfredo Pareto's observation, 80% of the land in Italy was owned by 20% of its population, where 20 % of something is always responsible for 80 % of the results. He recognized a universal principle he called the vital few and trivial many and placed it into writing.

Partial Functional Failure is when the asset is still functioning, but it is below the performance standards that the user wants the asset to function.

Parts per Million (ppm) is used to measure the concentration of a contaminant in soils and sediments where 1 ppm equals 1 mg of substance per kg of solid (mg/kg). It is a unit used to describe small concentrations in water, where 1 ppm is equivalent to 1 mg/liter because a liter of water weighs approximately 1000 grams. It is also used to describe the concentrations of contaminants in lubricants.

Patent is the granting of a right by an authority to the originator of the invention. Just like in publishing this book, copyright is placed at the beginning of this page, which grants the author the right to publish this book. In the case of a patent, this provides the inventor the exclusive rights to the patent process whether this is a design or a new invention.

Penetration is the measure of the grease's consistency and depends on whether the consistency had been altered by handling. ASTM D217, and D1403 test, measures the penetration of unworked and worked greases.

Performance Rate is sometimes referred to as the throughput or equipment Performance Efficiency. It reflects whether the equipment is running at its full capacity or speed for individual products. Performance Efficiency measures Speed Losses, Idling, and Minor Stoppages. To determine the performance efficiency of the machine, the ideal or design UPH should be known.

Physical Cause is the first level of Root Cause Failure Analysis that refers to the physical cause of why the part or component failed. This is the technical explanation of why things broke or failed. This is said to be the metallurgical aspect as to why the failure occurred. It is also referred to as the Failure Analysis. The analysis will end on the component or part level.

Physiological Factors are a type of human error that refers to environmental stress affecting human performance. These stresses can include high or low temperatures, loud or irritating noise, excessive humidity, excessive or high vibration, exposure to toxic chemicals, radiation, or working too long without adequate break time, not to mention the day-to-day pressure from the boss.

Planned Maintenance is one of the TPM pillars that aims to improve the current maintenance activities, whose goal is to zero out all unplanned breakdowns of the equipment or asset. They are also the mentors of Autonomous Maintenance. This should be the strongest pillar in any TPM implementation, since if Planned Maintenance is weak, then Autonomous Maintenance will collapse.

Planning is defined as the total process set up to ensure that the right resources and materials arrive at the right place, at the right time, and do the right job in the right way. Planning also has a macro meaning when reference is made to an array of scheduled shutdowns for the year, the business plan, the marketing plan, the budget, and others.

PLCC (Plastic Leaded Chip Carrier) is a plastic, square, surface mount chip IC package that contains leads on all four sides and is manufactured by semiconductor industries. The lead pins extend down and back under and into tiny indentations in the housing. PLCC is a type of integrated circuit package that can be used to enable us to be mounted on a printed circuit board directly soldered either to the board or within a socket.

P-M Analysis is a method that analyzes chronic losses according to the inherent principles and natural laws that govern them. P stands for Phenomena and Physical. A phenomenon is a deviation from a normal to an abnormal state. M stands for Mechanism and the 4Ms (man, machine, method, and materials).

PM Compliance is a maintenance indicator that will measure how many PM tasks were actually completed and closed divided by the total number of tasks to be accomplished on or before the due date.

Potential Failure is defined as an identifiable physical condition that indicates that a functional failure is about to occur or is in the process of occurring.

Poka-Yoke also referred to as Mistake Proofing, is mostly used by manufacturing industries, especially to help an equipment operator avoid errors and mistakes during operations. Its purpose is to eliminate product defects by preventing, correcting, or drawing attention to human errors, which were developed by Shigeo Shingo, as part of the Toyota Production System. It was originally termed as Baka-Yoke, which was called foolproofing or idiot-proofing. Later on, Shigeo Shingo changed it to a milder name Poka-Yoke.

Pour Point is an oil analysis test that refers to the temperature at which oil will solidify due to extreme cold temperatures making the oil no longer capable of flowing. This oil analysis test only applies to countries with winter or extremely cold seasons. This refers to the W rating for engine oils. Note that W stands for winter.

Pour point Depressants are critical substances added to the additive compound to prevent the wax formation in the base oil from forming large crystal networks that can inhibit the flow of lubricating oil at cold temperatures, usually during the winter season.

Precision Maintenance involves performing maintenance work in a consistent, precise, and industry-accepted way. If properly implemented, maintenance should yield exactly the same results no matter who is performing the tasks whether the person is the most experienced or least experienced in the craft.

Predictive means to declare or indicate something in advance, especially foretell based on observation, experience, or scientific reason. Predictive comes from the Latin word "pre," meaning before, and "diction," from the Latin word "dicare," meaning to proclaim.

Predictive Maintenance is a type of maintenance performed on the equipment based on the equipment's actual condition with specialized diagnostic monitoring tools or Predictive Maintenance instruments to determine potential failures. Predictive Maintenance is a maintenance activity geared to indicating where a piece of equipment is on the critical wear curve and predicting its remaining useful life.

Preventive Maintenance is a type of maintenance performed on a fixed schedule or time-based interval. Preventive Maintenance is used when the parts have worn-out rates and are directly related to their age.

Primary function explains why the asset was purchased. It has something to do with the volume, output, speed, and quality.

Proactive-Corrective-Maintenance, which I can define as any activity done on the equipment such as overhauling, replacement, or other means resulting from Predictive Maintenance tasks before the failure is about to happen.

Proactive Maintenance is about understanding the root cause of why a part keeps on failing and performing corrective measures to eliminate the problem completely and avoid the recurrence of failure. In Proactive Maintenance, maintenance is ahead of failure.

Productive Work includes those works that are scheduled by the planner such as routine Preventive Maintenance activities or any future works, which are planned and scheduled.

Profit refers to the amount realized by a company after subtracting the expenses it incurred when providing a service or goods from the total revenue.

Protective Devices refer to those devices placed in the equipment and assets to protect something against failure. Samples of protective devices include led, beacons, sensors, alarms, emergency stops, lighting arresters, and so on. The protective device's function is to indicate that the protected function is still working and is functional. These are the defenses of the equipment against failures.

Psychological Factors can be grouped according to those, which are unintended, and those errors that are intended. Unintended errors can be grouped into slips and lapses, while intended errors can be grouped according to mistakes and violations.

Pumpability is the ability of the grease to be pumped or pushed through a system, which is similar to the viscosity of the oil. This indicates how easy or hard the grease can flow through lines, nozzles, and grease dispensing units.

Purchasing Cost is also called the Ordering Cost. This is the sum of the fixed cost that is incurred each time an item is ordered. This includes the physical activities required to process an order. Usually, the purchasing cost can include the purchaser's salary, telephone costs, receiving clerk salary, account payable costs, internet costs, electricity, and other costs.

Quality Rate measures the loss incurred due to rejected, reworked, and return products. Quality Rate is the product of the following two factors, the number of products scrapped, reworked, or rejected and the number of items produced.

Radio Frequency Identification (RFID) refers to a technology where the digital data encoded in RFID tags or smart labels are captured by a reader through radio waves. RFID systems consist of three components, an RFID tag, or smart label, an RFID reader, and an antenna. For industries, this is usually used for MRO Spare parts.

Random Failures are failures that can occur at any given period. This means that the probability that an item will fail in any one period is the same as it is in any other period. This means that the conditional probability of failure is not constant. This is also referred to as chance failures and wear out is not identifiable which means that the item can fail at any given time or period.

Raw Materials Inventory: These are inventory parts and items used by industries to produce their finished products. If you work in the automotive industry, this will refer to the different parts needed to assemble a car.

RCFA Logic Tree diagram is a tool that uses deductive logic to guide the thought process used to draw correct conclusions. Therefore, a logic tree is a disciplined methodology that prompts the user to answer questions that will eventually identify the root cause of a failed event.

RCM Decision Diagram is an algorithm that allows the users to make a decision-making process to identify the most appropriate and feasible maintenance tasks to address a particular failure mode.

RCM Default Tasks are RCM tasks available when it is not feasible to perform a Preventive or an On-Condition task to address a particular failure mode. Default tasks on RCM include

Failure Finding Tasks for hidden failures, Run to Fail, Switching intervals for redundant functions, and redesign or modification.

RCM Streamline Version is a shortcut or streamlined RCM version for improving the effectiveness of current maintenance programs and strategies. It starts with the existing maintenance program used within the plant and counter check as to what failure modes these tasks addressed.

Reactive-Corrective-Maintenance is the activity of repairing or troubleshooting the equipment after a failure occurs. This will be done after the failure happens, similar to the original definition of corrective maintenance.

Reactive Maintenance is a type of maintenance strategy that simply means fixing it only when it fails. As the saying goes, when it ain't broke, don't fix it; when it fails, then we come and fix it. Other terms used to designate Reactive Maintenance are run to fail, run to destruction, corrective maintenance, fire-fighting mode, and stop the bleeding syndrome. RCM termed this as no-scheduled maintenance.

Recurrence Prevention includes a list of things and corrective actions that need to be done to indicate that the failure or defect has not repeated itself. This will usually be done in the case of the physical, human, and system cause only.

Redesign or Modification means that if the rest of the maintenance tasks are not feasible and worth addressing the failure mode, then the last option for maintenance will be to resort to redesign or modification, especially when the failure mode will have safety or environmental consequences. The goal here is to eliminate or reduce the consequences of failure from happening.

Redundancy or Standby means duplicating the system or component, which is not affected if failures and breakdowns occur. Failures are allowed or tolerated through some forms of redundancy, standby, or when the asset has some form of duplicated function.

Redundant tasks are the same task done by different groups of people, which can either be the PM crew or a third-party contractor. These are considered intrusive and should be deleted from the Preventive Maintenance lists since this will just end up in a waste of manpower and resources. What is important is to decide who is the best person or group to perform the task.

Reforming is a process designed to increase the amount of gasoline that can be produced from crude oil. Hydrocarbons in the naphtha stream have almost the same number of carbon atoms as gasoline, but their structure is generally more complex. Reforming rearranges naphtha hydrocarbons into gasoline molecules. The other products of reforming are light gases and high-octane gasoline blending components called reformate.

Reliability is the probability that an item will operate without failure throughout a specified interval and that the item will perform its intended function under specified operational and environmental conditions under a given and specified time.

Reliability-Centered Maintenance is a process used to determine the most feasible maintenance requirements in its present operating context or state that it is being operated.

This strategy is used to define the correct maintenance tasks for equipment or asset with the aid of an RCM decision diagram or algorithm.

Reordering point is the sum of the minimum stock, and in certain cases, we add the safety stock. Others called this the buffer stock. The Reordering Point is equal to the minimum stock plus the safety stock.

Repair are the activities that are done after an asset or equipment failed. These are the activities to be done to restore the equipment back to its operating state.

Replacement Asset Value (RAV) is also called the Estimated Asset Value (EAV). It is the cost of maintaining the asset, which is measured against the value of the asset. It can be said as the percentage of the cost to replace the asset. The lower the value of RAV, the more effective we are in maintaining and preserving the asset.

Return on Investment (ROI) directly measures the amount of return for a given investment, relative to the investment's cost. In calculating the Return on Investment (ROI), the benefit or return gained from an investment is divided by the cost of the investment. The result is expressed as a percentage or a ratio.

Revenue refers to the income a company earns by providing services or selling products in a given financial year.

Risk is something or someone that can cause a problem, harm, injury, damage, danger, hazard, or loss to the industry. Risk is the potential for uncontrolled loss that is of definite value to an organization. It is an intentional interaction with uncertainty.

Risk Management is the identification, evaluation, and prioritization of risks as the effect of uncertainty on objectives followed by coordinated and economical application of resources to minimize, monitor, and control the probability or impact of unfortunate events to maximize the realization of opportunities.

Risks Priority Number or RPN in FMEA is the sum of severity multiplied by its detection and its occurrence. This is also the sum of the consequences multiplied by its criticality, confidence, and probability.

RNM Cost refers to the total repair and maintenance cost incurred on the equipment or system for a given period. The trend we want should be that the lower the RNM cost, the better. RNM costs should be tracked and reported either weekly or every month.

Rockwell Hardness Test refers to the hardness of metals and alloys. With this method, a hardness number is determined by the difference in the penetration resulting from applying an initial minor load followed by a major larger load.

Roll Stability of Grease: ASTM D1831 is a grease test that assesses how stable the grease is when it is subject to operating conditions. This test allows us to determine if the grease will handle the intended load and for how long before it begins to fail.

Root Cause Analysis, both RCFA, and RCA is performed by determining the physical, human, system, and latent cause of the problem. The distinction is on the word failure, for equipment related we used the term RCFA. Hence, Root Cause Analysis can be used for safety, accidents and near-miss investigations, and quality defects for industries.

Root Cause Failure Analysis is a basic investigative tool that allows us to understand why things went wrong by identifying the problem's basic source or origin. Root Cause Failure Analysis is the investigation process or probing of a problem, which is entirely based on the evidence unfolded on equipment-related failures. The level of root causes will include the physical, human, system, and the latent cause of the problem.

Rooticians are people who are knowledgeable in conducting a Root Cause Analysis or Root Cause Failure Analysis investigation. They are the fact-finding group deployed to find the cause of a problem or incident.

Rotable Spare Parts are reusable spares or components that can be reconditioned and reused, such as motors and engines. These can be component or inventory items that can be repeatedly restored or refurbished to be used once again on the equipment into a fully serviceable condition after it failed. The value of rotable spare parts depends on the remaining useful life of the production equipment it supports.

Ruled-Based Mistakes usually occur when people believe that they are following the correct course of action when doing specific tasks based on a specific rule or procedure, but the course of action is inappropriate.

RULER (Remaining Useful Life Evaluation Routine ASTM D-6971-04 3 and ASTM D-6810-02) is an oil analysis test that measures the remaining useful life of the lubricant. This test can be used to determine whether the oil needs to be changed or can be extended. The RULER oil analysis test will measure the remaining level of antioxidant additive in the lubricating oil.

Run to fail is a maintenance strategy, which tells us that it is time to repair the failure when a machine fails. The failure will happen first, and then maintenance reacts by repairing it. This is the very essence of reactive maintenance. Maintenance is done at a point when there is repair or actual breakdown of the equipment. This occurs when repair action is taken on a problem only when the problem results in a machine failure or breakdown. Run to fail is not valid if the consequences of failure may have safety or environmental consequences.

SAE JA1011 is also known as the Evaluation Criteria for Reliability-Centered Maintenance (RCM) Processes. This document describes the criteria to indicate that any process or analysis to derive the tasks is compliant with the classical RCM process requirements. This also separates the classical RCM from the Streamlined RCM.

Safety Consequences: A failure consequence where a failure mode can cause injury or kill someone else. Failure modes that can cause accidents or even near-misses, as a result, will be included in this category.

Safety stock is an addition of items, which is applied mostly to critical parts for the storekeeper to increase their confidence in avoiding stock out. This is also called the buffer stock.

Safety Supplies Inventory refers to items regularly supplied to employees by the EHS, such as gloves, hard hats, goggles, safety shoes, ear protectors, and overhaul outfits used as part of the plant's uniform in conducting their regular work routine provided to operators and maintenance people working in the plant.

Saponification is the process used to develop the grease thickeners where fatty acids are used to react with an alkali to form a chemical soap.

Schadenfreude means the pleasure we gained from another person's distress, especially if they perceived the person deserved it because they engaged themselves in a bad deed.

Scheduled Discard Tasks refer to removing an item or spare to be replaced at a specified time interval before the part wears out completely. This is included as a Preventive Maintenance task. Also termed as scheduled replacement of parts. The frequency of scheduled discard tasks is governed by the age at which the item or component shows a rapid increase in the conditional probability of failure.

Scheduled Restoration Tasks are Preventive Maintenance tasks that entail re-manufacturing a single component or overhauling an entire assembly on or before a specified age limit, regardless of its condition at the time, also termed as Scheduled Rework. The frequency of a scheduled restoration task is governed by the age at which the item or component shows a rapid increase in the conditional probability of failure.

Scheduled Maintenance Tasks, also called Preventive Maintenance tasks, are scheduled tasks that need to be done on the equipment or asset at scheduled intervals. In RCM, these refer to as scheduled discard and scheduled restoration tasks.

SDE Analysis classification of spares is based on the lead time to acquire the part. This classification is carried out based on the lead time required to procure the spare part. Scarce (S) are those items that are imported and those items, which require more than 6 months of lead time. Difficult (D) are Items that require more than a fortnight but less than 6 months of lead time. Note that one fortnight means 2 weeks or 14 days. Easily Available are easily available items, usually with less than a fortnight's lead time.

Seal Swell is one of the functions of the oil in which the oil must be compatible with the seals and must not cause the seal to crack, shrink, or degrade; rather, it must cause the seals to expand slightly when contact with the oil to ensure proper sealing.

Secondary function refers to the other functions of the asset besides the primary function. Most assets are expected to fulfill more than one function besides their primary functions

Secondary Damage refers to the damage to the equipment that is caused by the primary failure mode. For example, a bearing seizes that it caused a fracture on the shaft or other parts due to excessive vibration.

Set-Up Loss, Conversion, or Changeover in several manufacturing industries, equipment is not dedicated, producing different product types with the same equipment. Hence, when it is time to change one product to another, it will require the time required to remove dies, and jigs for one product, clean up, prepare dies and jigs for the next product, reassemble the equipment, adjust the equipment, perform trial runs and make further adjustments until the product of acceptable quality is now obtained from the equipment.

Severity in FMEA/FMECA refers to the impact of the failure mode and its effects. Severity considers the worst scenario that can happen after a failure or breakdown takes place. It is a rating corresponding to the seriousness of the effects of a potential failure mode. This is also the consequence of the failure mode.

Shear force is the sum of the effect of shear stress over a surface that results in a shear strain. The important thing to consider is that the force acting on an object is parallel.

Shear Stress is also called tangential stress. Unlike both tensile and compressive stress, the force applied is not perpendicular but parallel to the area. Once a force is applied in parallel in opposite directions, it can cause the object to deform. Shear stress is a force acting parallel to a surface or to a planar cross-section of an object.

Shutdown Loss is the time when equipment is shut down for scheduled maintenance. However, shut down related work generally affects the operating time of the equipment. Shutdown-related work must be regarded as a planned downtime loss and reduction of shutdown work time, which must be reduced.

Single Minute Exchange of Dies (SMED) is a process developed by Shigeo Shingo to reduce set-up time and changeover for manufacturing industries. The success of this system was illustrated in 1982 at Toyota when the die punch setup time in the cold-forging process was reduced over three months from one hour and forty minutes to three minutes.

Situation is a set of things happening and the conditions that exist at a particular time and place. Merriam-Webster dictionary states that a situation is how something is placed concerning its surroundings. The outcome is something that follows as a result or consequence of the situation. It is a result or effect of an action, situation, or event.

Skill is the product of personal motivation and thorough training. The end result is mastery, which is used to perform one's job correctly and efficiently, developed over time.

Soft skills are training that improves our personal well-being that shapes how we work and interact with others. An example of this training includes Supervisory Skills, Leadership, Team-Building, Work Ethics, Effective Communication, Teamwork, Time-Management, and the like.

Soot is a by-product of the combustion process and can escape the piston rings through blow-by. This may be caused either by low compression, or too much idling due to stop and start, such as in traffic jams, or air-fuel mixture. Soot can range from 3 microns and above,

Sour Crude Oil is a crude oil that contains a high amount of sulfur, which is one of the main impurities in crude oil. It is very common to find crude oil containing sulfur impurities. When the crude oil's total sulfur content is more than 0.5%, the oil is called sour.

Spare Parts Management is the science of managing parts inside the storeroom when operations and maintenance required them in case of a breakdown that seeks replacement or Preventive Maintenance routine and replacement activities.

Speed Operating Rate is the ratio of the Theoretical Cycle Time divided by the Actual Cycle Time of the equipment.

Squirrel Stores are maintenance secret hiding places for their own spare parts. Squirrel Stores are unofficial stores in which maintenance keeps their own spare parts for their own need and benefit.

Squirrelling is the act where maintenance keeps the spare parts themselves. Why does maintenance do this? It is because there are a lot of stock-outs, which means that the part reflects in the system, but physically it is out of stock.

Slip occurs when somebody does something incorrectly or does something in the wrong sequence. If a person will perform something in sequences such as step 1, step 2, step 3, step 4, and step 5, and what the person did was step 1, step 2, step 4, step 3, and step 5, the person committed a slip. This is when a human action takes place that is not intended. The occurrence of a lack of one's trains of thought is derived mostly from unconscious behavior.

Snake Oil is a quack remedy or panacea. This is a euphemism for deceptive marketing that promises its customers a silver bullet solution to all their problems. As the word implies, these are considered too good to be true products. Examples of these are lubricant additives commercialized on Home TV shopping performing tests on oil beyond your wildest imagination but only to be recognized by the Federal Trade Commission as a fraud.

Spalling is a fracture of the running surface and subsequent small, discrete material particle removal. It is the pitting or flaking away of bearing material. Spalling can occur in the inner ring, the outer ring of the bearing's rolling elements.

Spectroanalysis (ASTM D5185 and ASTM D4951) is an oil analysis of metal content and additive package. This test will check 19 elements and reports them in ppm or percentage. This oil analysis tests limit its test from 10 microns and below, and its disadvantage is not being able to detect particles larger than 10 microns.

Sporadic Failures are failures that often indicate sudden large and abrupt deviations from the norm. This refers to catastrophic failures and is the opposite of chronic failures. Usually, their cause is easy to detect since there are just one or a couple of causes.

Stakeholders are a group of people in Root Cause Failure Analysis that will include any of the following; the person being accused, the person whose behavior needs to change, the person who has to spend money, the person accusing, or anyone that the Principal Investigator and evidence gathering team think is involved in the incident.

Start-Up Loss is a type of loss that occurs during starting up the equipment. Start-up loss means that the material loss is caused at the initial stage of product launching, namely the loss caused during the period from a start-up of production to the stabilized production stage.

Strategic Asset Management Plan (SAMP) is a document that specifies how the organizational objectives can be converted into asset management objectives. Included in the SAMP document is the relationship between organizational objectives and asset management objectives.

Strategy is a careful plan or method for achieving a particular goal, usually over a long period. It also includes the carrying out of plans to achieve the goal. It is particularly a long-term plan for success. The strategy also means to maintain and build a competitive advantage over their competition.

Static friction is the force that keeps a motionless object from being pushed or pulled across a surface. The only force that is acting on this block will be the weight of the object and its resisting force.

Strain is the change of shape of the object after compressive or tensile stress is applied, which can also be equal to the Modulus of Elasticity (E), which is equal to the stress divided by the strain of the object. For tensile and compressive stress, the force applied is perpendicular or at 90 degrees to the area. It is also the quantity that describes the amount of deformation that can occur within a material body whenever a load is applied.

Strength in the metallurgical sense is the metal part's property to resist the part's stress imposed upon the part.

Strength of Material is its ability to withstand an applied load without failure or plastic deformation. The field of strength of materials deals with forces and deformations that result from acting on a material.

Stress is defined as the force per unit area, often considered as the force acting through a small area within a plane.

Stock-Out is when maintenance requests a part in the storeroom where there is an actual inventory in the system, but the actual or physical inventory is zero.

Storeroom's space utilization is equal to the actual space consumed by the storage and spares divided by the total space of the storeroom.

Sweetening is replacing a certain amount of oil so that new additives can be mixed and added to the existing used oil inside the equipment. Unlike changing oil, where the oil's total volume is replaced. Around 25 to 30% of the used oil will be replaced with new fresh oil during the sweetening process so that new additives will be mixed with the existing oil.

Synthetic oil will fall on Group IV and V Base Stocks for lubricating oil. Synthetic oil is the result of a chemical reaction called synthesis, in which the end result is a uniformly shaped molecules that are more resistant to heat and impossible to achieve through the crude oil refining process compared to mineral-based oil, whose molecules are uneven in size.

TAN / TBN Crossover is where the value of both TAN and TBN becomes equal. When fresh oil enters the service, the TAN will be low, while TBN will be high. As the engine runs, TAN will increase, and TBN will fall. When the TAN value becomes equal to the TBN, the TAN/TBN crossover had been reached which means that it is time to change the oil.

Tangible Measurements or quantitative measurements are those measurements that are easy to quantify and have a direct impact on the bottom line performance of the plant.

Tasks Duplication means that the same task is performed by different people or groups. Decide on who is the best and most qualified to perform the task.

Tensile Stress is when you have an object in a vertical position such as a cylinder, and a downward force is applied to an object, there will always be an opposite force, which will be equal to the downward force. Both the upward and downward forces will cause the object to be under tensile stress.

Theoretical Cycle Time also called the Ideal or Design Cycle Time is the Ideal production rate of the machine and is sometimes referred to as the Design Speed. The reciprocal of Theoretical cycle time is the units per hour or UPH.

Tool Algo is an instrument that monitors and records the number of strokes and is stored on the main computer so that tools, dies, punches, or whatever is monitored will be replaced after reaching its dictated stroke or cycles on the machine.

Tool Cutting Blade Change is a loss in the equipment or machine incurred on swapping or changing any consumable tooling item when it has become worn-out, ineffective, or severely damaged.

Torsional Stress is the twisting of an object caused by the force acting on the object's longitudinal axis. The twisting effect is known as torsion. Torsional stress can lead to deformation. The twisting is called torque, which is also called the twisting moment.

Total Acid Number (TAN) is a measure of the total acid concentration present in a lubricant. Occasionally, the decrease of an additive package may cause an initial increase in the TAN of the fresh oil.

Total Base Number (TBN) is a measure of alkaline concentration present in a lubricant. Engine oil is formulated with alkaline additives to combat the build-up of acids in a lubricant as it breaks down. Once alkaline additives are depleted, the lubricant no longer performs its function, and the engine is at risk of oxidation, corrosion, sludge, varnish, and other problems.

Total Functional Failure occurs when there is a total loss of function on the equipment. This can happen both in the primary and secondary functions of the equipment.

Total Productive Maintenance is a plant improvement methodology system that enables continuous and rapid improvement of the manufacturing process through employee involvement, employee empowerment, and closed-loop measurement of results. TPM is a production-driven improvement methodology designed to optimize equipment reliability and ensure efficient management of its assets. TPM aims to build up a corporate culture that thoroughly pursues production system efficiency, improvement, and Overall Equipment Effectiveness. This methodology originated in Japan.

Training Gap is defined as the difference between the skills required to complete the job and the existing skillset of any particular team member. Focusing all our training based on the needs may or may not still contribute to the bottom line results.

Training Needs Analysis (TNA) is the process in which the company identifies the training and development needs of its employees so that they can do their job effectively, depending on the specific needs of that function in the organization.

Treating in oil refineries has several options for their treating processes, but the primary purpose of the majority of them is the elimination of unwanted sulfur compounds. A variety of intermediate and finished products, including gasoline, kerosene, jet fuel, and gases, are dried and sweetened. Sweetening is a major refinery treatment of gasoline that treats sulfur compounds to improve the color, odor, and oxidation stability.

Tribology was coined by Peter Jost in 1966, derived from the Greek word tribos, meaning rubbing, and translating the word literally on the rubbing science. The word tribology was not widely used until the Oxford English Dictionary defines tribology as the branch of science and technology concerned with interacting surfaces in relative motion and associated matters such as friction, wear, lubrication, and design bearings.

Trim is the cutting of the dambar that shortens the leads together. The form is the forming and bending of the leads into the correct shape and position. Singulation is the cutting of the tie bars attached in the individual units to the lead frame resulting in the individual separation of each unit from the lead frame at the end of line process in the semiconductor industry.

Turnaround is a scheduled event where an entire system or process of an industrial plant such as an oil and gas refinery, petrochemical plant, power plant, pulp and paper mill, and similar plants will be shut down. This is usually done at a longer interval.

Unplanned Breakdowns are unexpected failures occurring on the equipment in which the operator will generate a work request where the role of the maintenance or technician is to troubleshoot the equipment until it becomes operational once again.

Uptime is the time in which the equipment and asset are operating. It is also the time during which the equipment and asset are working without failure. Other terms to designate uptime includes operating time, productive time, running time, or time the equipment is utilized.

Utilization is the proportion of time that the equipment is utilized for its intended function and purpose. It is also the time the equipment is operating and running.

Varnish is an oil contaminant made of oxidized or carbonaceous adhesive material covering the engine's internal surfaces. As the oil is oxidized, these compounds form a consistent, hard, and shiny substance. Varnish occurs because of many factors, such as oxidization, friction, and heat. These are the products of the evaporation of the oil.

VED Analysis indicates that the classification is based on the criticality of the spare. Vital (V) means that a spare part will be termed vital if its non-availability will be a very high loss due to production downtime or a high cost will be involved in the part is procured on an emergency basis. Essential (E) will be a moderate loss incurred due to the non-availability of the part. Desirable (D) are those that are desirable if the production loss is not very significant due to its non-availability. Most of the parts that will fall under this category are equipment that has some redundancy making it less critical in the plant.

Velocity is the average speed at which the fluid can move from a given point. The velocity is measured in either mile per hour, feet per second, or in SI or International System of Units. This is an important factor in sizing the hydraulic lines.

Vertical set-up time will include the time the equipment was installed and the commissioning time until the time it is now ready for the start of operation.

Vertical Start-up time usually starts when the purchaser generates a Purchase Order for additional new equipment or a repeat order of new equipment, which will be the same model that they are currently using.

Vibration can be defined by these common terms, such as swinging back and forth, oscillating, unbalanced, and shaking. The vibration occurs when a machine or machine component moves from its neutral or normal position to a lower and upper extreme limit of travel. This happens because the force is applied to the rotating equipment and components. Vibration can also be a cyclic or pulsating motion of a machine or machine components from its point of rest.

Vibration analysis is one of the most common types of Predictive Maintenance techniques that analyze these plotted vibrations signals to diagnose abnormal vibrations. This is the most used technique for analyzing the condition of rotating machinery.

Violation occurs when someone knowingly and deliberately commits an error. Violations fall under three categories: routine violation, exceptional violation, and acts of sabotage.

Viscosity is the measurement of a fluid's resistance to flow. This indicates the oil's capacity to lubricate by providing a thin film to separate metal-to-metal contact. An increase or decrease in viscosity can indicate many things in the oil and the equipment. Viscosity is not a perfect Oil Analysis test as we still need to perform other tests to diagnose the exact problem of why the viscosity of the oil changes.

Viscosity Index (VI) of a lubricant describes how the oil's viscosity will change depending on the temperature change. This means that if the temperatures increase, the viscosity will decrease, and vice versa. The viscosity index is a characteristic used to indicate variations in the viscosity of the lubricating oil concerning temperature changes.

Visual controls are any means or devices used to provide ease in inspection and expose early problems and deviations to the operators. It easily alerts operators of problems in the equipment. It is a business management technique that originated in Japan, where information is communicated using visual signals instead of texts or other written instructions. The design is deliberate in allowing quick recognition of the information being communicated to increase efficiency and clarity. This is also used to spot the anomaly or abnormality on the equipment more easily. Also termed visual management.

Volatility is the property that describes the degree and rate at which the oil will vaporize under given pressure and temperature. This means that the higher the volatility rate, the higher the oil lost due to evaporation.

Waddington Effect means the less invasive the Preventive Maintenance is, the better the maintenance's outcome can be as more maintenance can lead to early failures. This can be said to be the original concept of infant mortality failures.

Wallodi Weibull was born on June 18, 1887, in Sweden and died on Oct 12, 1979, in Annecy, France. He has written many books on strength of materials, fatigue, the rupture in solids, bearings, and the Weibull Distribution, which he invented in 1937. Weibull distribution is about life data analysis, which is used to determine the probability and capability of parts, components, or systems to perform their desired functions over a given time without failure.

Water Vapor Transmission Rate (WVTR) is a measure of water vapor passage through a substance. It is a measure of the permeability of vapor barriers.

Water Washout is a test on grease that refers to ASTM D 1264, which covers evaluating the resistance of lubricating grease to water washout from a bearing.

Wear is defined as damage to a solid surface caused by the removal or displacement of materials by any means of mechanical action of a contacting solid, liquid, or gas. It may cause significant surface damage, and the process is usually gradual. The damage is usually thought of as gradual deterioration, correlated with the equipment's operating age.

Wear debris can be defined as metal particles produced from the breakdown of surfaces within a machine. These particles can range from submicron size to chunks of metal as large as those seen by the naked eye.

Wear Metal Debris Analysis Tests is an oil analysis test instrument that indicates the number of metal contaminants present in the oil. By knowing this information, maintenance can understand what parts inside the equipment are on the verge of wearing out. Each of these elements will have its own specific limits, as specified in the oil analysis tables.

Wear-Out Failures or Age-Related Failures are parts that will eventually survive and reach a specific age before they fail. Age specified may be in the form of running hours, time, number of strokes, calendar days, number of revolutions, number of stress applied, or any other form.

Weibull Distribution is often used to describe the lifetime of parts. It is used to analyze and predict failure rates and describe the failure of parts and equipment.

Wire-bond is a semiconductor machine responsible for adhering or welding a thin wire to a bare semiconductor die-chip pad and the other end of the wire to a conductive pad on a substrate. The wire is usually made of 99% pure gold and is usually very thin, often 0.001 to 0.0013 of an inch.

Work-in-process (WIP) Inventory comprises of all the materials, components, parts, assemblies, and subassemblies that are being processed in production or are still waiting to be processed within the system.

Work Order for Preventive Maintenance is a document containing all the information needed to perform and accomplish the maintenance task and provides the process for completing that task. PM Work Orders include the scope of the work to be done, who will perform the task, and other relevant details.

World Class is the ability to compete anywhere globally to meet and beat any competitor anywhere in the world in terms of product, price, quality, and on-time delivery. (Extracted from Terry Wireman's book on World Class Maintenance)

World Class Maintenance Management is the art and science of managing maintenance resources performed by the best-in-class industries worldwide.

Wrench Time is an indicator that is used to measure how much time the maintenance people actually spend on doing actual maintenance work on the equipment. Others refer to this as the tool time or when maintenance is actually holding a tool, which of course is not only the wrench.

Yield Losses can be defined as the total loss between the input of raw material and the output of finished goods. This may also include the losses incurred due to reject products or those that had been scrapped.

RSA Maintenance Books Collection in Series

I have written these books inspired by the millions of maintenance mankind from different industries who want only the correct maintenance strategies for their equipment and assets to function the way we want. These books are written in series, explaining each discipline's maintenance in detail based on its original concept on World Class Maintenance, The 12 Disciplines.

- Volume 1: World Class Maintenance Management–The 12 Disciplines
- Volume 2: Maintenance–Roadmap to Reliability
- Volume 3: Reliability–A Shared Responsibility for Both Operators and Maintenance
- Volume 4: Cutting–Edge Maintenance Management Strategies
- Volume 5: Problems and Solutions on MRO Spare Parts and Storeroom
- Volume 6: Lubrication Tactics for Industries Made Simple
- Volume 7: Decoding Reliability-Centered Maintenance Process for Manufacturing Industries
- Volume 8: RSA Reliability and Maintenance Newsletter Vault Collection, Subscribers Edition
- Volume 9: Investigating Equipment Failures through Root Cause Failure Analysis
- Volume 10: Maintenance Indices–Meaningful Measures of Equipment Performance
- Volume 11: Implementing Preventive Maintenance for Industries the Right Way
- Volume 12: Extending Equipment's Life Cycle – The Next Challenge for Maintenance

Rolly Angeles Reliability and Maintenance Encyclopedia

Available at https://www.amazon.com/Rolly-Angeles/e/B07B3T1TXC/ref=ntt_dp_epwbk_0
RSA Reliability Website: https://rsaonlinebookstore.company.site/
Draf2Digital: https://books2read.com/ap/RWJydN/Rolly-Angeles

Index